The Alkaloids
Volume 13

A Specialist Periodical Report

The Alkaloids
Volume 13

A Review of the Literature Published
between July 1981 and June 1982

Senior Reporter
M. F. Grundon *School of Physical Sciences, New University of Ulster, Coleraine, North Ireland*

Reporters
W. A. Ayer *University of Alberta, Canada*
K. W. Bentley *University of Technology, Loughborough*
A. S. Chawla *Panjab University, Chandigarh, India*
R. Dharanipragada *University of West Virginia, U.S.A.*
G. Fodor *University of West Virginia, U.S.A.*
H. Guinaudeau *University of Limoges, France*
D. M. Harrison *New University of Ulster*
R. B. Herbert *University of Leeds*
A. H. Jackson *University College, Cardiff*
J. A. Lamberton *C.S.I.R.O., Melbourne, Australia*
J. R. Lewis *University of Aberdeen*
S. W. Page *University of Georgia, U.S.A.*
S. W. Pelletier *University of Georgia, U.S.A.*
A. R. Pinder *Clemson University, South Carolina, U.S.A.*
D. J. Robins *Glasgow University*
J. E. Saxton *University of Leeds*
M. Shamma *Pennsylvania State University, U.S.A.*

The Royal Society of Chemistry
Burlington House, London W1V 0BN

ISBN 0-85186-367-1
ISSN 0305-9707
Copyright © 1983
The Royal Society of Chemistry

All Rights Reserved
No part of this book may be reproduced or transmitted in any form
or by any means – graphic, electronic, including photocopying, recording,
taping, or information storage and retrieval systems – without
written permission from The Royal Society of Chemistry

Printed offset by
J. W. Arrowsmith Ltd., Bristol, England

Made in Great Britain

Foreword

Once again an annual review of the alkaloid literature is provided and this time there is a two-year coverage of Lycopodium Alkaloids. In an effort to cope with rising costs of production this Volume has been produced from camera-ready copy and I am grateful to the authors for their willing co-operation.

It is now apparent, however, that the Specialist Periodical Reports on "The Alkaloids", "Biosynthesis","Terpenoids and Steroids" and "Aliphatic and Related Natural Product Chemistry" are not financially viable in their present book format and in 1984 they are to be replaced by a review journal to be called "Natural Product Reports". The new journal is to be published every two months and will continue to give comprehensive annual surveys of the four areas of natural product research; additional articles on subjects not covered by existing Specialist Periodical Reports on topics such as chemotaxonomy and enzymology and on advances in physical techniques will be included periodically and author and subject indices are to be provided. The flexible production timetable should ensure more rapid publication of manuscripts compared with S.P.R.'s. In order to maintain continuity the S.P.R. Senior Reporters are members of the Editorial Board of Natural Products Reports and in the case of the alkaloid chapters it is expected that the present authors will contribute to the Journal. We all hope that the support we have received over the years from alkaloid chemists will be extended to this new and exciting venture. As always comments and suggestions will be much appreciated.

May 1983   M.F. GRUNDON

Contents

Chapter 1 Biosynthesis   1
     By R.B.Herbert
   1 Pyrrolidine and Piperidine Alkaloids   1
     1.1 Nicotine   1
     1.2 Cocaine and Cuscohygrine   3
     1.3 β-Pyrazol-1-ylalanine   3
     1.4 Pyrrolizidine Alkaloids   4
     1.5 Anabasine   4
     1.6 The Early Stages of Alkaloid Biosynthesis   5
   2 Phenethylamine and Isoquinoline Alkaloids   7
     2.1 Norlaudanosoline Synthase   8
     2.2 Hordenine and Normacromerine   9
     2.3 Hasubanonine, Protostephanine, and Laurifinine   9
     2.4 Aporphine Alkaloids   10
     2.5 Bisbenzylisoquinoline Alkaloids   11
     2.6 Aristolochic Acid   12
   3 Alkaloids Derived from Tryptophan   13
     3.1 β-Carboline Alkaloids   13
     3.2 Terpenoid Indole Alkaloids   14
     3.3 Ergot Alkaloids   15
     3.4 Cyclopiazonic Acid   18
     3.5 Streptonigrin   18
     3.6 Penitrem   20
     3.7 Roquefortine   20
     3.8 Hinnuliquinone   21
   4 Miscellaneous   22
     4.1 Geldanomycin, Rifamycin, and Antibiotic A23187   22
     4.2 Phenazines and Phenoxazinones   25
     4.3 Pseurotin A   29
     4.4 Cytochalasins   30
     4.5 Cycloheximide   31
     4.6 Streptothricin   31
     4.7 β-Lactam Antibiotics   33
     4.8 Acridone Alkaloids   37

    4.9 Malonomicin 37
    References 39

Chapter 2 Pyrrolidine, Piperidine, and Pyridine
    Alkaloids   44
    By A.R. Pinder
    1 Pyrrolidine Alkaloids 44
        1.1 *Sceletium* Alkaloids 46
    2 Piperidine Alkaloids 46
        2.1 Decahydroquinoline Alkaloids 49
        2.2 Spiropiperidine Alkaloids 50
    3 Pyridine Alkaloids 50
    References 53

Chapter 3 Tropane Alkaloids   55
    By G. Fodor and R. Dharanipragada
    1 Occurrence and Structure of New Alkaloids 55
    2 Synthesis and Chemical Transformations 57
    3 Pharmacology 62
    4 Analytical Studies 63
    References 63

Chapter 4 Pyrrolizidine Alkaloids   65
    By D.J. Robins
    1 Synthesis of Macrocyclic Pyrrolizidine Alkaloids 65
    2 Synthesis of Necine Bases 69
    3 Alkaloids of the Boraginaceae 72
    4 Alkaloids of the Compositae 73
    5 Alkaloids of the Leguminosae 74
    6 Alkaloids in Micro-organisms 75
    7 Alkaloids in Insects 75
    8 General Studies 76
    9 Pharmacological and Biological Studies 77
    References 79

Chapter 5 Indolizidine Alkaloids   82
    By J.A. Lamberton
    1 Swainsonine 82
    2 *Prosopis* Alkaloids 82
    3 *Elaeocarpus* Alkaloids 83
    4 *Dendrobates* Alkaloids 84
    5 Other Syntheses 85
    References 86

Chapter 6 Quinolizidine Alkaloids — 87
By M.F. Grundon

1 The Lupinine-Cytisine-Sparteine-Matrine-*Ormosia* Group — 87
  1.1 Occurrence — 87
  1.2 Structural and Stereochemical Studies — 87
  1.3 Synthesis — 92
2 Sesquiterpenoid Alkaloids — 93
3 Alkaloids of the Lythraceae — 95
References — 97

Chapter 7 Quinoline and Acridone Alkaloids — 99
By M.F. Grundon

1 Quinoline Alkaloids — 99
  1.1 Occurrence — 99
  1.2 Non-terpenoid Quinolines — 101
  1.3 Prenylquinolinones and Hemiterpenoid Tricyclic Alkaloids — 102
  1.4 Furoquinoline Alkaloids — 109
  1.5 Dimeric Quinolinone Alkaloids — 112
2 Acridone Alkaloids — 114
  2.1 Occurrence — 114
  2.2 New Alkaloids — 116
References — 119

Chapter 8 β-Phenylethylamines and the Isoquinoline Alkaloids — 122
By K.W. Bentley

1 β-Phenylethylamines — 122
2 Isoquinolines — 122
3 Benzylisoquinolines — 123
4 Bisbenzylisoquinolines — 125
5 Pavines and Isopavines — 128
6 Analogues of Cularine — 129
7 Berberines and Tetrahydroberberines — 129
8 Secoberberines — 134
9 Protopines — 135
10 Phthalideisoquinolines — 135
11 Spiro-benzylisoquinolines — 138
12 Indanobenzazepines — 139
13 Rhoeadines — 142
14 Emetine and Related Alkaloids — 143

15 Morphine Alkaloids   144
  16 Benzophenanthridines   155
  17 Colchicine and Related Bases   157
     References   160

Chapter 9 Aporphinoid Alkaloids   172
     By M. Shamma and H. Guinaudeau
  1 Proaporphines   172
  2 Aporphines   172
  3 Dimeric Aporphines   179
  4 Oxoaporphines   180
  5 4,5-Dioxoaporphines   181
  6 Phenanthrenes   182
  7 Aristolochic Acids and Aristolactams   182
     References   184

Chapter 10 Amaryllidaceae Alkaloids   187
     By M.F. Grundon
  1 Isolation and Structural Studies   187
  2 Synthesis   189
     References   195

Chapter 11 *Erythrina* and Related Alkaloids   196
     By A.S. Chawla and A.H. Jackson
  1 Isolation and Structure Determination   196
  2 Synthesis   198
     References   204

Chapter 12 Indole Alkaloids   205
     By J.E. Saxton
  1 Introduction   205
  2 Simple Alkaloids   205
     2.1 Non-tryptamines   205
     2.2 Non-isoprenoid Tryptamines   207
  3 Isoprenoid Tryptamine and Tryptophan Derivatives   211
     3.1 Non-terpenoid Alkaloids   211
     3.2 Ergot Alkaloids   214
  4 Monoterpenoid Alkaloids   220
     4.1 Alkaloids of *Aristotelia* and *Borreria* Species   220
     4.2 Corynantheine-Heteroyohimbine-Yohimbine Group, and Related Oxindoles   221
     4.3 Sarpagine-Ajmaline-Picraline Group   237

4.4 Strychnine-Akuammicine-Ellipticine Group 240
    4.5 Aspidospermine-Aspidofractine-Eburnamine Group 244
    4.6 Catharanthine-Ibogamine-Cleavamine Group 254
  5 Bisindole Alkaloids 259
  6 Biogenetically Related Quinoline Alkaloids 266
    6.1 *Cinchona* Group 266
    6.2 Camptothecin 267
    References 269

Chapter 13 *Lycopodium* Alkaloids   277
    By W.A. Ayer
    References 280

Chapter 14 Diterpenoid Alkaloids   281
    By S.W. Pelletier and S.W. Page
  1 Introduction 281
  2 Structural Elucidations and General Studies 282
    Configuration of the C(1)-Oxygen Function of the Lycoctonine Alkaloids 282
    High-Performance Liquid Chromatographic Methods for the Determination of Aconitine Alkaloids 285
    Alkaloids of *Aconitum carmichaeli* 285
    Alkaloids of *Aconitum crassicaule* 286
    Alkaloids of *Aconitum delavayi* 289
    Alkaloids of *Aconitum episcopale* Levl. 289
    Alkaloids of *Aconitum finetianum* Hand-Mazz. 289
    Alkaloids of *Aconitum flavum* Hand-Mazz. 292
    Alkaloids of *Aconitum japonicum* Thumb. 292
    Alkaloids of *Aconitum jinyangense* W.T. Wang 292
    Alkaloids of *Aconitum karakolicum* Rapcs. 292
    Alkaloids of *Aconitum koreanum* [*coreanum*] 292
    Alkaloids of *Aconitum leucostomum* 294
    Alkaloids of *Aconitum monticola* Steinb. 294
    Alkaloids of *Aconitum nagarum* 294
    Alkaloids of *Aconitum paniculatum* Lam. 296
    Alkaloids of *Aconitum pendulum* 296
    Alkaloids of *Aconitum saposhnikovii* B. Fedtsch. 298
    Alkaloids of *Aconitum sinomontanum* Nakai 298
    Alkaloids of *Aconitum subcuneatum* Nakai 298
    Alkaloids of *Aconitum talassicum* M. Pop. 298
    Alkaloids of *Aconitum yesoense* Nakai 298
    Alkaloids of *Atragene sibirica* L. 300

　　　　　Alkaloids of *Consolida orientalis* 300
　　　　　Alkaloids of *Delphinium cardiopetalum* DC. 301
　　　　　Alkaloids of *Delphinium dictyocarpum* DC. 301
　　　3 Chemical Studies 302
　　　　　Transformation Products from Lycoctonine 302
　　　　　The Origin of Oxonitine 304
　　　　　Diterpenoid Alkaloid Synthetic Studies 305
　　　　　References 306

Chapter 15 Steroidal Alkaloids　　　　　　　　　　　　309
　　　　　By D.M. Harrison
　　　1 Alkaloids of the Apocynaceae 309
　　　2 *Buxus* Alkaloids 311
　　　3 *Solanum* Alkaloids 312
　　　4 *Fritillaria* and *Veratrum* Alkaloids 314
　　　5 Miscellaneous Steroidal Alkaloids 319
　　　　　References 319

Chapter 16 Miscellaneous Alkaloids　　　　　　　　　　322
　　　　　By J.R. Lewis
　　　1 Muscarine Alkaloids 322
　　　2 Imidazole Alkaloids 322
　　　3 Oxazole and Isoxazole Alkaloids 323
　　　4 Peptide Alkaloids 324
　　　5 Alkaloid-containing Sources and Unclassified
　　　　　Alkaloids 329
　　　　　5.1 *Amanita muscaria* 329
　　　　　5.2 *Coix lachryma-jobi* 329
　　　　　5.3 *Dulacia guianensis* 329
　　　　　5.4 *Streptomyces aureofaciens* 329
　　　　　5.5 *Zanthoxylum arborescens* 330
　　　　　References 330

1
Biosynthesis

BY R. B. HERBERT

Continuity with previous Reports in this series is maintained. Background information for new work appearing here is, as usual, obtainable through earlier Reports to which reference is given; two comprehensive reviews are also cited.[1,2]

1 Pyrrolidine and Piperidine Alkaloids

1.1 Nicotine. — The biosynthesis of nicotine (6) is well established[1,2] to be from ornithine (1), sequentially through putrescine (2), its N-methyl derivative (3), and (4) (cf. Vol. 12, p.1; Vol. 11, p.1). New results very usefully allow deduction of

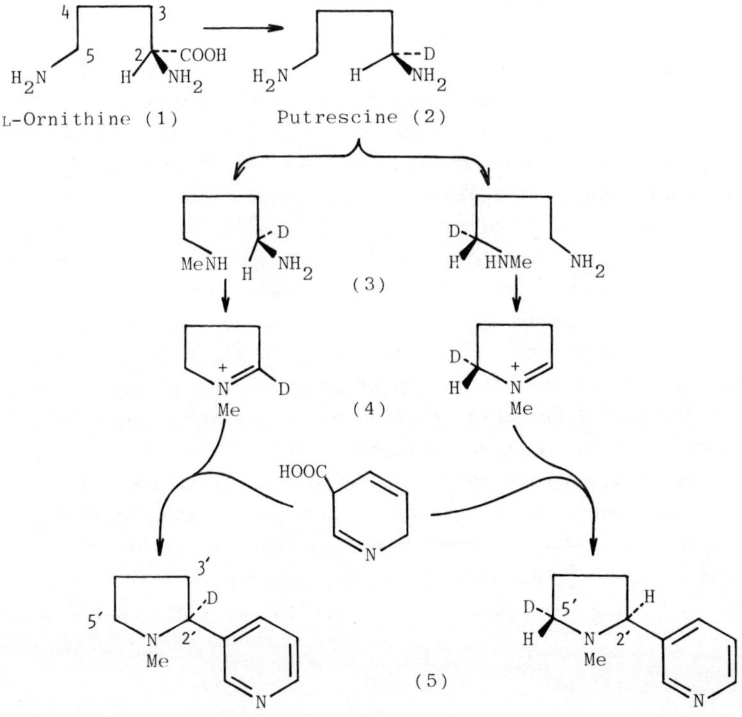

Scheme 1

stereochemistry involved in each of the biosynthetic steps.³

(R)-[1-²H]Putrescine (2) was well incorporated into nicotine (6) in tobacco plants. The nicotine [see (5)] showed deuterium n.m.r. signals, of similar height, corresponding to the 2'-proton and the 5'-pro-R proton in (6). If the putrescine labels are traced through Scheme 1, the reader will see that this observation accords with stereospecific removal of the pro-S proton from a -CH$_2$NH$_2$ group of putrescine (7) in the oxidation of (3). [Removal of the 1-pro-R hydrogen atom of putrescine {see (7)} would have given nicotine labelled only at C-5']. This stereochemistry is the same as that found for other reactions catalysed by diamine oxidase (cf. Vol. 10, p.9).

Levels of tritium retention in nicotine (6) using DL- and L-[(R S)-5-³H]ornithine samples as precursors showed that L-ornithine (1) rather than the D-isomer is the preferred precursor. Tritium retention was measured relative to DL-[5-¹⁴C]-ornithine administered at the same time. (For earlier application of the method used here, see Vol. 5, p.7.)

DL-[2-³H, 5-¹⁴C]Ornithine gave nicotine with loss of half of the tritium present in the precursor. This is consistent with decarboxylation of (1) occuring with retention of tritium and subsequent loss of half of it during the conversion of (3) into (4). Since the result with the [²H]putrescine is that the oxidation of (3) results in loss of the 1-pro-S proton of putrescine (7), the putrescine derived from L-[2-³H]ornithine (the usable part of the DL-precursor) must have had tritium in the (S)-configuration. Therefore decarboxylation of L-ornithine proceeds with retention of configuration, as is the case with bacterial ornithine decarboxylase.[4,5] Indeed all the amino-acid decarboxylases so far studied catalyse decarboxylation in the same stereochemical sense (see ref. 6 and refs. cited in ref. 5).

The stereochemistry of the last step in nicotine biosynthesis (Scheme 1) follows from the known stereochemistry of the alkaloid.

It has been found that treating callus of Nicotiana tabacum with urea leads to an increase in the level of nicotine production.[7] By correlation with this, the content of ornithine, citrulline, and arginine, which are urea cycle intermediates and nicotine precursors, was higher in treated callus than in untreated callus.

Good evidence[8] has been obtained that N'-isopropylnornicotine is produced during air-curing of tobacco leaves and is not formed in

Biosynthesis

intact plants. It was deduced to be formed from nornicotine.

1.2 Cocaine and Cuscohygrine. — Preliminary results[9] (cf. Vol. 12, p.3), which showed that cocaine (8) is derived in part from ornithine (1) in Erythroxylon coca, have been published in full.[10] New, and most interesting, information is that the label from DL-[5-^{14}C]ornithine appeared equally divided between the two bridgehead carbon atoms (C-1 and C-5) in cocaine (8). This indicates that this alkaloid, in contrast to the structurally similar tropane alkaloids,[1,2] is derived from ornithine by way of a *symmetrical* intermediate (putrescine). It is interesting to note that [5-^{14}C]ornithine incorporation into cuscohygrine (9) in E. coca was also by way of a symmetrical intermediate,[9,10] by contrast with the biosynthesis of cuscohygrine in other plants which does not involve any symmetrical intermediate[1,2] (cf. Vol. 12, p.3). The key intermediate in the biosynthesis of these pyrrolidine alkaloids is (4). It appears quite simply that its biosynthesis from ornithine in some plants (e.g. E. coca and Nicotiana species) is via putrescine (7); in others it is not.

1.3 β-Pyrazol-1-ylalanine. — The biosynthesis of β-pyrazol-1-yl--L-alanine (10), which contains an unusual N-N linkage, has been studied.[11] The clear evidence, using whole plants and cell-free extracts of cucumber (<u>Cucumis sativus</u>), is that 1,3-diaminopropane is a precursor for the heterocyclic ring of (10). Pyrazole can also act as a precursor;[11] it is enzymically condensed with O-acetylserine to give (10).[12]

1.4 Pyrrolizidine Alkaloids. —

The origins of the pyrrolizidine ring system [as (11)] which is found in these alkaloids has been receiving recent, well-merited attention (<u>cf</u>. Vol. 12, p.4; Vol. 11, p.2; Vol. 10, p.13). Work relating to the incorporation principally of [^{14}C]ornithine and [^{14}C]putrescine, which was previously published in preliminary form,[13] (<u>cf</u>. Vol. 10, p.13) is now available in a full paper.[14]

A careful analysis of the difficulties of unravelling pyrrolizidine alkaloid biosynthesis without ambiguity is contained in a full paper[15] which is now available from one of two groups to use putrescine as a precursor labelled with ^{13}C and ^{15}N (<u>cf</u>. Vol. 12, p.5). Additional results, which relate to the incorporation of radioactive ornithine, putrescine, and spermidine, support the most recent results; Δ^1-pyrroline, a possible alkaloid precursor, was not incorporated.

There is a cautionary tale relating to the use of mixed ^{3}H and ^{14}C labels.[15] It was found with some samples of retronecine (11), where tritium was on carbon next to nitrogen, that there could be enough of a difference in pK_a compared to when protium was present for partial separation of ^{14}C-labelled and ^{3}H-labelled species to occur on chromatography, with consequent disastrous change in isotope ratio.

1.5 Anabasine. —

The specific incorporation of lysine (13) into the piperidine ring of anabasine (12) has been re-examined,[16] with confirmation of earlier results.[1,2] A mixture of DL-[4,5-^{13}C$_2$]- and DL-[6-^{14}C]-lysine was used as precursor. The ^{13}C n.m.r. spectrum of the derived anabasine (12) showed satellites for C-4' and C-5' due to the presence of the two contiguous ^{13}C lysine labels in the alkaloid. This neatly confirms the previously deduced orientation of the lysine skeleton in the piperidine ring of (12). (For other applications of the approach used here, see Vol. 12, p.1; Vol. 11, pp.1 and 19.)

Degradation of the anabasine gave results showing that over 98% of the [6-^{14}C]lysine label was located at C-6'. This confirms

Biosynthesis

that lysine is incorporated into (12) without the intervention of any symmetrical intermediate. This excludes cadaverine (17) as an intermediate formed from lysine (13). In the above experiment, inactive cadaverine was added during isolation. When reisolated it was found to be essentially devoid of radioactivity, which indicates that cadaverine was not formed from the radioactive lysine fed. (For further discussion, see Section 1.6 below).

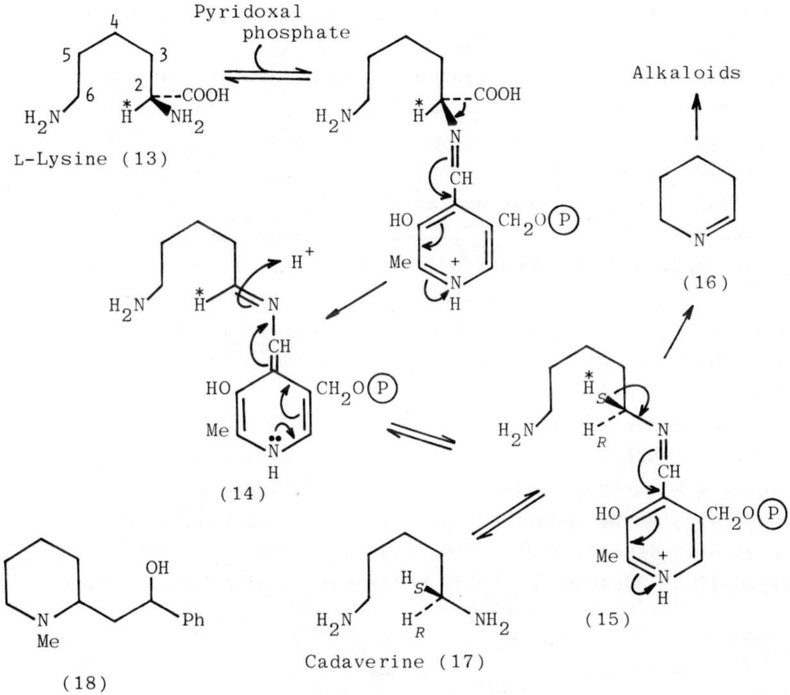

Scheme 2

1.6 The Early Stages of Alkaloid Biosynthesis. — It is well established that L-lysine (13) is incorporated into some piperidine alkaloids by way of a symmetrical intermediate; it is accepted that this symmetrical intermediate is cadaverine (17), which is also an alkaloid precursor. Lysine, however, is incorporated into other alkaloids, e.g. anabasine (12) (Section 1.5) and sedamine (18), without the intervention of any symmetrical intermediate. Cadaverine (17), although able to act as an alkaloid precursor, cannot be an intermediate formed from the lysine fed, because of its symmetry (cf. Vol. 10, p.9; refs. 1 and 2).

An ingenious model has been developed which accounts for the biosynthesis of all piperidine alkaloids (cf. Vol. 10, p.9). The key idea is that those alkaloids which are formed with symmetrization of a lysine label are biosynthesized by way of free cadaverine (17); those which are formed without symmetrization of label are biosynthesized by way of cadaverine, which remains unsymmetrical by being (co)enzyme-bound [as (15)] until oxidation occurs to give (16) (Scheme 2).

The decarboxylation by lysine decarboxylase of L-lysine (13) to give cadaverine (17) occurs with retention of configuration [protonation occurs on the α-face of the imine (14)]. The oxidation of cadaverine (17) occurs with loss of the 1-pro-S proton, which is the proton originally sited at C-2 in L-lysine (13) (cf. Section 1.1). It follows that L-[2-^{3}H]lysine should give alkaloids such as sedamine (18) with loss of the tritium label. But such a label is known to be retained on formation of sedamine (18).[17] This observation is difficult to reconcile with the model shown in Scheme 2 (cf. Vol. 10, p.9) but may be accommodated in a modified version.

If instead of protonation of the imine function in (14), in the lysine decarboxylase reaction, nucleophilic attack by the δ-amino-group of lysine [see (19)] occurs, (16) is obtained directly and independently of cadaverine, and without loss of the C-2 proton of lysine. This "modified" decarboxylase would function, it is suggested, for the biosynthesis of alkaloids such

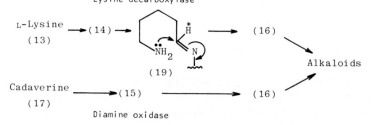

Scheme 3

Scheme 4

as sedamine (18), see Scheme 3; lysine and cadaverine serve, on this model, as essentially independent alkaloid precursors (cf. Section 1.5 for supporting evidence). For those alkaloids, which are biosynthesized from lysine via cadaverine, the course of biosynthesis is straightforward [with normal protonation on the α-face of (14)] (Scheme 4). If the hypothesis is correct then it follows that all alkaloids formed unsymmetrically from lysine would retain the C-2 proton of the amino-acid. Those formed symmetrically would lose half of the label from a L-[2-^{3}H]lysine precursor [complete retention with the formation of (17) and loss of half the label on oxidation of (17)]. This has not yet been tested but in the analogous case of nicotine biosynthesis (cf. Schemes 1 and 4) half of the tritium from [2-^{3}H]ornithine is lost on formation of the alkaloid (Section 1.1). It may be noted that in the original hypothesis (Scheme 2) a similar fate would be predicted for the lysine C-2 proton in the biosynthesis of both groups of piperidine alkaloids.

2 Phenethylamine and Isoquinoline Alkaloids

The seminal ideas [18] relating to the oxidative coupling of phenols have found very widespread application, and yet the actual mechanism of coupling in vivo remains obscure. In new work, it has been found that extracts of Papaver somniferum which had peroxidase activity were unable to catalyse the conversion, by phenol oxidative coupling, of reticuline into salutaridine.[19]

The establishment of alkaloid—producing tissue cultures of Berberis species,[20] of Papaver bracteatum,[21] of Corydalis ophiocarpa,[22] and of P. somniferum [23] have been reported. Some alkaloid interconversions with cultures of the last two species were also reported.

All of the twenty-seven known spirobenzylisoquinoline alkaloids, e.g. (20), contain a methylenedioxy-group on ring D. It has been suggested that the formation of these groups from an ortho-methoxy-phenol is encouraged by steric compression around a C-9 methoxy-substituent which is relieved very substantially upon the formation of a methylenedioxy-group in vivo.[24]

2.1 Norlaudanosoline Synthase.

Strong evidence, including some arising from the use of enzyme preparations,[25] has been obtained by several groups of workers that isoquinoline alkaloids are formed from a phenethylamine [as (21)] and an α-keto-acid [as (22)]; condensation affords the amino-acid [as (24)], which upon decarboxylation and reduction gives a typical isoquinoline base, exemplified by norlaudanosoline (25)[1,2] (cf. Vol. 2, p.10; Vol. 6, p.17; Vol. 7, p.10; Vol. 9, p.8; Vol. 10, p.15). An analogous pathway has been established for β-carboline alkaloids (this Report, Section 3.1). New evidence from another reputable group of workers is a stark contradiction.[26]

An enzyme that synthesises norlaudanosoline (25) has been isolated, and purified, from several plant species which normally produce isoquinoline alkaloids. Substrates for the enzyme were dopamine (21) and, most surprisingly, 3,4-dihydroxyphenylacetaldehyde (23), and not 3,4-dihydroxyphenylpyruvic acid (22). 4-Hydroxyphenylacetaldehyde was a substrate for the enzyme but not 4-hydroxyphenylpyruvic acid or phenylpyruvic acid. The product of the clearly enzyme-catalysed reaction between (21) and (23) was norlaudanosoline (25) [predominantly the (S)-isomer]. No doubt the question of the normal intermediacy of (22) versus (23) in isoquinoline biosynthesis will receive urgent attention. In particular, it would be useful to discover if amino-acid precursors [as (24)] are required to have a particular chirality. If so, one could conclude that their utilization in biosynthesis is by a normal enzyme-catalysed reaction.

2.2 Hordenine and Normacromerine. — The metabolism of hordenine (26) in Hordeum vulgare plants has been studied.[27] The alkaloid is ultimately degraded to C_6-C_1 intermediates that are incorporated into polymeric material.

Further information[28] on the biosynthesis of normacromerine (27) (cf. Vol. 11, p.8; Vol. 10, p.15; Vol. 9, p.7) is that it can be formed in Coryphantha macromeris from normetanephrine (28), which is in turn formed from norepinephrine (29). Normetanephrine was shown to be a natural constituent of the cactus. Octopamine (30) was a poor normacromerine precursor.

2.3 Hasubanonine, Protostephanine, and Laurifinine. — Full papers on the biosynthesis of hasubanonine (31) and protostephanine (32), which are most interesting benzylisoquinoline variants, have been published:[29] an epic piece of research. (Preliminary accounts[30] were reviewed in Vol. 8, p.8; Vol. 6, p.26).

Protostephanine (32) and laurifinine (33) have closely related structures, and the latter has been shown to derive from (+)-N-norprotosinomenine (34) in Cocculus laurifolius.[31] [The biosynthesis of (32) stands in marked contrast.] The precursor (34) and norlaudanosoline (25) were utilized for biosynthesis, but not three other isoquinolines with methylation patterns different from those of (34). (±)-N-Norprotosinomenine [as (34)] was incorporated without loss of its O-methyl groups or the proton at C-1 (that is, of the (+)-isomer, the (−)-isomer not being utilized).

C. laurifolius also produces coccuvine (35), an alkaloid of the Erythrina type. The biosynthesis of this alkaloid had previously been deduced to proceed from (34) by way of the N-desmethyl derivative of laurifinine (33)[32] (cf. Vol. 11, p.14). The new results are complementary then to the old ones.

2.4 *Aporphine Alkaloids.* — The biosynthesis of N-methylcrotsparine (36), N-methylcrotsparinine (38), and N-methylsparsiflorine (37) in Croton sparsiflorus has been studied.[33] The key precursor is N-methylcoclaurine (39), each enantiomer serving specifically as a precursor for either (36) and (37), on the one hand, or (38) on

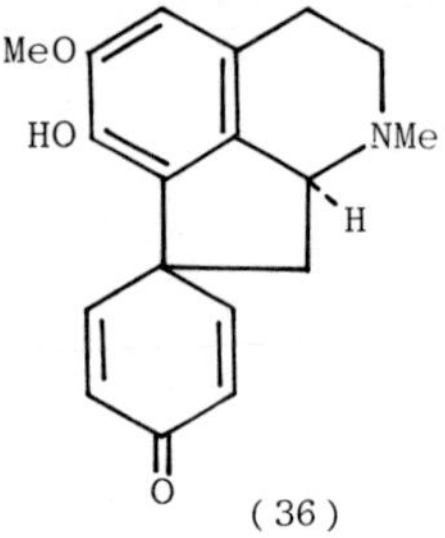

(36)

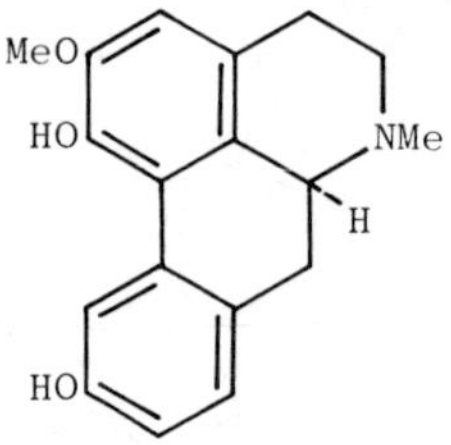

(37)

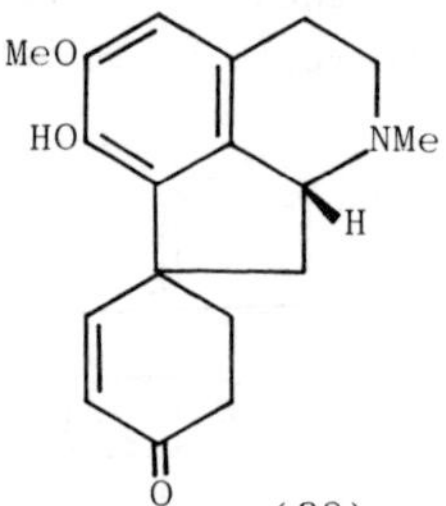

(38)

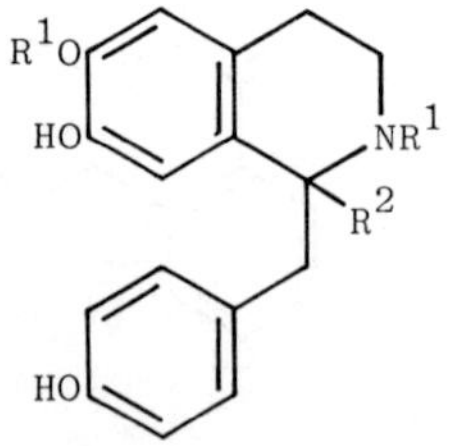

(39) R^1 = Me, R^2 = H
(40) R^1 = H, R^2 = COOH

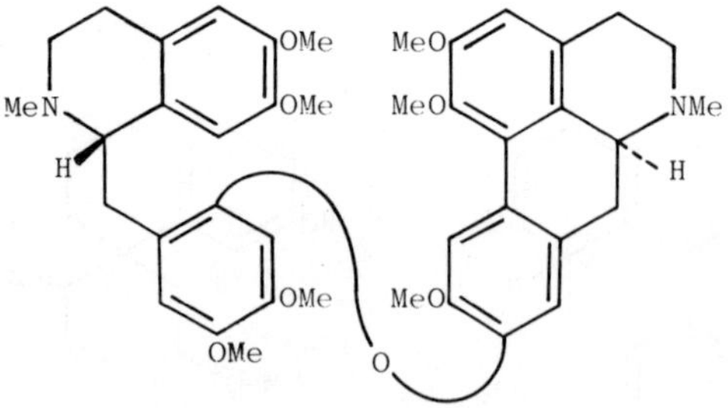

(41)

the other. The specific incorporation of (40), dopamine, and 4-hydroxyphenylpyruvic acid is to be noted. The results are entirely complementary to others obtained with this plant (cf. Vol. 6, p.19; Vol. 11, p.10).

2.5 Bisbenzylisoquinoline Alkaloids. — The biosynthesis of a number of bisbenzylisoquinoline alkaloids has been investigated (cf. Vol. 12, p.11; Vol. 10, p.16; Vol. 9, p.11). One of these investigations concerned tiliacorine and tiliacorinine. The results, previously published in preliminary form,[34] and reviewed in Vol. 9, are now available in a full paper [35] (with the loss of two coworkers).

(42) R = Me
(43) R = H

(44) R^1 = Me, R^2 = H
(45) R^1 = H, R^2 = Me

(46)

Preliminary results using Thalictrum minus which showed that thalicarpine (41) was formed from two molecules of reticuline (42)[36] (cf. Vol. 12, p.13) have been included in a full paper.[37] The results are supported by those of other workers obtained with Cocculus laurifolius.[38] Of several related isoquinolines tested, reticuline (42) was the best precursor; norreticuline (43) was also satisfactorily incorporated,[37,38] and it is specifically the (S)-isomer of reticuline which is used in biosynthesis.[38] The expectation that the aporphine moiety is formed before elaboration of the bisbenzylisoquinoline skeleton is supported by the incorporation into (41) of tritiated isoboldine (44),[37,38] and, at a lower level, boldine (45).[38] Both isoboldine (44) and reticuline (42) were found to be present in T. minus.[37] However, norreticuline (43) was transformed into (41) with loss of the 4'-O-methyl which is expected to appear in the aporphine half of (41), but with retention of the 4'-O-methyl group which should be present in the other half. It follows that a demethylation occurs at some point in the course of biosynthesis involving the aporphine half.[38] This could occur to provide a free hydroxy-group necessary for phenol oxidative coupling within a bisbenzylisoquinoline precursor. This would mean that isoboldine (44) is not a normal intermediate in the biosynthesis of thalicarpine (41).

A possible biosynthesis for some bisbenzylisoquinoline alkaloids has been advanced.[39]

2.6 Aristolochic Acid. — The pattern and specificity of the incorporation of tyrosine[40] and norlaudanosoline (25)[41] into aristolochic acid (50) demonstrate that this acid is derived by degradation of an aporphine alkaloid; the nitro-group is derived from the amino-group of tyrosine.[40] Results of further experiments establish that the aporphine intermediate is stephanine (49).[42]

Tritiated samples of stephanine (49) and (at a lower level) prestephanine (48) were incorporated into (50) in <u>Aristolochia bracteata</u>. Orientaline (47) was shown to be a specific precursor, with the (R)-isomer preferred over the (S)-isomer. Nororientaline (46) was shown to be a specific precursor but, since its level of incorporation was approximately ten times lower than that of (47),

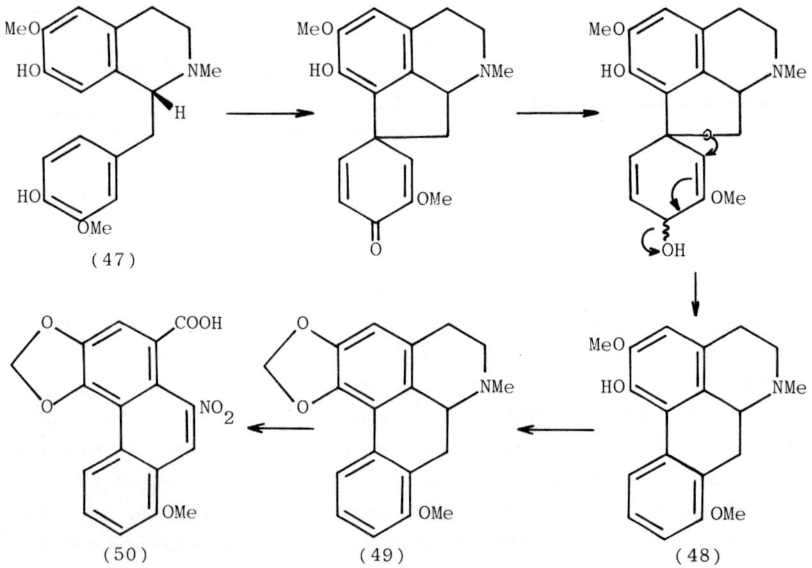

Scheme 5

the major pathway from (25) may involve N-methylation before completion of the O-methylation pattern. The precursors (46), (47), and (49) were isolated from A. bracteata in radioactive form after feeding [3-^{14}C]tyrosine, thus helping to establish their normal intermediacy in aristolochic acid (50) biosynthesis. It may be concluded that the biosynthesis of (50) takes the course shown in Scheme 5.

In accord with normal experience in experiments with isoquinoline alkaloids (cf. Vol. 11, p.12), administered dopa was found to provide only that part of (50) which was the phenethylamine moiety in (47)[42] (cf. ref. 40). The methylenedioxy-group of (50) was found to arise from a methoxy-phenol,[42] again in accord with other work.[1,2]

3 Alkaloids Derived from Tryptophan

3.1 β-Carboline Alkaloids.—

Preliminary results[43] (cf. Vol. 12, p.14) which show that the amino-acid (51) is an efficient precursor for harman (52) in Passiflora edulis and is a natural plant constituent, have been published in full.[44] The amino-acid (51) is also an efficient precursor for eleagnine (53) in Eleagnus angustifolia and is a natural constituent of this plant too.[44] Attention is drawn to the very sensitive monitoring of intact precursor incorporation obtainable with mixed ^{14}C and ^{3}H labels, which the results illustrate.

(55)

3.2 Terpenoid Indole Alkaloids.

Loganin (62) is a key early intermediate in the biosynthesis of terpenoid indole alkaloids, e.g. catharanthine (54) and vindoline (55).[1,2] It is known[1,2] to be formed from geraniol (56)/nerol (57) by way of the hydroxy-derivatives (58). The steps beyond (58) leading to loganin (62) have been probed by testing tritiated samples of the isomeric pairs (59), (60), and (61) as precursors for catharanthine (54), vindoline (55), and loganin (62) in Catharanthus roseus plants.[45] Satisfactory, and similar, incorporations were recorded for each pair, indicating that biosynthesis involves oxidation of the hydroxy-groups in (58) to aldehyde functions prior to cyclisation to give the loganin (62) skeleton. The similar levels of incorporation of the precursors meant that it was not possible to conclude whether the C-1 or the C-10 alcohol group in (58) normally undergoes oxidation first.

The conclusion (cf. Vol. 12, p.16) that hydroxyloganin (63) is probably not an intermediate in the bioconversion of loganin (62) into secologanin (64) has been supported by the results of others.[46] Both (63) and its C-6 epimer were fed to C. roseus. Negligible incorporations into vindoline (55) and catharanthine (54) were obtained.

Several carotenoid-inducers have been found to be effective in promoting the production of terpenoid indole alkaloids in C. roseus cell cultures. This possibly results from the induction of terpenoid precursors.[47] More papers have appeared concerning the production of alkaloids by C. roseus tissue cultures.[48,49]

The chemical conversion of alkaloids of the vobasine type into those of the ervatamine type can readily be effected, using a 'modified Polonovski' reaction. This involves formation of an N-oxide and its rearrangement, using trifluoroacetic anhydride as reagent. Dregamine (65) has been converted into 20-epi-ervatamine

(66) in this way. It has now been found that cytochrome-P-450-dependent mono-oxygenases from rat liver are able to carry out the conversion of (65) into (66) in the presence of NADPH and oxygen, arguably by a similar mechanism.[50] It could thus be that the 'modified Polonovski reaction' corresponds closely in mechanism in appropriate cases with normal alkaloid biosynthesis. In incubations with liver microsomes, in addition to the formation of (66), simple demethylation of (65) occured as the major reaction.

Anhydrovinblastine (70)[51] and 20'-deoxyleurosidine (71)[52] can both act as late precursors for vinblastine (72) (cf. Vol. 12, p.15; Vol. 11, p.19; Vol. 10, p.22). A hypothetical biogenetic grid has been put forward which allows for (70) and (71) to be alternative intermediates in the biosynthesis of (72) via (69).[52] Formation of (69) from the N-oxide (67) of (71), or equivalent, should result in anti-elimination and loss of the 21'α-proton. This has been examined with [21'α-^{3}H, methyl-^{14}C]anhydrovinblastine (73).[53]

The (73) was incubated with cell-free extracts of C. roseus leaves. Radioactive vinblastine (72) was isolated which had the same ^{3}H:^{14}C ratio as the precursor. This proves that anhydrovinblastine is incorporated intact into (72) and shows that tritium is not lost in this transformation. The fact that the tritium label is retained argues against (69), formed from (70), being an intermediate in vinblastine biosynthesis.

3.3 Ergot Alkaloids. — There is good evidence that ergot alkaloids, represented by elymoclavine (76), are biosynthesized by way of chanoclavine-I (74) and the aldehyde (75) (cf. ref. 2; Vol. 5, p.27). It has now been shown, with cultures of a Claviceps strain and using samples of chanoclavine-I (74) chirally tritiated at C-17, that bioconversion of (74) into (76) involves removal of the 17-pro-R and retention of the 17-pro-S proton of (74).[54] Proton removal occurs in the same stereochemical sense as for yeast and liver alcohol dehydrogenases,[55] which further supports the conclusion that (75) is an intermediate in the bioconversion of (74) into (76). More support is gained from the isolation of (75), which accumulated in a Claviceps mutant blocked in the synthesis of tetracyclic ergot alkaloids.[56]

Secondary metabolism leading to the ergot alkaloids begins with tryptophan. It has been found that the naphthylalanines

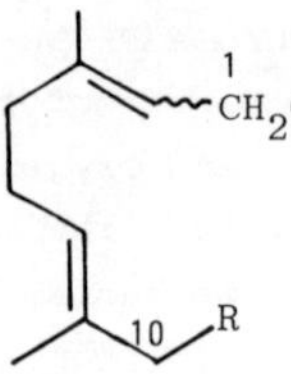

(56) R = H; *E*-isomer
(57) R = H; *Z*-isomer
(58) R = OH; *E/Z*-isomers

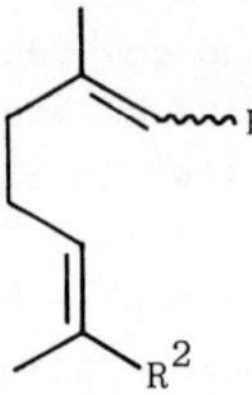

(59) R^1 = CHO, R^2 = CH_2OH; *E/Z*-isomers
(60) R^1 = CH_2OH, R^2 = CHO; *E/Z*-isomers

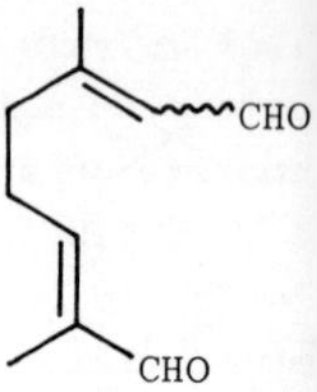

(61) *E/Z*-isomers

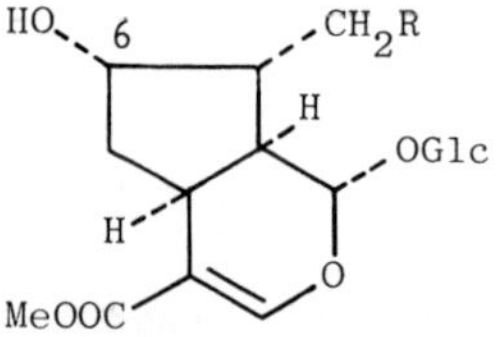

(62) R = H
(63) R = OH

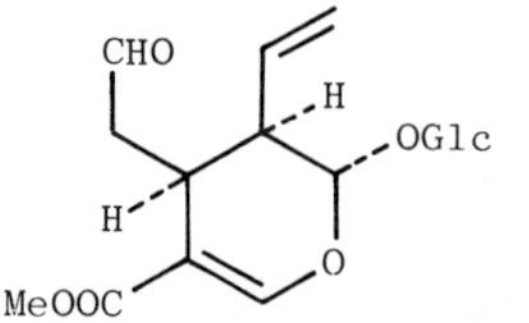

(64)

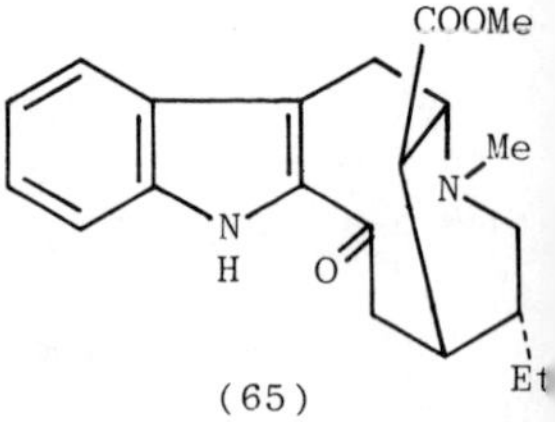

(65)

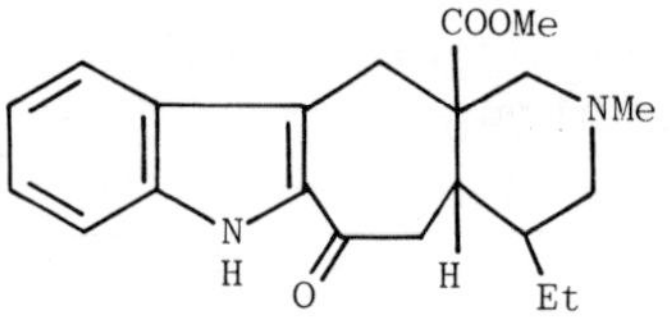

(66)

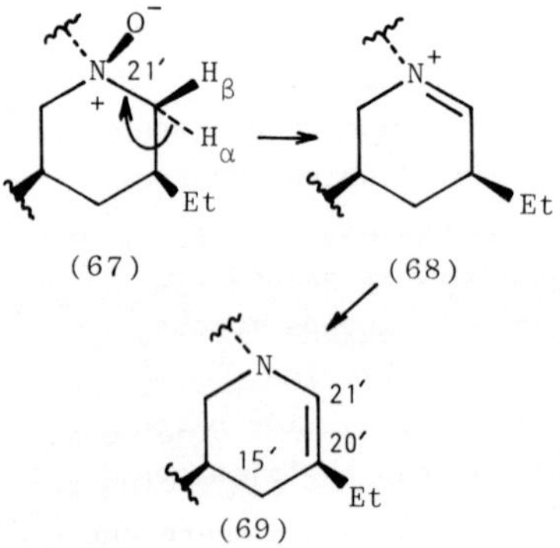

(67) (68)

(69)

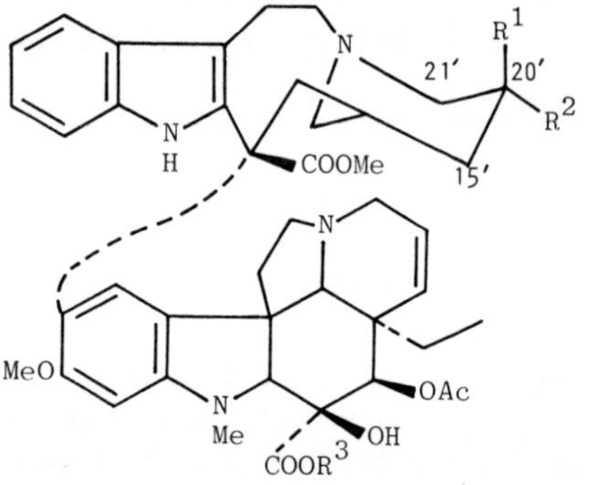

(70) R^1 = Et, $\Delta^{15'(20')}$, R^3 = Me
(71) R^1 = Et, R^2 = H, R^3 = Me
(72) R^1 = OH, R^2 = Et, R^3 = Me
(73) R^1 = Et, $\Delta^{15'(20')}$, R^3 = $^{14}CH_3$; 21'α-^{3}H

(77) and (78) are able to induce alkaloid production, presumably by mimicking tryptophan and inducing the synthesis of the first enzyme on the pathway, DMAT synthetase.[57]

Sphacelia sorghi cultures have been found to be able to utlize hydroxyproline in place of proline for the synthesis of the peptidic moiety of dihydroergosine, 9'-hydroxydihydroergosine being produced instead.[58] But, S. sorghi would not accept isoleucine or norleucine instead of leucine. (For related examples, where one amino-acid is replaceable by an analogue, see Vol. 11, p.23; Vol. 10, p.26).

3.4 Cyclopiazonic Acid.

Results pertaining to the incorporation into α-cyclopiazonic acid (79) of tryptophan chirally tritiated at C-3, previously published in a preliminary communication,[59] are now available in a pull paper[60] (cf. Vol. 6, p.30). In addition it has been shown that (2R S)-[2-^{3}H, 3-^{14}C]-tryptophan was incorporated into (79) with retention of approximately a half of the tritium.[60] It may be reasonably assumed that this loss is from the (2R)-tryptophan in the mixture fed when it is converted into the (2S)-isomer by transamination via indolepyruvic acid and the (2S)-[2-^{3}H]tryptophan in the mixture then is incorporated intact. This retention accords with other work using β-cyclopiazonic acid (80), which was incorporated into (79) without tritium loss from C-5 (= C-2 in tryptophan)[61] (cf. Vol. 6, p.30).

3.5 Streptonigrin.

A review has been published[62] on the biosynthesis of some heterocyclic compounds containing nitrogen, which have been studied using stable isotopes. A resumé of the current knowledge of the biosynthesis of streptonigrin (85) is included. Although this Streptomyces flocculus metabolite is known to derive from tryptophan, the origin of the quinoline portion has remained obscure (cf. Vol. 11, p.24; Vol. 10, p.23; Vol. 9, p.24). In new experiments, [U-^{13}C]glucose (81) has been used as a nice probe for the origins of this portion of streptonigrin.[63] (For related applications of [U-^{13}C]glucose in biosynthetic studies see Section 4.1; Vol. 12, p.24).

The glucose labelling of the tryptophan (86) half of (85) was that expected for normal biosynthesis along the shikimate pathway, and served as an internal control [see Scheme 6; glucose carbons which remain attached throughout biosynthesis are indicated with thickened bonds]. A similar biosynthesis by way of the shikimate pathway was apparent for the quinoline portion in (85), and 4-aminoanthranilic acid (82) has been singled out as a likely key intermediate. Excitingly, related intermediates are involved in the biosynthesis of other metabolites (see this Report, Section 4.1; Vol. 12, p.21). Notably, one of the amination reactions occurs here onto C-6 of shikimic acid (87) (or equivalent), i.e. the same site as in anthranilic acid (84) biosynthesis. Three molecules of erythrose 4-phosphate (83) are implicated in this

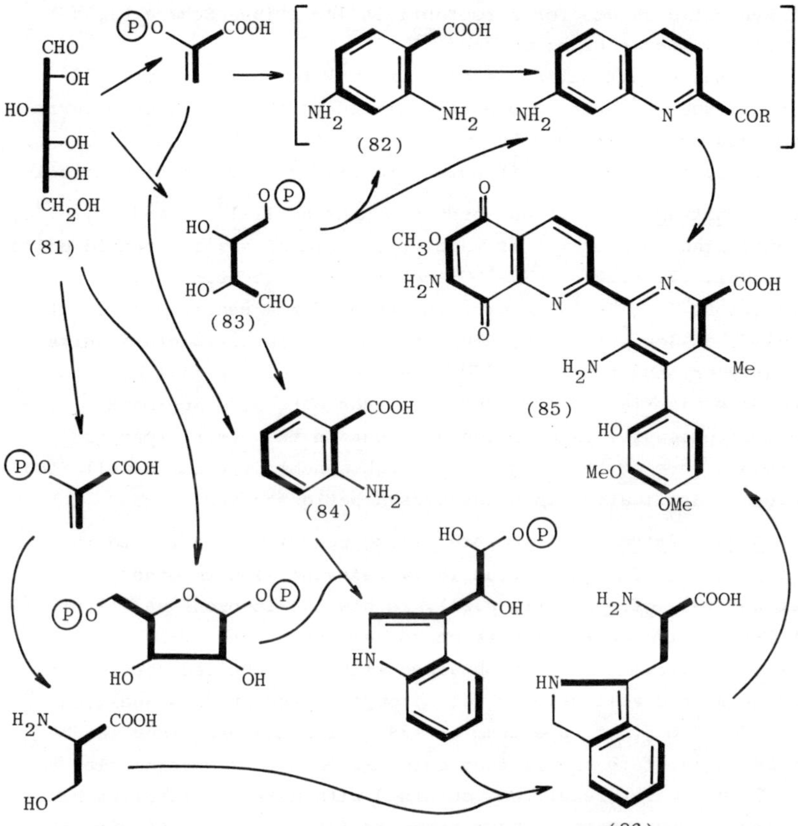

Biosynthesis

Scheme 6

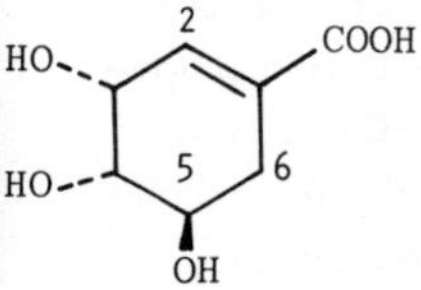

(81) (82) (83) (84) (85) (86)

(87) 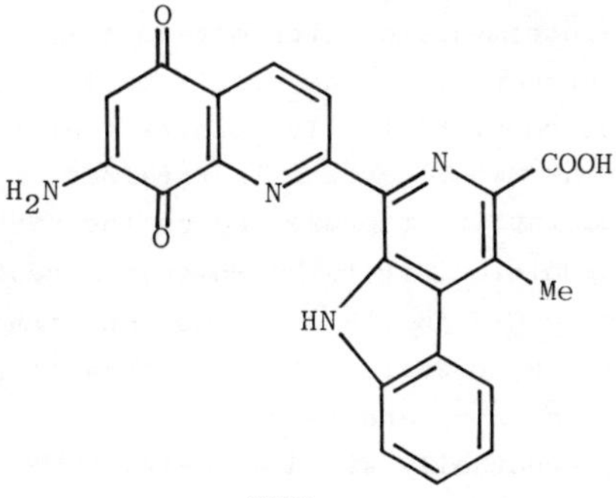 (88)

biosynthetic scheme for streptonigrin (85) (see Scheme 6), a finding which is being checked.[62]

Finally, attention has been drawn[62] to the structure (88) of lavendamycin.[64] It is clearly related to that of streptonigrin and this metabolite may be an intermediate in the biosynthesis of (85), or, at the least, it points to likely steps of biosynthesis.

3.6 Penitrem. — It has been shown, using $[1-^{13}C]$- and $[1,2-^{13}C_2]$-acetate, that penitrem A is made up in part of six isoprenoid units (cf. Vol. 12, p.20). $[2-^{13}C]$Acetate has also been used as a precursor.[65] The ^{13}C n.m.r. spectrum of the penitrem A isolated showed, at low intensity, coupling between intact acetate units, clearly arising from $[1,2-^{13}C_2]$acetate which must have been formed from the material fed; other couplings were also apparent. The sort of extensive information apparent in the n.m.r. spectrum obtained here at 125.76 MHz with high sensitivity could well find valuable application in other biosynthetic studies.

3.7 Roquefortine. — An interesting question relating to the biosynthesis of roquefortine (89), relevant also to other metabolites (cf. Vol. 11, p.20), is how the isoprene unit arrives at C-14. (For earlier work on roquefortine see: Vol. 11, p.22; Vol. 10, p.25.) Results obtained with ^{13}C-labelled acetic acid and mevalonic acid help in the attempt to answer this question.[66]

The ^{13}C n.m.r. spectrum of (89) which had been labelled by $[1,2-^{13}C_2]$acetate showed that both C-26 and C-27 were coupled to C-23; a similar result was obtained with material labelled by $[2,3-^{13}C_2]$mevalonate. This means that the regiospecificity of the label in the methyl groups of dimethylallyl pyrophosphate, formed as an intermediate after mevalonate, is lost at a later stage of biosynthesis. (The results actually show that randomization is not complete.) It appears that the distinction between isoprene methyl groups is retained in the related metabolite echinulin, represented by the part-structure (91).[67] In addition, it is known that a deuterium label at C-2 of tryptophan [= C-6 in (89)] is lost on formation of roquefortine (89) (cf. Vol. 10, p.25). It is reasonable to associate this proton loss with entry and rearrangement of the reversed isoprene unit which is eventually sited at C-14. Therefore, (91), with regiospecificity of methyl labels maintained, is a likely intermediate. Rearrangement would then occur with necessary loss of

this regiospecificity to give (92). It is suggested that (91) is
formed either by direct alkylation or by an aza-Claisen-type
rearrangement on (90) (cf., however, Vol. 11, p.20).

As a result of turns through the Krebs' cycle, the tryptophan
part of roquefortine was found[66] to be labelled to some extent,
and in a predictable way, by [1,2-$^{13}C_2$]acetate.

The effect of L-tryptophan, L-histidine, and DL-mevalonic
acid on the production of roquefortine and ergot alkaloids by
P. roqueforti has been studied.[68]

3.8 Hinnuliquinone. — This quinone (93) is produced by the
fungus Nodulisporium hinnuleum. Reasonable incorporations of
[U-^{3}H]tryptophan and [2-^{14}C]mevalonate give support to the expected
origins of this metabolite (the specificity of the labelling was
not established).[69]

4 Miscellaneous

4.1 Geldanomycin, Rifamycin, and Antibiotic A23187. — It is now clear that 3-amino-5-hydroxybenzoic acid (103) is the long-sought, key C_7N intermediate involved in the biosynthesis of rifamycins and mitomycins. On the other hand, the C_7N intermediate involved in the biosynthesis of the 3-aminoacetophenone residue in pactamycin (101) is 3-aminobenzoic acid (100) (cf. Vol. 12, p.21).

Geldanomycin (94)

The C_7N unit found in each of these metabolites has its origins in the shikimate pathway. The intermediate on this pathway from which diversion occurs, leading to the appropriate C_7N intermediate, is believed to be earlier than shikimic acid (87) itself and is possibly 3-dehydroquinic acid (98) or DAHP (97) (cf. Vol. 11, p.28; Vol. 9, p.34).

The incorporation of $[U-^{13}C]$glucose into geldanomycin (94) in Streptomyces hygroscopicus has been examined.[70] (For other, similar applications of $[U-^{13}C]$glucose as a biosynthetic probe, see Section 3.5; Vol. 12, p.24). Because of the low enrichments obtained at first, ^{13}C-depleted glucose (99.9% ^{12}C) was used as a carbon source in the production medium on which the S. hygroscopicus was grown; this glucose was also used as a diluent when the $[U-^{13}C]$glucose was administered. The ^{13}C n.m.r. spectrum of the derived geldanomycin clearly showed coupled signals, which arose from the label in the precursor, set against a low background of natural-abundance ^{13}C singlets. It was clear that C-15, C-16, and C-21 of the benzoquinone ring of (94) constitute an intact C_3 unit derived from glucose and C-17 through C-20 an intact C_4 unit.

Biosynthesis

This corresponds, respectively, to biosynthesis from a unit of phosphoenolpyruvate (96) and a unit of erythrose 4-phosphate (95) along the shikimate pathway. Most importantly, the "meta"-amino-group in (94) has become attached to a different carbon atom of a shikimate pathway precursor than in the biosynthesis of pactamycin (101)[71] (cf. Vol. 12, p.24). It may be noted that it had earlier been deduced that the C_7N intermediate(s) leading to the mitomycins and rifamycins bears an amino-group on the same carbon atom as in geldanomycin (cf. Vol. 11, p.28; Vol. 6, p.45). 3-Amino-5-hydroxybenzoic acid (103), a known C_7N intermediate in the biosynthesis of mitomycins and rifamycins, is thus probably an intermediate in geldanomycin biosynthesis too. From the foregoing it is clear that 3-aminobenzoic acid (100) is biosynthesized along a different pathway to (103) (see Scheme 7). In support, 3-aminobenzoic acid (100) was found not to be a precursor for geldanomycin

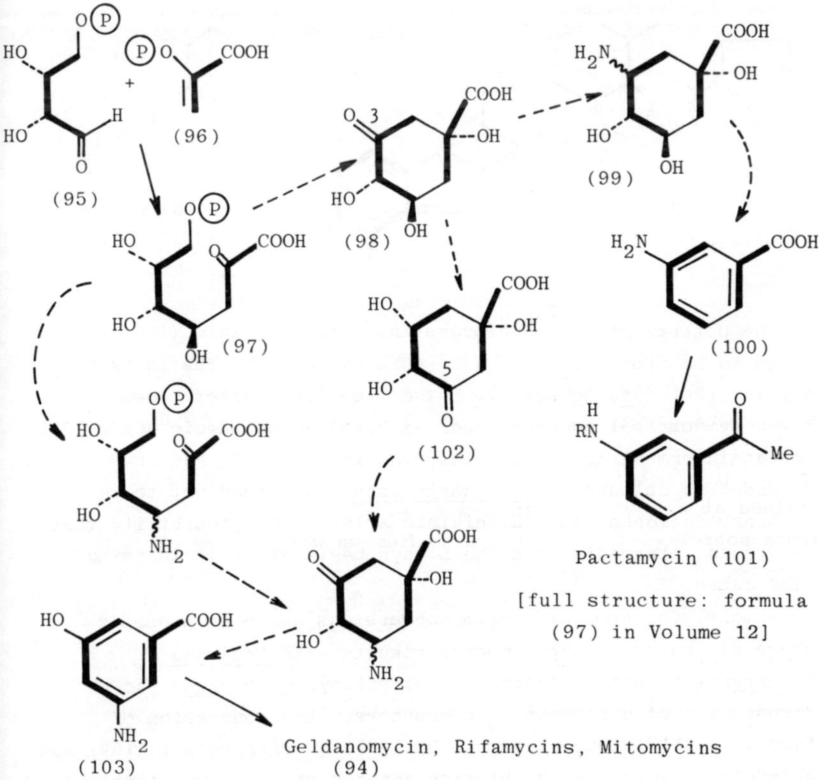

Pactamycin (101)

[full structure: formula (97) in Volume 12]

Geldanomycin, Rifamycins, Mitomycins (94)

Scheme 7

(94).[70] The combined results obtained may be summarized as shown in Scheme 7 (cf. the biosynthesis of streptonigrin, Scheme 6).

The biosynthesis of antibiotic A23187 (104), produced by Streptomyces chartreusis, has been studied with [U-^{13}C]glucose[72] and metabolites of the shikimate pathway.[73] Unlike the secondary metabolites discussed above, shikimic acid (87) was deduced to act as a precursor for the C_7N_2 unit of (104). Neither anthranilic acid nor tryptophan was incorporated. Therefore the C_7N_2 unit in (104) is formed from an intermediate on the shikimate pathway lying after shikimic acid but before anthranilic acid (84). It may be noted that one of the nitrogen atoms in this unit in (104) is located on the same shikimate carbon atom as it is in anthranilic acid (84). (See above for similar conclusions on streptonigrin biosynthesis.)

(104)

(105) R = H
(106) R = OH

The pattern of [U-^{13}C]glucose incorporation into (104) was deduced to be from a C_3 plus a C_4 unit as shown by the thickened bonds in (104) (cf. Scheme 7). The labelling pattern does not allow a symmetrical compound such as 2,6-diaminobenzoic acid (105) to be an intermediate in the biosynthesis of (104), and indeed (105), when added to cultures of S. chartreusis, was found not to reduce the incorporation of [U-^{14}C]shikimic acid. The possibility that (106) is an intermediate in the biosynthesis of (104) is being investigated.

A model for some rifamycin conversions has been proposed[74] (Scheme 8), based on experimental results with Nocardia mediterranei mutants, conversions of rifamycin in vivo, and the incorporation of different C_3 precursors; the conversion of rifamycin S (107) into rifamycin B (108) and rifamycin L (109) was completely inhibited by a thiamine antagonist. It was concluded

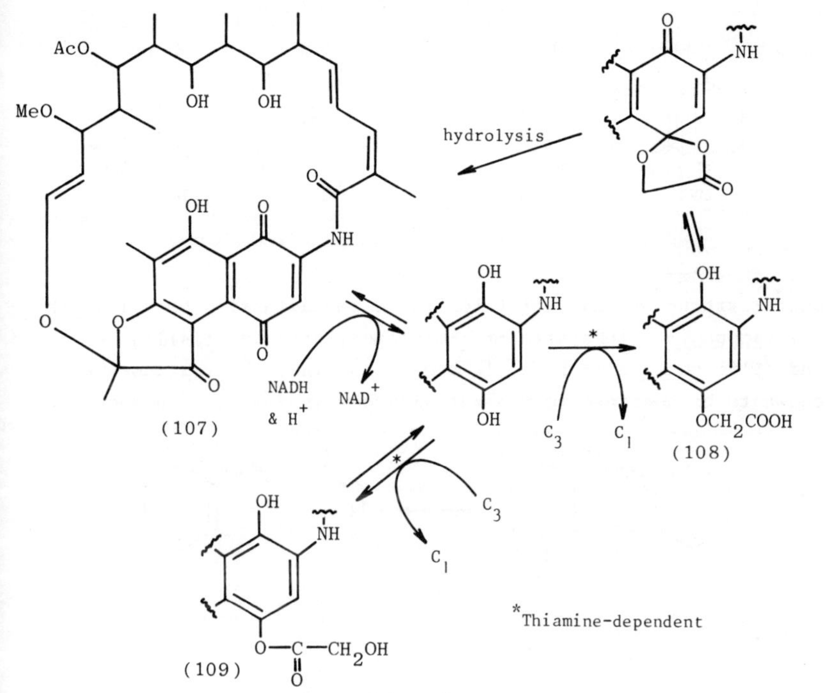

Scheme 8

that the C_3 unit involved in the biosynthesis of (108) was related to glycerol and of (109) was related to pyruvate.

4.2 Phenazines and Phenoxazinones.

A preliminary report[75] (cf. Vol. 9, p.28) concerning the incorporation of phenazine-1,6-dicarboxylic acid (113) and its methyl ester (114) into phenazine-1-carboxylic acid (115) in Pseudomonas aureofaciens, and of (113) into lomofungin (112) in Streptomyces lomodensis, has appeared in full[76] (cf. Vol. 10, p.28; Vol. 12, p.29). In addition,[76] (111) has been isolated from an extract of S. lomodensis cultures after the extract had been treated with excess diazomethane. The (111) incorporated label from radioactive (113). It is reasonable to conclude from this that (110) (or a methyl ester?) is produced from (113) as an intermediate in the biosynthesis of lomofungin (112).

Phenazine-1,6-dicarboxylic acid (113) has been found to be an apparently specific precursor for the phenazines (117) and (118) in S. luteoreticuli.[77] The dimethyl ester (114) was generally a

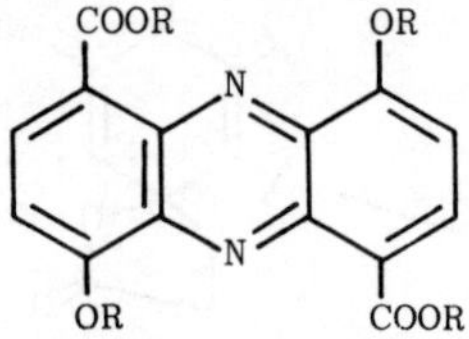

(110) R = H
(111) R = Me

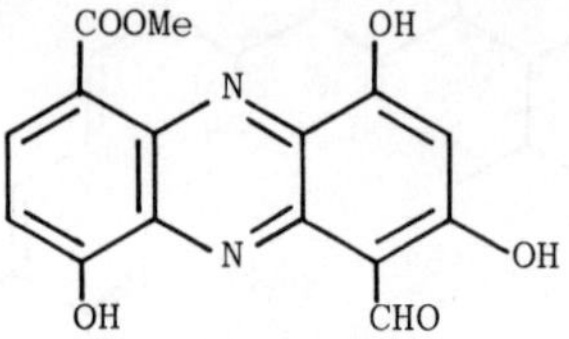

(112)

better precursor than (113) for these phenazines. Methyl [methyl-^{3}H, carbonyl-^{14}C]phenazine-1,6-dicarboxylate [as (114)] gave (117) and (118) containing both ^{14}C and ^{3}H labels; the tritium was shown to be confined to the methyl group in (117). In some

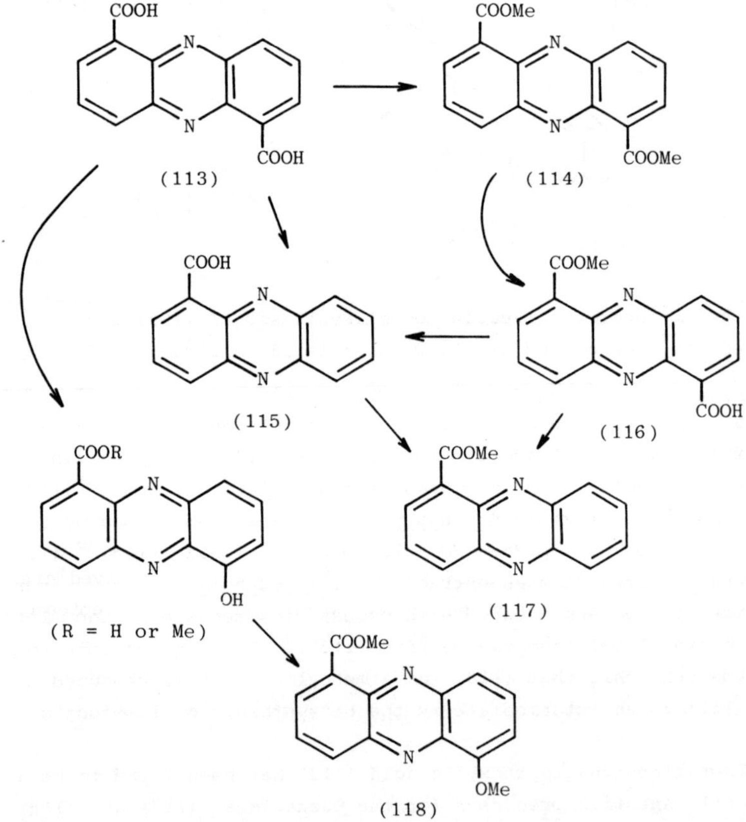

(113)

(114)

(115)

(116)

(R = H or Me)

(117)

(118)

Scheme 9

experiments the extent of tritium labelling was enhanced over that of ^{14}C. It was concluded that the methoxycarbonyl groups in (117) and (118) can arise intact from those in (114) but that methyl groups removed from (114) in the course of biosynthesis to give, e.g., (116) prior to decarboxylation to give (117) are somehow retained and introduced into endogenously synthesized (unlabelled) material with consequent enhancement of tritium relative to ^{14}C.

Phenazine-1-carboxylic acid (115) was identified as a normal intermediate in the formation of (117). It was not significantly utilized, however, for the biosynthesis of (118), which correlates with other observations.[78] The pathways deduced for phenazine biosynthesis in S. luteoreticuli are summarized in Scheme 9; some of the steps may be reversible.

Two molecules of shikimic acid (87) are used for the construction of the phenazine ring system; diversion from the shikimate pathway to phenazine biosynthesis seems to occur at shikimic acid (87) or at chorismic acid (119), or at a compound in between.[2] In order to cast light on the nature of the aminated intermediate which must be formed from one of these compounds en route to the phenazines, the source of the phenazine nitrogen atoms has been examined.[79]

(S)-[$CO^{15}NH_2$]glutamine was as efficiently incorporated as shikimic acid into iodinin (121) in Brevibacterium iodinum. It served as a significantly better source of phenazine nitrogen than did [^{15}N]ammonium sulphate, which was better utilized than (S)-[^{15}N]glutamic acid. Significant dilabelling of iodinin by the glutamine proves that both phenazine nitrogens can derive from the amide nitrogen of this amino-acid. As in phenazine biosynthesis, glutamine is the primary nitrogen source in anthranilic acid biosynthesis. It was suggested that, as in anthranilic acid biosynthesis, the substrate on which amination occurs in phenazine biosynthesis is chorismic acid (119) and the enzymes involved are similar.[79] One wonders if shikimate metabolites like streptonigrin and the ones discussed in the section immediately above will also turn out to have their nitrogen atoms derived primarily from glutamine.

A striking and consistent feature of the glutamine and ammonium sulphate incorporations into iodinin (121) was that the latter gave a statistical distribution of label between $^{15}N_1$ and $^{15}N_2$ species whereas the former gave a significantly higher than

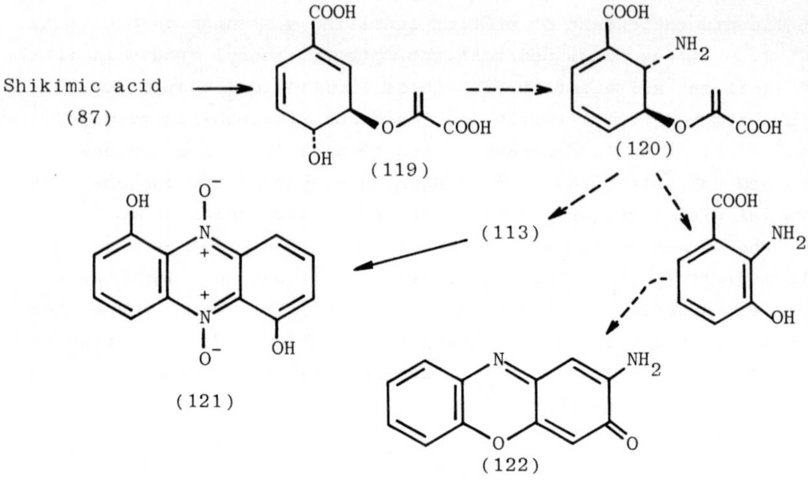

Scheme 10

statistical amount of $^{15}N_2$ compared to $^{15}N_1$ species.[79] This has been shown[80] to be due to swamping by labelled glutamine, but not ammonium sulphate, of the endogenous glutamine pool for a period after each administration of the precursor.

B. iodinum cultures produce 2-aminophenoxazinone (122) in addition to iodinin (121). Although, actinomycin (124) produced by Streptomyces antibioticus derives from tryptophan[2] (cf. Vol. 11, p.26; Vol. 6, p.42) neither this amino-acid nor anthranilic acid served as precursors for (122).[79] Shikimic acid (87), however, was shown to be a precursor for (122) and glutamine was the primary source of the nitrogen atoms. Indeed there was a close relationship between the incorporations of ^{15}N from labelled glutamine, ammonium sulphate, and glutamic acid into (121) and (122). It follows from all the evidence that the biosyntheses of (121) and (122) are closely related and these metabolites probably derive from a common intermediate. This was suggested to be (120), which could also be an intermediate in the biosynthesis of other metabolites. The biosynthesis of (122) and of phenazines was suggested to be as shown in Scheme 10.[79]

Purification has been reported[81] of phenoxazinone synthase, the enzyme which catalyses phenoxazinone formation from 4-methyl-3-hydroxyanthranilic acid (123) in the biosynthesis of actinomycin (124). Two forms of the enzyme, one of high and one of low molecular weight, were isolated. The relative amount of the

Biosynthesis

higher molecular weight protein increased with the age of the culture. Both forms consumed the same amount of oxygen per mole of phenoxazinone produced and, from antibody studies, it was apparent that they have structural features in common.

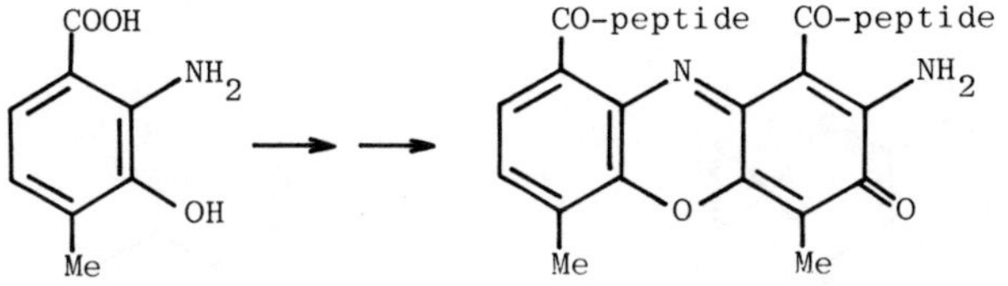

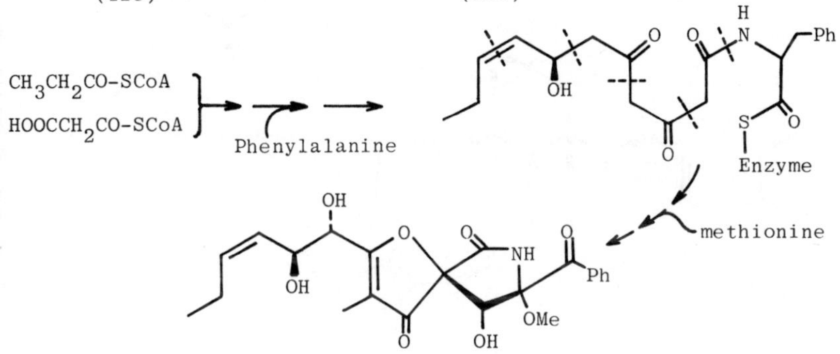

Scheme 11

4.3 Pseurotin A.

The biosynthetic origins of pseurotin A (125), which is produced by Pseudeurotium ovalis, have been determined (see Scheme 11).[82] The $\underline{C}$- and $\underline{O}$-methyl groups originate from the methyl group of methionine without loss of any of the hydrogen atoms. More than half of the deuterium in [2-^{13}C, ^{2}H$_{2}$]propionic acid used as a precursor was retained on formation of pseurotin A (125). Thus pseurotin E (126), a minor metabolite of P. ovalis, cannot be an intermediate in the biosynthesis of (125), because its formation would require complete loss of deuterium.

As a monitor for the intact utilization of phenylalanine in the biosynthesis of (125), the incorporation of [2-^{13}C, ^{15}N]phenylalanine was compared to that of [^{15}N]phenylalanine, analysis being particularly by mass spectrometry. Although extensive exchange of ^{15}N had occurred as expected by transamination, it was apparent that more than a statistical amount of ^{15}N was attached to ^{13}C in (125), i.e. transamination was not complete. This excludes nitrogen-free intermediates lying after phenylalanine in the biosynthesis of pseurotin A (125). An outline biosynthetic pathway to (125) has been suggested.

(126)

(127) R = H
(128) R = Ac

(129)

4.4 Cytochalasins.

— The cytochalasins so far examined are made up of a polyketide plus phenylalanine (cf. Vol. 7, p.28; Vol. 6, p.44). A similar origin for chaetoglobosin A (127) is apparent with tryptophan replacing phenylalanine. Results of experiments with ^{13}C- and ^{14}C-labelled precursors confirm this.[83] The metabolites, (127) and (128), are biosynthesized in Chaetomium globosum from nine and ten acetate/malonate units, respectively, one unit of tryptophan, and three C_1 units derived from methionine (C-12 and the methyl groups at C-18 and C-16).

$[2-^{13}C, ^2H_3]$Acetate was incorporated into (128) with retention of significant amounts of deuterium only in the starter acetate unit (C-11) (cf. Vol. 12, p.31 for similar observations with other cytochalasins) and the O-acetyl group. On the other hand, the C_1-units derived from methionine each retained all three of the protons from the precursor, as expected. Results obtained with (2S)-[1-^{14}C]-, (2S)-[3-^{14}C]-, and (2RS)-[3-^{14}C]tryptophan, each of which was fed in a mixture with (2S)-[5'-^3H]tryptophan, show that tryptophan is incorporated intact and that naturally predominant (2S)-tryptophan is the preferred precursor for (128). Results of further experiments show that (2S)-tryptophan is utilized without loss of the proton or the amino-group at C-2.

4.5 Cycloheximide. — It has been shown, using $[1,2,3-^{13}C_3]$-malonate, that a C_3 unit derived from malonate is stereospecifically the source for one side of the glutarimide ring of cycloheximide (129) in cultures of Streptomyces naraensis[84] (cf. Vol. 12, p.31). Results obtained using $[1-^{13}C]$-, $[2-^{13}C]$-, and $[1,2-^{13}C_2]$-acetate and $[^{13}C]$bicarbonate, with cultures of S. griseus, also lead to the conclusion that the glutarimide ring of (129) is formed stereospecifically.[85] However, the first group of workers, supported by earlier work,[86] concluded that C-4 through C-6 derive from malonate while C-2 and C-3 derive from acetate whereas the second group concluded that C-4, C-2, and C-3 derive from malonate and the other two carbons derive from acetate. The conflict between the two conclusions seems to turn on the interpretation of ^{13}C n.m.r. spectra.

4.6 Streptothricin. — Current knowledge on the biosynthesis of streptothricin F (134), a Streptomyces antibiotic, has been reviewed.[62] Preliminary, yet ingeniously predictive, experiments with $[1,2-^{13}C]$acetate gave a labelling pattern in the streptolidine moiety [as (133)] of streptothricin F (134), consistent with biosynthesis from acetyl-CoA via the citric acid cycle, α-ketoglutarate (130), and arginine (131) (Scheme 12).[87] The acetate labelling of the β-lysine moiety in (134) was consistent with biosynthesis for this fragment occurring through α-lysine formed by way of the diaminopimelic acid pathway.[2,88] Further results confirm the intermediacy of arginine and lysine in streptothricin F biosynthesis.

L-[Guanido-^{13}C, $^{15}N_2$]arginine (135; labels: ●) and DL-[guanido-^{13}C; 2-^{15}N]arginine (135; labels: ■) were incorporated into streptothricin F as shown (134; ● and ■, respectively).[89] It is clear that all three nitrogens in the heterocyclic moiety [as (133)] of (134) have their origin in arginine. Coupling was seen between the nitrogen and carbon marked ■ in the ^{13}C n.m.r. spectrum of (134) derived from the second of the two arginine precursors. This very nicely establishes the intact incorporation of arginine and strongly supports the validity of biosynthesis as shown in Scheme 12,

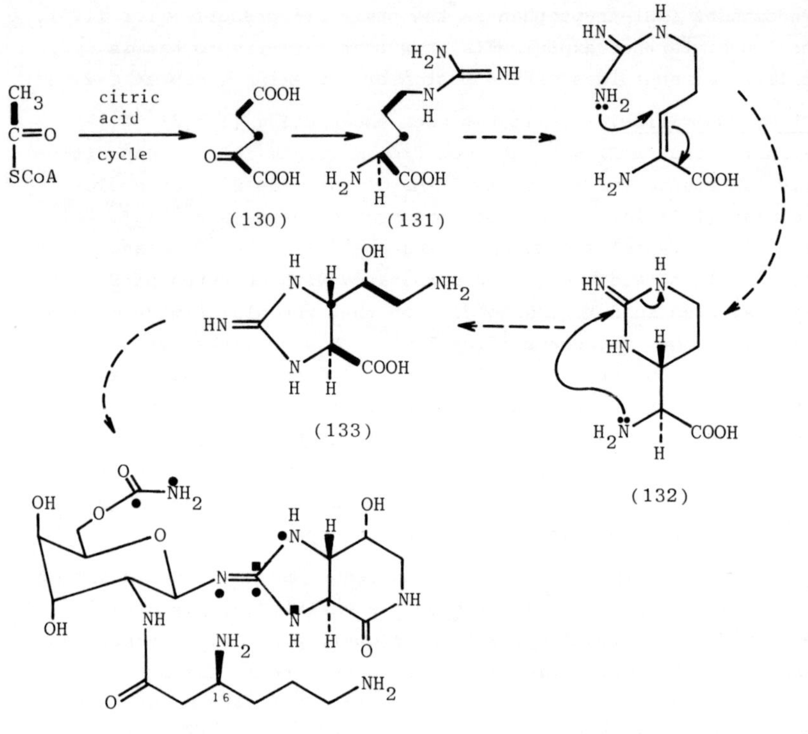

Scheme 12

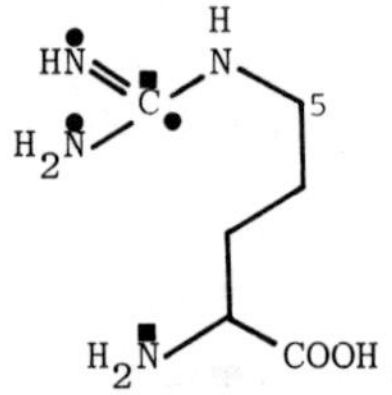

(135)

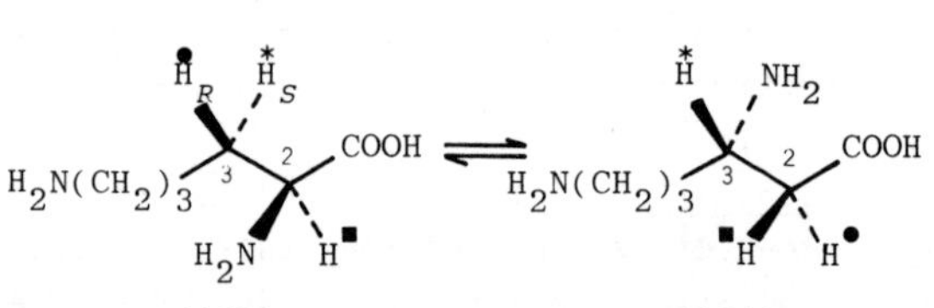

(136) (136)

particularly the key step shown in (132).

DL-[5-^{14}C]arginine [as (135)] was used as a precursor together with the L-arginine sample labelled with stable isotopes. The latter was incorporated twice as efficiently as was the DL-precursor, thus establishing that L-arginine, and not its enantiomer, is used for streprothricin F biosynthesis.[89]

Lysine-2,3-aminomutase catalyses the reversible interconversion of L-α-lysine (136) and L-β-lysine (137). It has been shown, using a cell-free extract of <u>Clostridium</u> species, that, in the rearrangement of (136) which affords (137), the 3-<u>pro-R</u> hydrogen of α-lysine is transferred to C-2, and the 3-<u>pro-S</u> hydrogen is retained at C-3 of (137); inversion of configuration occurs in this reaction at both C-3 and C-2.[90] Further examination of the biosynthesis of streptothricin F (134) provides additional information on the mutase reaction.[91] A mixture of DL-[3-^{13}C, 2-^{15}N]- and L-[U-^{14}C]-lysine was used as a precursor. The relative level of incorporation of the two species was in good agreement with utilization only of the L-isomer of lysine in biosynthesis. The ^{13}C n.m.r. spectrum of the streptothricin F obtained showed a high-intensity doublet for C-16 which arose from coupling to the adjacent nitrogen, which must therefore have been ^{15}N isotope. The origin of the β-lysine moiety in (134) is clearly from α-lysine (136) and the rearrangement reaction must involve intramolecular migration of nitrogen

4.7 β-Lactam Antibiotics. —

Using cell-free preparations of eucaryotic organisms, <u>e.g.</u>, <u>Cephalosporium acremonium</u>, it has very clearly been shown that the biosynthesis of penicillins involves the cyclisation of an intact molecule of the tripeptide (138) to give isopenicillin N (139) (<u>cf</u>. Vol. 12, p.25). <u>Streptomyces clavuligerus</u> is a procaryotic organism which produces β-lactam antibiotics. A cell-free preparation of this organism has been obtained which would convert (138) into (139); penicillin N (140) was also isolated as a result of isomerase action.[92] Similarities and differences in properties compared to those of cell-free preparations of <u>C</u>. <u>acremonium</u> were noted. The results obtained here complement those obtained hitherto with eucaryotic fungi.

Oxygen isotopes, associated with tripeptide (138), have been added to the battery of isotopic labels used to probe the mechanism of penicillin biosynthesis.[93] The tripeptide (138) enriched with ^{17}O/^{18}O labels on each of the oxygen atoms was transformed into isopenicillin N (139) by a cell-free extract of <u>C</u>. <u>acremonium</u> without any loss of label. In a complementary experiment, (138) was transformed into (139) without any incorporation of label from ^{17}O/^{18}O-enriched water. It follows that

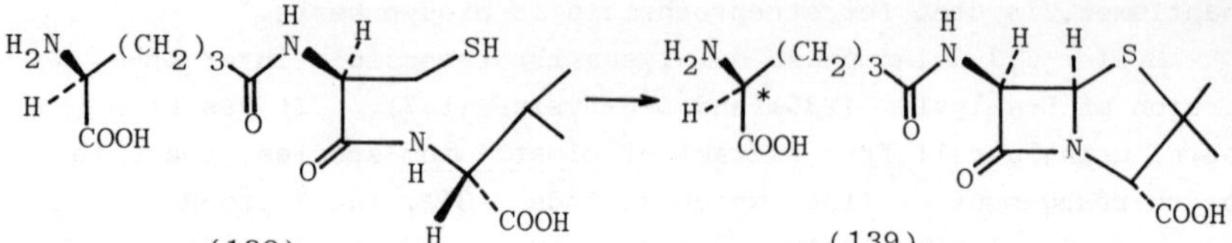

(138)

(139)

(140) epimeric at *

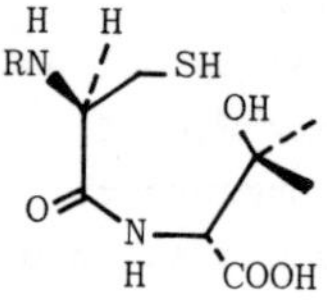

(141) 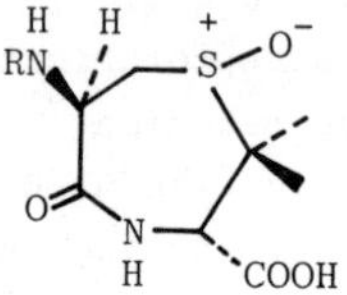 (142) 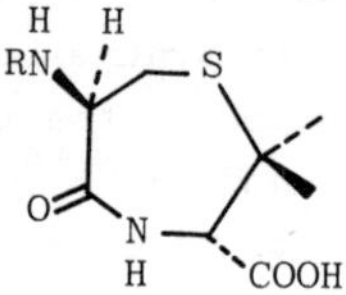 (143)

(R = L-α-amino-δ-adipyl)

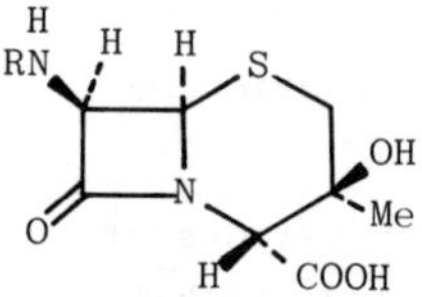

(144) 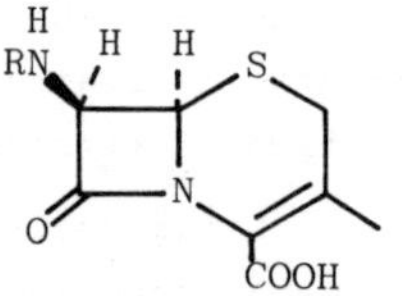 (145) 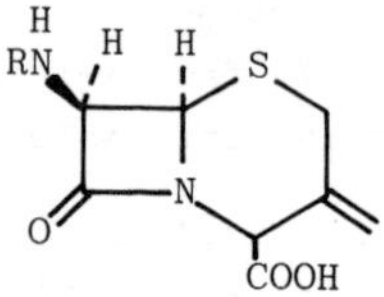 (146)

(R = D-α-amino-δ-adipyl)

no dehydration-hydration step can occur in the conversion of (138) into (139), and any mechanism involving such a step is excluded.

[^{18}O]Valine is incorporated into penicillin V with loss of one oxygen atom (cf. Vol. 12, p.27). This logically occurs before formation of (138).

Using a β-lactam-negative mutant of C. acremonium and (3S,3S)-[4-^{13}C]valine as substrate, it has been demonstrated[94] that the conversion of valine into the tripeptide (138) occurs with retention of configuration, in agreement with other results (cf. Vol. 12, p.27). That the proton at C-3 in valine is lost during the biosynthesis of the penicillin nucleus has been confirmed.[95]

The compounds (141), (142), and (143), considered as possible intermediates in penicillin biosynthesis, have been incubated with the C. acremonium cell-free system.[96] No antibiotic formation was observed from any of them, thus excluding them as biosynthetic intermediates. Neither (142) nor (143) inhibited the transformation of (138) into (139); weak inhibition was observed with (141).

The cepham (144), which had been isolated from C. acremonium broth, has been found not to be converted into deacetoxycephalosporin C (145) by a cell-free preparation of C. acremonium.[97] Using the same system, penicillin N (140) was transformed into (145). It appears that (144) is an end-product of metabolism, and its formation, together with that of (145) from (140), has been rationalized schematically.[97]

The cepham (146) has also been found not to be converted into (145) using C. acremonium preparations.[97,98] Further, (146) inhibits the conversion of (140) into (145).

(R^1 = L-α-amino-δ-adipyl)

(147) R^2 = Me, R^3 = Et
(148) R^2 = Et, R^3 = Me
(149) R^2 = H, R^3 = Me

(150) R^2 = Me, R^3 = Et
(151) R^2 = Et, R^3 = Me
(152) R^2 = H, R^3 = Me

The adaptability towards modified substrates of the cyclase enzyme which is responsible for the conversion of the LLD-tripeptide (138) into (139) has been tested.[99] The LLD-analogues (147), (148), and (149) were incubated with a cell-free preparation of C. acremonium and were found to be converted into isopenicillin N analogues but with lower efficiency than for (138) into (139), which is virtually quantitative (respectively, 36, 4 and 10% yield). All three tripeptide analogues inhibited the transformation of (138) into (139).

The products obtained from (147) and (148) were, respectively, (150) and (151). The analogue (149) gave a mixture of C-2 epimers of demethylisopenicillin N; the major epimer was (152). From this and the yield of the penicillin analogues obtained it is apparent that cyclization to a penicillin is favoured with the larger group at C-2 in the β-configuration.

In spite of the multitude of experiments now carried out, the mechanism for the cyclisation of (138) to give (139) remains quite obscure. Indeed, with so many experiments done, it is high time that something definitive on the mechanism of cyclisation was discovered. Some support for a radical mechanism involving an organo-iron intermediate has been provided by a model chemical reaction;[100] the cyclase enzyme is known to be dependent on the presence of Fe^{2+}.

L-Glutamic acid and some non-metabolizable analogues have been found to stimulate penicillin production by <u>Penicillium chrysogenum</u>.[101] Similar stimulation was not observed with glutamine or ammonium chloride as alternative nitrogen sources. At the enzymic level, glutamic acid caused an increase in δ-(L-α-aminoadipyl)-L-cysteine synthetase concentration. It was concluded that an increase in glutamic acid concentration at the end of the logarithmic phase of growth of <u>P. chrysogenum</u> normally induced this synthetase and thereby penicillin biosynthesis.

The validity of previous predictions (<u>cf</u>. Vol. 10, p.29) that C-10, C-2, C-3, C-8, and C-9 of clavulanic acid (154) derive from a molecule of glutamic acid (153) have been established as correct.[102] DL-[3,4-$^{13}C_2$]glutamic acid (153) gave appropriately labelled (154) (labels in each: ●). Coupling was observed between the labelled carbon atoms in the ^{13}C n.m.r. spectrum of a derivative of (154), showing that the glutamic acid was utilized intact. Labelling of other carbon atoms in (154) was observed consistent with metabolism of the (153) to other clavulanic acid precursors by way of the citric acid cycle.

The biosynthetic origins of nocardicin A (156) have been established (cf. Vol. 12, p.27; Vol. 9, p.33). Serine (155) is one of the precursors and provides the carbon atoms of the β-lactam ring of (156). Its incorporation is without change of oxidation level at the hydroxymethyl group [= C-4 in (156)]. It follows as likely that β-lactam formation is through simple (assisted) displacement of the hydroxy-group by amide nitrogen within some seryl intermediate.

Further results show that L-serine is a very much better precursor than the D-isomer for nocardicin A (156). The L-serine was incorporated with extensive, but not complete, loss of a tritium label located at C-2 in (155). Any cyclisation mechanism which involves loss of the C-2 proton is thus excluded.[103]

Serine samples chirally deuteriated at C-3 were incorporated into (156). The chirality of the deuterium at C-4 of (156) was determined by ^{2}H n.m.r. spectroscopy. The results showed that serine is incorporated into (156) with inversion of configuration at its hydroxymethyl group, which is consistent with cyclisation occurring by a simple nucleophilic displacement of the hydroxy-group.[103]

4.8 Acridone Alkaloids. — The partial derivation of these alkaloids from anthranilic acid[2] has been confirmed for rutacridone (157) in tissue cultures of Ruta graveolens.[104] The remainder of (157) should derive from three molecules of acetate and one of mevalonate. Evidence for the former could be obtained but not for the latter: labelled mevalonate appeared to be utilized with at least partial randomization of label (cf. ref. 2).

4.9 Malonomicin. — Malonomicin (158) is produced by Streptomyces rimosus; its name reflects the presence of an almost unique aminomalonic acid residue. Experiments with ^{13}C-labelled precursors and detailed analysis of the labelling patterns in the derived metabolite show that (158) is biosynthesized from one unit each of 2,3-diaminopropionic acid (C-3, -4, and -5), acetic acid (C-1 and -2), succinic acid (C-6, -7, -8, and -9/10), carbon dioxide (C-10/9), and L-serine (C-11, -12, and -13).[105] The specificity of the succinic acid incorporation has been confirmed with [2,3-^{2}H$_4$]succinic acid; one deuterium atom was retained and was found to be located at C-7 in the derived (158).[106] Fumaric acid was used for biosynthesis, but apparently by way of succinic acid. [Related biosynthetic origins for other tetramic acids, i.e. tenuazonic acid (cf. Vol. 5, p.49) and cyclopiazonic acid (cf. Vol. 5, p.26), may be noted.]

Results of further experiments[106] with deuteriated precursors

(159) R = H
(160) R = -CCH$_2$CH$_2$COOH
           ‖
           O

(161) R = H
(162) R = -CCH$_2$CH$_2$COOH
           ‖
           O
(163) R = -CCH$_2$C(COOH)$_2$
           ‖ |
           O   NH$_2$

established that (162) is a precursor for malonomicin (158). The incorporation of the precursor was low but clearly intact. No clear evidence could be obtained on the intermediacy of (160), which should be a precursor for (162). Neither (159) nor (161) was incorporated into (158). Later potential intermediates, such as (163), were not incorporated into (158).

References

1 R. B. Herbert, in 'Comprehensive Organic Chemistry'. ed. D. H. R. Barton and W. D. Ollis, Pergamon, Oxford, 1978, Vol. 5, p. 1045.

2 R. B. Herbert, in 'Rodd's Chemistry of Carbon Compounds', 2nd edn., ed. S. Coffey, Elsevier, Amsterdam, 1980, Vol. IV, Part L, p. 291.

3 I. D. Wigle, L. J. J. Mestichelli, and I. D. Spenser, J. Chem. Soc., Chem. Commun., 1982, 662.

4 J. C. Richards and I. D. Spenser, reported at the 62nd Canadian Chemical Conference, C.I.C., Vancouver, 1979; quoted in ref. 3.

5 G. R. Orr and S. J. Gould, Tetrahedron Lett., 1982, 23, 3139.

6 A. R. Battersby, R. Murphy, and J. Staunton, J. Chem. Soc., Perkin Trans. 1, 1982, 449.

7 G. A. Ravishankar and A. R. Mehta, Experientia, 1981, 37, 1143.

8 E. Leete, Phytochemistry, 1981, 20, 1037.

9 E. Leete, J. Chem. Soc., Chem. Commun., 1980, 1170.

10 E. Leete, J. Am. Chem. Soc., 1982, 104, 1403.

11 E. G. Brown, K. A. M. Flayeh, and J. R. Gallon, Phytochemistry, 1982, 21, 863.

12 I. Murakoshi, H. Kuramoto, J. Haginiwa, and L. Fowden, Phytochemistry, 1972, 11, 177.

13 D. J. Robins and J. R. Sweeney, J. Chem. Soc., Chem. Commun., 1979, 120.

14 D. J. Robins and J. R. Sweeney, J. Chem. Soc., Perkin Trans. 1, 1981, 3083.

15 G. Grue-Sørensen and I. D. Spenser, Can. J. Chem., 1982, 60, 643; preliminary communication: G. Grue-Sørensen and I. D. Spenser, J. Am. Chem. Soc., 1981, 103, 3208.

16 E. Leete, J. Nat. Prod., 1982, 45, 197.

17 R. N. Gupta and I. D. Spenser, Phytochemistry, 1970, 9, 2329.

18 D. H. R. Barton and T. Cohen, in 'Festschrift Dr. A. Stoll', Birkhaüser, Basle, 1957, p. 117.

19 T. Robinson and W. Nagel, Phytochemistry, 1982, 21, 535.

20 H. Hinz and M. H. Zenk, Naturwissenschaften, 1981, 68, 620.

21 E. J. Staba, S. Zito, and M. Amin, J. Nat. Prod., 1982, 45, 256.

22 K. I. Wasa and N. Takao, Phytochemistry, 1982, 21, 611.

23 W. J. H. Tam, W. G. W. Kurz, F. Constabel, and K. B. Chatson, Phytochemistry, 1982, 21, 253.

24 G. Blaskó, S. F. Hussain, and M. Shamma, J. Am. Chem. Soc., 1982, 102, 1599.

25 A. I. Scott, S.-L. Lee, and T. Hirata, Heterocycles, 1978, 11, 159.

26 M. Rueffer, H. El-Shagi, N. Nagakura, and M. H. Zenk, FEBS Lett., 1981, 129, 5.

27 C. A. Russo and E. G. Gros, Phytochemistry, 1982, 21, 609; ibid., 1981, 20, 1763.

28 W. J. Keller, Phytochemistry, 1981, 20, 2165.

29 A. R. Battersby, R. C. F. Jones, R. Kazlauskas, A. P. Ottridge, C. Poupat, and J. Staunton, J. Chem. Soc., Perkin Trans. 1, 1982, 2010; A. R. Battersby, R. C. F. Jones, R. Kazlauskas, C. W. Thornber, S. Ruchirawat, and J. Staunton, ibid., p. 2016; A. R. Battersby, R. C. F. Jones, A. Minta, A. P. Ottridge, and J. Staunton, ibid., p. 2030.

30 A. R. Battersby, R. C. F. Jones, R. Kazlauskas, C. Poupat, C. W. Thornber, S. Ruchirawat, and J. Staunton, J. Chem. Soc., Chem. Commun., 1974, 773; A. R. Battersby, A. Minta, A. P. Ottridge, and J. Staunton, Tetrahedron Lett., 1977, 1321.

31 D. S. Bhakuni and S. Jain, Tetrahedron, 1981, 37, 3171.

32 D. S. Bhakuni and S. Jain, Tetrahedron, 1980, 36, 2153.

33 D. S. Bhakuni and S. Jain, Tetrahedron, 1981, 37, 3175.

34 D. S. Bhakuni, A. N. Singh, S. Jain and R. S. Kapil, J. Chem. Soc., Chem. Commun., 1978, 226.

35 D. S. Bhakuni and S. Jain, J. Chem. Soc., Perkin Trans. 1, 1981, 2598.

36 N. L. Marekov and A. K. Sidjimov, Tetrahedron Lett., 1981, 22, 2311.

37 A. K. Sidjimov and N. L. Marekov, Phytochemistry, 1982, 21, 871.

38 D. S. Bhakuni and S. Jain, Tetrahedron, 1982, 38, 729.

39 I. R. C. Bick, Heterocycles, 1981, 16, 2105.

40 F. Comer, H. P. Tiwari, and I. D. Spenser, Can. J. Chem., 1969, 47, 481.

41 H. P. Schütte, U. Orban, and K. Mothes, Eur. J. Biochem., 1967, 1, 70.

42 V. Sharma, S. Jain, D. S. Bhakuni, and R. S. Kapil, J. Chem. Soc., Perkin Trans. 1, 1982, 1153.

43 R. B. Herbert and J. Mann, J. Chem. Soc., Chem. Commun., 1980, 841.

44 R. B. Herbert and J. Mann, J. Chem. Soc., Perkin Trans. 1, 1982, 1523.

45 A. R. Battersby, M. Thompson, K.-H. Glüsenkamp, and L.-F. Tietze, Chem. Ber., 1981, 114, 3430.

46 A. R. Battersby, N. D. Westcott, K.-H. Glüsenkamp, and L.-F. Tietze, Chem. Ber., 1981, 114, 3439.

47 S.-L. Lee, K.-D. Cheng, and A. I. Scott, Phytochemistry, 1981, 20, 1841.

48 W. Kohl, B. Witte, and G. Höfle, Z. Naturforsch., Teil. B, 1981, 36, 1153.

49 J. P. Kutney, L. S. L. Choi, P. Kolodziejczyk, S. K. Sleigh, K. L. Stuart, B. R. Worth, W. G. W. Kurz, K. B. Chatson, and F. Constabel, J. Nat. Prod., 1981, 44, 536; Helv. Chim. Acta, 1981, 64, 1837; F. Constabel, P. Gauder-LaPrairie, W. G. W. Kurz, and J. P. Kutney, Plant Cell Reports, 1982, 1, 139.

50 C. Thai, M. Dufour, P. Potier, M. Jaouen, and D. Mansuy, J. Am. Chem. Soc., 1981, 103, 4956.

51 R. L. Baxter, C. A. Dorschel, S.-L. Lee, and A. I. Scott, J. Chem. Soc., Chem. Commun., 1979, 257; K. L. Stuart, J. P. Kutney, T. Honda, and B. R. Worth, Heterocycles, 1978, 9, 1391.

52 F. Guéritte, N. V. Bac, Y. Langlois, and P. Potier, J. Chem. Soc., Chem. Commun., 1980, 452.

53 R. L. Baxter, M. Hasan, N. E. Mackenzie, and A. I. Scott, J. Chem. Soc., Chem. Commun., 1982, 791.

54 S. B. Hassam and H. G. Floss, J. Nat. Prod., 1981, 44, 756.

55 J. B. Jones and J. F. Beck, in 'Applications of Biochemical Systems in Organic Chemistry', ed. J. B. Jones, C. J. Sih, and D. Perlman, Wiley-Interscience, New York, 1976, Part 1, Ch. 4.

56 W. Maier, D. Erge, and D. Gröger, Planta Med., 1980, 45, 104.

57 J. E. Robbers, S. Srikrai, H. G. Floss, and H. G. Schlossberger, J. Nat. Prod., 1982, 45, 178.

58 S. M. Atwell and P. G. Mantle, Experientia, 1981, 37, 1257.

59 P. S. Steyn, R. Vleggaar, N. P. Ferreira, G. W. Kirby, and M. J. Varley, J. Chem. Soc., Chem. Commun., 1975, 465.

60 A. E. de Jesus, P. S. Steyn, R. Vleggaar, M. J. Varley, and N. P. Ferreira, J. Chem. Soc., Perkin Trans. 1, 1981, 3292.

61 C. W. Holzapfel and J. C. Schabort, S. Afr. J. Chem., 1977, 30, 233.

62 S. J. Gould, J. Nat. Prod., 1982, 45, 38.

63 S. J. Gould and D. E. Cane, J. Am. Chem. Soc., 1982, 104, 343.

64 T. W. Doyle, D. M. Balitz, R. E. Grulich, D. E. Nettleton, S. J. Gould, C. Tann, and A. E. Moews, Tetrahedron Lett., 1981, 22, 4595.

65 A. E. de Jesus, W. E. Hull, P. S. Steyn, F. R. van Heerden, R. Vleggaar, and P. L. Wessels, J. Chem. Soc., Chem. Commun., 1982, 837.

66 C. P. Gorst-Allman, P. S. Steyn, and R. Vleggaar, J. Chem. Soc., Chem. Commun., 1982, 652.

67 J. K. Allen, K. D. Barrow, and A. J. Jones, J. Chem. Soc., Chem. Commun., 1979, 280.

68 A. G. Kozlovsky, T. A. Reshetilova, and T. N. Medvedeva, Mikrobiologiya, 1982, 51, 48.

69 M. A. O'Leary and J. R. Hanson, Tetrahedron Lett., 1982, 23, 1855.

70 K. L. Rinehart, Jr., M. Potgieter, and D. A. Wright, J. Am. Chem. Soc., 1982, 104, 2649.

71 K. L. Rinehart, M. Potgieter, D. L. Delaware, and H. J. Seto, J. Am. Chem. Soc., 1981, 103, 2099.

72 M. J. Zmijewski, Jr., Tetrahedron Lett., 1982, 23, 1775.

73 M. J. Zmijewski, R. Wong, J. W. Paschal, and D. E. Norman, unpublished results quoted in ref. 72.

74 O. Ghisalba, R. Roos, T. Schupp, and J. Nüesch, J. Antibiot., 1982, 35, 74.

75 S. P. Gulliford, R. B. Herbert, and F. G. Holliman, Tetrahedron Lett., 1978, 195.

76 P. R. Buckland, S. P. Gulliford, R. B. Herbert, and F. G. Holliman, J. Chem. Res. (S), 1981, 362; (M), p. 4201.

77 P. R. Buckland, R. B. Herbert, and F. G. Holliman, J. Chem. Res. (S), 1981, 363; (M), p. 4225.

78 R. B. Herbert, F. G. Holliman, and P. N. Ibberson, J. Chem. Soc., Chem. Commun., 1972, 355.

79 R. B. Herbert, J. Mann, and A. Römer, Z. Naturforsch., Teil. C, 1982, 37, 159.

80 A. Römer and R. B. Herbert, Z. Naturforsch, in press.

81 H. A. Choy and G. H. Jones, Arch. Biochem. Biophys., 1981, 211, 55.

82 P. Mohr and Ch. Tamm, Tetrahedron, 1981, 37, Supplement No. 1, p. 201.

83 A. Probst and Ch. Tamm, Helv. Chim. Acta, 1981, 64, 2065.

84 H. Shimada, H. Noguchi, Y. Itaka, and U. Sankawa, Heterocycles, 1981, 15, 1141.

85 P. W. Jeffs and D. McWilliams, J. Am. Chem. Soc., 1981, 103, 6185.

86 Z. Vaněk, M. Puža, J. Cudlin, M. Vondrácek, and R. W. Rickards, Folia Microbiol., 1969, 14, 388; F. Johnson, in 'Fortschritte der Chemie Organische Naturstoffe', ed. W. Herz, H. Grisebach, and G. W. Kirby, Springer-Verlag, Wien, 1971, Vol. 29, p. 140.

87 S. J. Gould, K. J. Martinkus, and C.-H. Tann, J. Am. Chem. Soc., 1981, 103, 2871.

88 V. W. Rodwell, in 'Metabolic Pathways', 3rd edn., ed. D. M. Greenberg, Academic Press, New York, 1969, Vol. 3, p. 317; J. R. Kirkpatrick, L. E. Doolin, and O. W. Godfrey, Antimicrob. Agents, Chemother., 1973, 4, 542.

89 S. J. Gould, K. J. Martinkus, and C.-H. Tann, J. Am. Chem. Soc., 1981, 103, 4639.

90 D. J. Aberhart, H.-J. Lin, and B. H. Weiller, J. Am. Chem. Soc., 1981, 103, 6750. See this reference for a comprehensive list of supporting references.

91 S. J. Gould and T. K. Thiruvengadam, J. Am. Chem. Soc., 1981, 103, 6752.

92 S. E. Jensen, D. W. S. Westlake, and S. Wolfe, J. Antibiot., 1982, 35, 483.

93 R. M. Addington, R. T. Aplin, J. E. Baldwin, L. D. Field, E.-M. M. John, E. P. Abraham, and R. L. White, J. Chem. Soc., Chem. Commun., 1982, 137.

94 R. L. Baxter, A. I. Scott, and M. Fukumura, J. Chem. Soc., Chem. Commun., 1982, 66.

95 S. Wolfe, R. J. Bowers, D. A. Lowe, and R. B. Morin, Can. J. Chem., 1982, 60, 355.

96 G. Bahadur, J. E. Baldwin, T. Wan, M. Jung, E. P. Abraham, J. A. Huddleston, and R. L. White, J. Chem. Soc., Chem. Commun., 1981, 1146.

97 R. D. Miller, L. L. Huckstep, J. P. McDermott, S. W. Queener, S. Kukolja, D. O. Spry, T. K. Elzey, S. M. Lawrence, and N. Neuss, J. Antibiot., 1981, 34, 984.

98 J. E. Baldwin, B. Chakravarti, M. Jung, N. J. Patel, P. D. Singh, J. J. Usher, and C. Vallejo, J. Chem. Soc., Chem. Commun., 1981, 934.

99 G. A. Bahadur, J. E. Baldwin, J. J. Usher, E. P. Abraham, G. S. Jayatilake, and R. L. White, J. Am. Chem. Soc., 1981, 103, 7650.

100 J. E. Baldwin and A. P. Davis, J. Chem. Soc., Chem. Commun., 1981, 1219.

101 F. Lara, R. del Carmen Makeos, G. Vázquez and S. Sánchez, Biochem. Biophys. Research Commun., 1982, 105, 172.

102 S. W. Elson, R. S. Oliver, B. W. Bycroft, and E. A. Faruk, J. Antibiot., 1982, 35, 81.

103 C. A. Townsend and A. M. Brown, J. Am. Chem. Soc., 1982, 104, 1748.

104 A. Baumert, I. N. Kuzovkina, G. Krauss, M. Hieke, and D. Gröger, Plant Cell Reports, 1982, 1, 168.

105 D. Schipper, J. L. van der Baan, and F. Bickelhaupt, J. Chem. Soc., Perkin Trans. 1, 1979, 2017.

106 D. Schipper, J. L. van der Baan, N. Harms, and F. Bickelhaupt, Tetrahedron Lett., 1982, 23, 1293.

2
Pyrrolidine, Piperidine, and Pyridine Alkaloids

BY A. R. PINDER

1 Pyrrolidine Alkaloids

The ant venom alkaloids occurring in the genera <u>Solenopsis</u> and <u>Monomorium</u> have been reviewed briefly.[1] The venoms of <u>M. latinode</u> and <u>M. subopacum</u> contain an array of 2,5-dialkylpyrrolidines (1), and the pyrrolines (2) are also present. All were identified by gas-chromatographic separation followed by mass spectrometry.[1]

$Me(CH_2)_3$ —[pyrrolidine, N-R]— $(CH_2)_n Me$ $Me(CH_2)_3$ —[pyrroline]— $(CH_2)_n Me$

(1) n = 4 or 6, R = H   (2) n = 4 or 6
n = 6, R = Me

[pyrrolidine with N-CONH$_2$, 2-R^1, 2-R^2]

(3) R^1 = OMe, R^2 = H
(4) $R^1 R^2$ = O

The pyrrolidine <u>N</u>-amide (3) has been isolated from the bark of the West African tree <u>Hexalobus crispiflorus</u>. Its structure was settled by spectroscopic study. Squamolone (4) is also present.[2]

A second synthesis of jatropham (5), an antitumour alkaloid of <u>Jatropha macrorhiza</u>, has been reported, starting from succinimide. It is outlined in Scheme 1.[3]

The roots of certain <u>Achillea</u> spp. contain several new amides, two of which are the pyrrolidides, (6) and (7) respectively, of

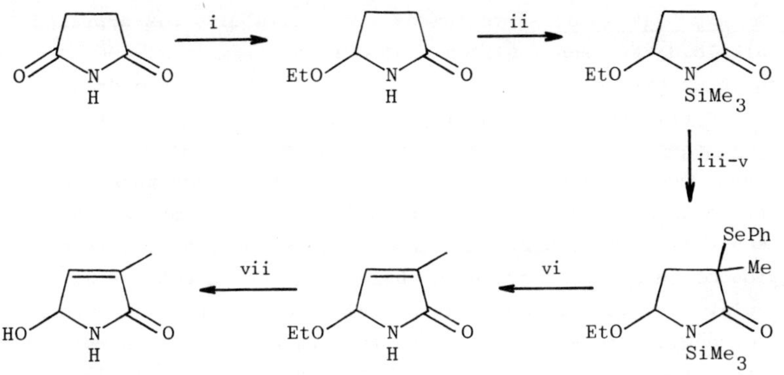

Pyrrolidine, Piperidine, and Pyridine Alkaloids

Reagents: i, NaBH$_4$, HCl, in EtOH-H$_2$O (9:1); ii, (Me$_3$Si)$_2$NH, heat; iii, LiNPr$^i_2$; iv, (PhSe)$_2$; v, MeI; vi, 30% H$_2$O$_2$ or mCPBA; vii, HOAc, H$_2$O, at 60°C

Scheme 1

2,3-dehydrolycaonic acid and lycaonic acid.[5] Another amide from A. tomentosa L. is the pyrrolideide (8), with a centrally placed triple bond,[6] all structures being deduced from spectral studies 4,5-Dihydro-okolasine (9) occurs in Piper guineense; it has been synthesised by a Wittig-Horner reaction between 3-(2-methoxy-4,5-methylenedioxyphenyl)propionaldehyde and 1-(diethoxyphosphonylacetyl)pyrrolidine, which leads stereospecifically to the required E-isomer.[7]

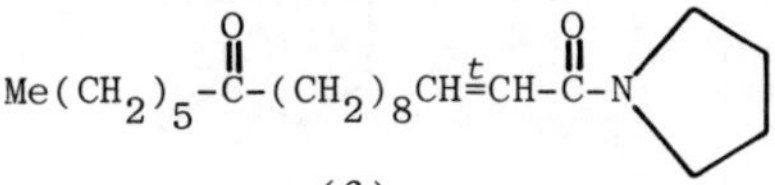

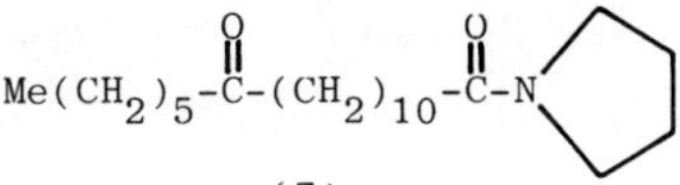

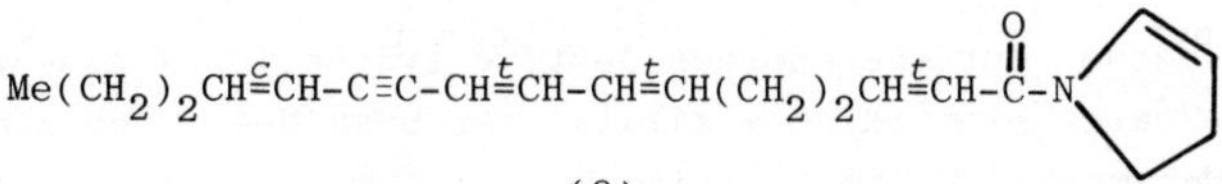

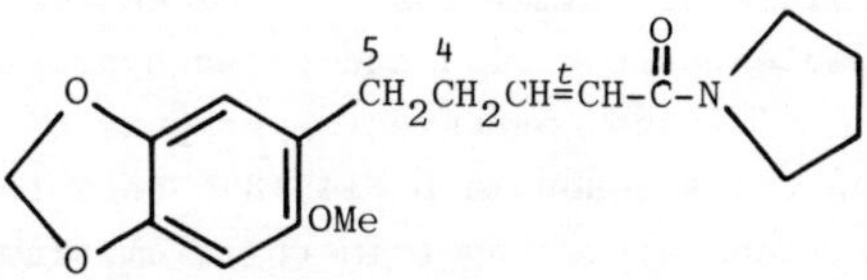

1.1 *Sceletium* Alkaloids.-A review of these alkaloids has appeared.[8] A new alkaloid has been isolated from Crinum oliganthum. On spectral evidence it has been assigned structure (10).[9] The dihydropyridone base (11), related to Sceletium alkaloid A_4, has been found in S. namaquense; its structure is based on spectral study.[10] Several new syntheses of mesembrine have been reported. One uses a cinnamonitrile as a synthon, with introduction of a "formyl anion" at the β-position, followed by Robinson annelation and finally elaboration of the cyanomethyl side-chain.[11] A second employs an intramolecular ene cyclisation of an acylnitroso-olefin to a hydroxamic acid, reducible to a lactam. Then follows a series of simple steps (N-methylation, hydroxylation via bromohydrin formation, debromination, oxidation, and finally lactam reduction) leading to (±)-mesembrine.[12] Another uses D-mannitol as a chiral template and furnishes (−)-mesembrine.[13] A fourth synthesis starts with a symmetrical phenylacetic acid derivative and leads to racemic mesembrine.[14] A synthesis of (±)-mesembranol has been described; it starts with a 1-arylcyclohexene, which is subjected to regio- and stereospecific heteroannelation to establish the octahydroindolone skeleton.[15]

(10)

(11)

2 Piperidine Alkaloids

A general method for the stereoselective synthesis of cis- and of trans-2,6-dialkylpiperidine alkaloids has been described and is outlined in Scheme 2.[16] The starting material was obtainable from 2-picoline, as a mixture of epimerides.[17] The scheme outlines the synthesis of (±)-solenopsin A (12) and (±)-dihydropinidine (13). Another stereoselective and convenient synthesis of (±)-solenopsin A (12) and also of (±)-solenopsin B (14) has been reported. It involves, as the critical step, the trimethylaluminium-catalysed Beckmann rearrangement of an oxime mesylate (Scheme 3). In the final reduction of the C=N bond it proved possible to control the stereochemical outcome.[18]

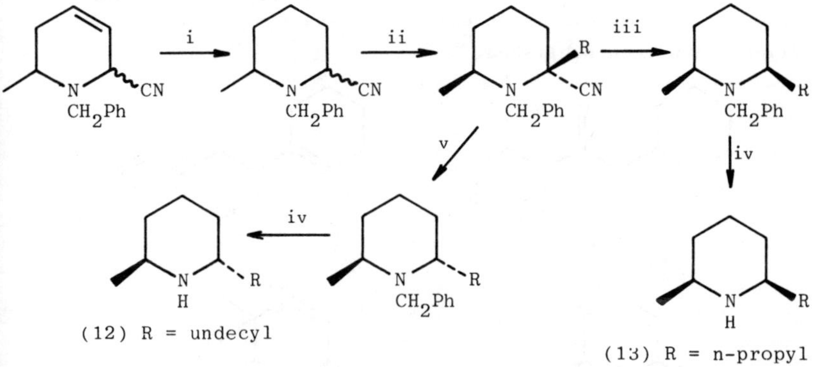
i
ii
iii
iv
v
iv
CN
R
CH2Ph
N
H
(12) R = undecyl
(13) R = n-propyl

Reagents: i, H_2, Pd/C, MeOH; ii, $LiNPr^i_2$, then RX; iii, Na, liquid NH_3, at -78^oC; iv, debenzylation; v, $NaBH_4$, MeOH

Scheme 2

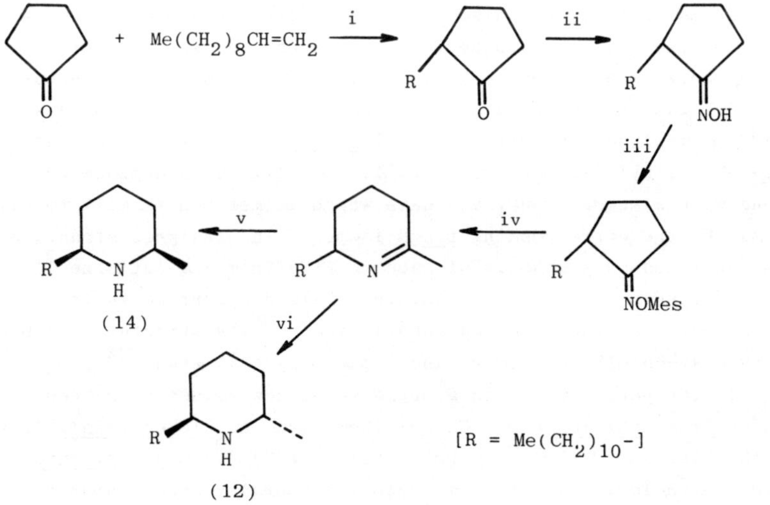
+ Me(CH2)8CH=CH2
i
ii
iii
iv
v
vi
O
R
NOH
NOMes
N
H
(14)
(12)
[R = Me(CH2)10-]

Reagents: i, Ag_2O, at 135^oC for 5 hours; ii, NH_2OH, NaOAc, MeOH; iii, $MeSO_2Cl$, Et_3N, CH_2Cl_2; iv, $AlMe_3$, PhMe, at -78^oC, then NaF, H_2O; v, Bu^i_2AlH or $LiAlH_4$; vi, $LiAlH_4$ + $AlMe_3$

Scheme 3

2-(3,4-Dimethoxyphenacyl)piperidine (15) occurs in <u>Boehmeria cylindrica</u> and is in part responsible for the intense antimicrobial action of extracts of the plant against <u>Candida albicans</u>.[19] Full details of a total synthesis of (±)-cassine, reported briefly earlier, have been published.[20]

A brief review of the chemistry and biology of ant venom alkaloids, some of which are piperidines, has been published.[1] The venoms of four unstudied <u>Solenopsis</u> (fire ant) species contain the expected 2,6-dialkylpiperidines, and one, <u>S</u>. sp. A (Puerto Rico), contains 2-(4-penten-1-yl)-1-piperideine(16), the structure of which was settled by n.m.r. and mass spectrometry.[1] The Australian mealybug ladybird (<u>Cryptolaemus montrouzieri</u>) secretes <u>cis</u>-1-(6-methyl-2-piperidyl)-2-propanone (17) as a defence weapon, along with a second, unstable base which isomerized readily to (17); it may be the corresponding <u>trans</u> isomer. The assigned structure was confirmed by synthesis of racemic (17) from 2,6-lutidine.[21]

Acalyphin (18) is a glucosidic alkaloid occurring in the aerial parts of the weed <u>Acalypha indica</u> L. Its structure has been arrived at chiefly by p.m.r. and c.m.r. spectral study.[22]

In the pepper alkaloid group, several new amides have been isolated from <u>Achillea</u> spp. One of them, from <u>A</u>. <u>biebersteinii</u> Afan., is the piperideide(19).[6] Others include (20) (from <u>A</u>. <u>lycaonica</u> Boiss. et Heldr.), and the hitherto structurally unencountered acetylenic amides (21) and (22) (from <u>A</u>. <u>spinulifolia</u> Fenzl ex Boiss.), and (23) and (24) (from <u>A</u>. <u>grandifolia</u> Friv.).[5] A one-step synthesis of piperine involves base-catalysed condensation of piperonal with <u>N</u>-crotonylpiperidine.[23] The 2-<u>trans</u>, 4-<u>cis</u> isomer (25) of all-<u>trans</u> wisaninehas been found in the root bark of <u>Piper</u>

$$Me(CH_2)_4CH\overset{t}{=}CH-CH\overset{t}{=}CHCON\diagup$$

(19)

(20) R = $Me(CH_2)_5CO(CH_2)_8CH\overset{t}{=}CH-$

(21) R = $Me(CH_2)_2CH\overset{c}{=}CHC\equiv C(CH_2)_2CH\overset{t}{=}CHCH\overset{t}{=}CH-$

(22) R = $MeC\equiv C-C\equiv C-C\equiv C(CH_2)_2CH\overset{t}{=}CHCH\overset{t}{=}CH-$

(23) R = $MeCH\overset{t}{=}CHC\equiv C-C\equiv C(CH_2)_2CH\overset{t}{=}CHCH\overset{t}{=}CH-$

(24) R = $MeCH\overset{t}{=}CHC\equiv C-C\equiv CCH\overset{t}{=}CHCH\overset{t}{=}CHCH\overset{t}{=}CH-$

(25)

guineense; it appears to be the first example of a "geometrically mixed" natural piperidide,[24] and proved to be identical with a synthetic product described earlier.[25] 4,5-Dihydrowisanine and 1-(2-methoxy-4,5-methylenedioxycinnamoyl)piperidine have been found in the same source and in P. peepuloides respectively; both have been synthesised by appropriate Wittig-Horner reactions, which led to the desired E-isomers.[7]

(26)   (27)   (28)

2.1 Decahydroquinoline Alkaloids.-A review entitled "Alkaloids of Neotropical Poison Frogs" includes cis-decahydroquinoline bases.[26] A new synthesis of (±)-pumiliotoxin C has been reported. It involves a Beckmann rearrangement-alkylation sequence brought about by organoaluminium reagents on the oxime tosylate (26), resulting in a ring-expansion to the racemic toxin (27).[27] The absolute configuration (28) assigned to gephyrotoxin on the basis of X-ray diffraction analysis[28] may have to be revised; a stereoselective synthesis starting with L-pyroglutamic acid led to (28), but the product proved to be (+)-gephyrotoxin whereas the natural product is (-).[29]

2.2 Spiropiperidine Alkaloids.-A review of histrionicotoxins has appeared,[26] and another on stereoselective syntheses of perhydro- and octahydrohistrionicotoxin.[30] New stereoselective routes to (±)-perhydrohistrionicotoxin[31,32,33] and to the natural (-)- and unnatural (+)- varieties[34] have been devised. It appears that nitramine and isonitramine, two alkaloids from Nitraria spp., must be reformulated as epimeric spiropiperidines (29) and (30) respectively, rather than decahydroquinolines.[35,36]

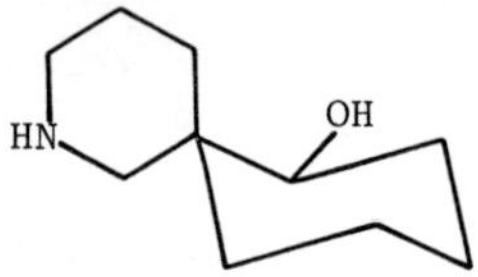

(29) 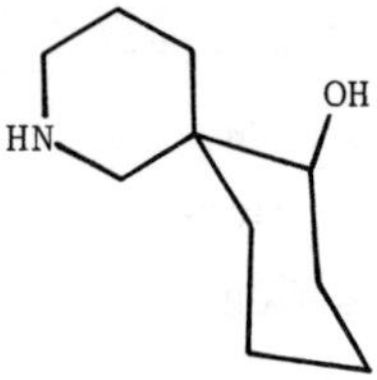   (30) 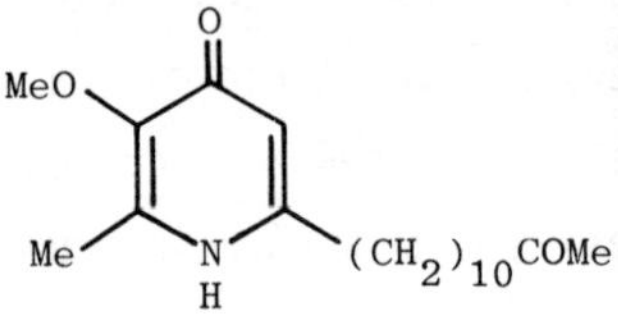   (31)

3 Pyridine Alkaloids

It has been pointed out that the ketonic Melochia alkaloid (31), hitherto called "melochinone," must be renamed melochininone, because the former name has been given to another alkaloid from the same source.[37] The syntheses of natural (-)-melochinine (32) and its enantiomer have been described (Scheme 4). The (S)- and (R)-

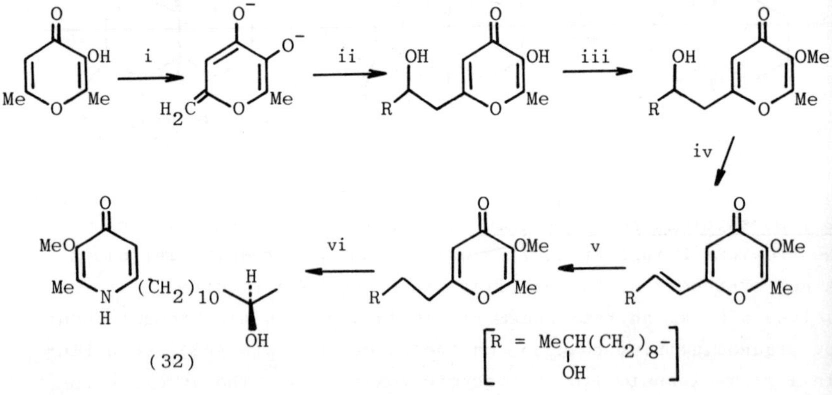

Reagents: i, LiN(SiMe$_3$)$_2$, at -70°C; ii, MeCH(CH$_2$)$_8$CHO, then HOAc, H$_2$O; iii, MeI;
iv, TsOH; v, H$_2$, Pd; vi, NH$_3$

Scheme 4

enantiomers of aldehyde (33) were synthesised; step ii using the former led to the natural base, and using the latter afforded the (+)-base.[38]

$$\text{Me}\overset{\text{OH}}{\underset{|}{\text{CH}}}(\text{CH}_2)_8\text{CHO}$$
(33)

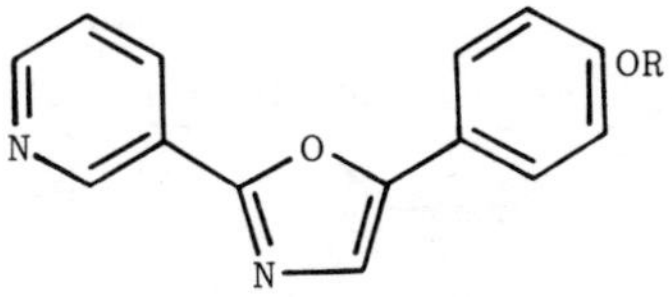

(34) R = Me
(35) R = Me$_2$C=CHCH$_2$-

The ripe fruits of <u>Aegle marmelos</u> contain O-methylhalfordinol (34) and O-isopentenylhalfordinol (35).[39] Actinidine (36) is a minor component of the defensive secretion of the Australian cocktail ant <u>Iridomyrmex nitidiceps</u>.[40] A new synthesis of (±)-actinidine has been described (Scheme 5).[41] Full details of an earlier briefly described synthesis of (±)-muscopyridine have been provid-

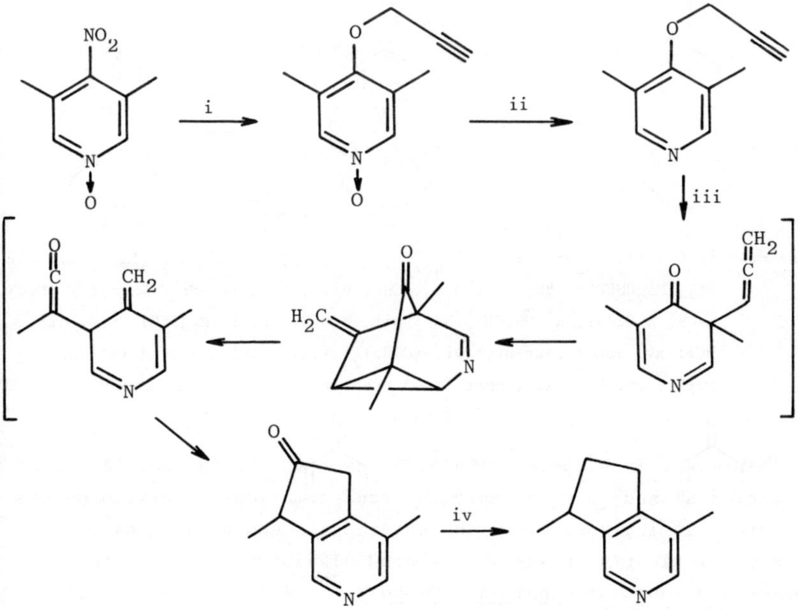

Reagents: i, HOCH$_2$C≡CH, K$_2$CO$_3$; ii, PCl$_3$; iii, 450°C; iv, Huang-Minlon reduction

Scheme 5

ed,[42] and two syntheses of the natural (R)-(+)-base (37) described (Schemes 6 and 7).[43]

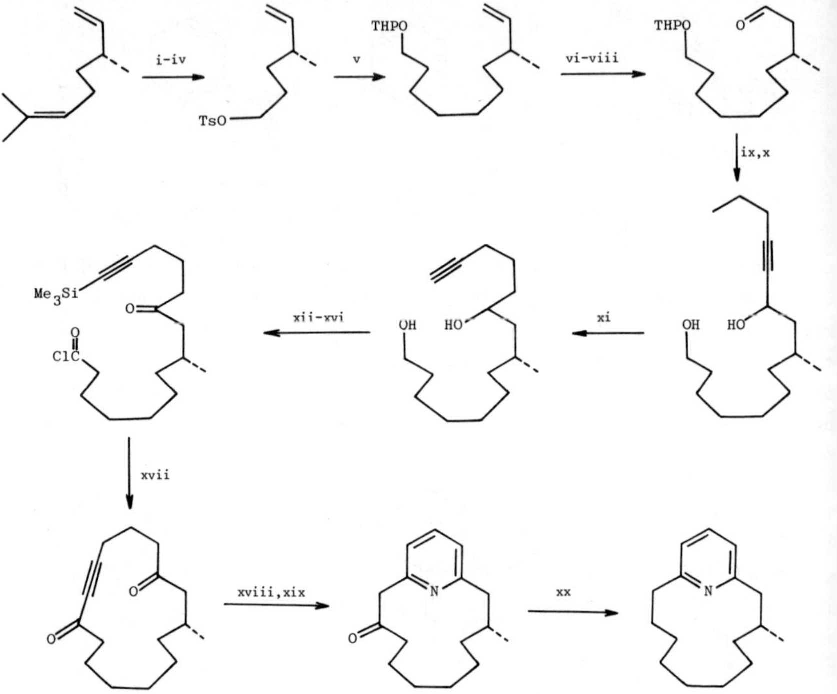

Reagents: i, O_3; ii, Me_2S; iii, $LiAlH_4$; iv, TsCl, py; v, $THPO(CH_2)_4MgBr$, Li_2CuCl_4; vi, $(Me_2CHCHMe)_2BH$; vii, H_2O_2, OH^-; viii, PCC; ix, $Pr^nC{\equiv}CMgBr$; x, TsOH, MeOH, H_2O; xi, $K^+ \; {}^-NH(CH_2)_3NH_2$; xii, EtMgBr; xiii, Me_3SiCl; xiv, dil. HCl; xv, Jones reagent; xvi, $(COCl)_2$; xvii, $AlCl_3$, high dilution; xviii, aq. NH_3; xix, heat, then preparative g.l.c.; xx, H_2NNH_2, H_2O, KOH

Scheme 6

Interest in nicotine alkaloids has been maintained. A review of microbial and in vitro enzymic transformations of alkaloids has appeared; it includes a section on nicotine and related bases.[44] A range of nicotine bases and related alkaloids occurs in the leaves and roots of Duboisia hopwoodii.[45] 4,4-Ditritio-(+)-nicotine has been synthesised; it is 10-30 times less potent than natural (-)-nicotine in the CNS.[46] Nicotine appears to show, in its reaction with simple alkyl halides, a marked preference for attack on the pyridine nitrogen.[47] t-Butyllithium reacts with nicotine to

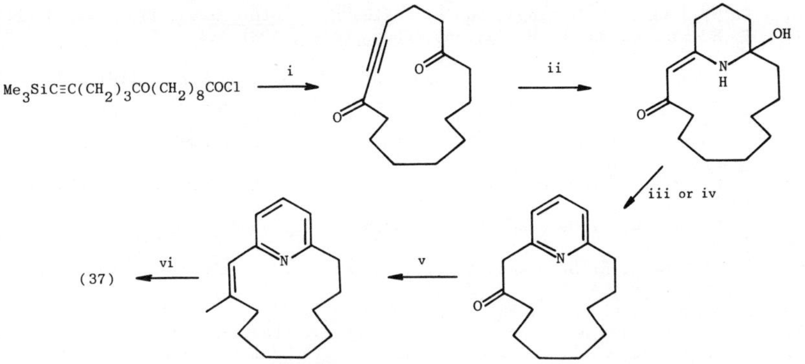

Reagents: i, AlCl$_3$, high dilution; ii, aq. NH$_3$; iii, heat, then preparative g.l.c.; iv, tetrachloro-o-benzoquinone; v, MeLi in THF; vi, mono(isopinocampheyl)-borane [from (−)-α-pinene]

Scheme 7

generate 6-t-butylnicotine along with two products resulting from the cleavage of the pyrrolidine ring. The recovered nicotine and the 6-t-butyl derivative are substantially racemised.[48] The quaternary salt (38), obtained by reaction of nicotine with O-mesylhydroxylamine, undergoes a triple ring expansion to the 1,2-diazocine (39) on treatment with sodamide in liquid ammonia.[49]

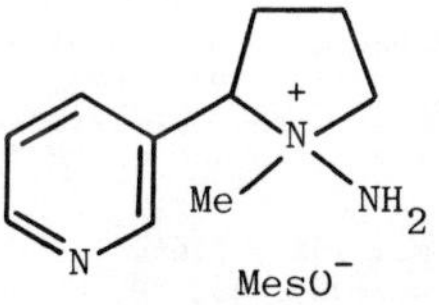

References

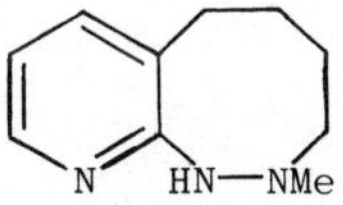

1. T. H. Jones, M. S. Blum, and H. M. Fales, Tetrahedron, 1982, 38, 1949.
2. H. Achenbach, C. Renner, and I. Addae-Mensah, Liebigs Ann. Chem., 1982, 1623.
3. T. Nagasaka, S. Esumi, N. Ozawa, Y. Kosugi, and F. Hamaguchi, Heterocycles, 1981, 16, 1987.
4. J. C. Hubert, J. B. P. A. Wijinberg, and W. N. Speckamp, Tetrahedron, 1975, 31, 1437.
5. H. Greger, M. Grenz, and F. Bohlmann, Phytochemistry, 1982, 21, 1071.
6. H. Greger, M. Grenz, and F. Bohlmann, Phytochemistry, 1981, 20, 2579.
7. S. Linke, J. Kurz, and H.-J. Zeiler, Liebigs Ann. Chem., 1982, 1142.
8. P. W. Jeffs, in 'The Alkaloids, Chemistry and Physiology,' ed. R.G.A. Rodrigo, Vol. XIX, Chapter 1, Academic Press, New York, 1981.
9. W. Döpke, E. Sewerin, Z. Trimiño, and C. Julierrez, Z. Chem., 1981, 21, 358.

10. P. W. Jeffs, T. M. Capps, and R. Redfearn, J. Org. Chem., 1982, 47, 3611.
11. I. H. Sanchez and F. R. Tallabs, Chem. Lett., 1981, 891.
12. G. E. Keck and R. R. Webb, J. Org. Chem., 1982, 47, 1302.
13. S. Takano, Y. Imamura, and K. Ogasawara, Tetrahedron Lett., 1981, 22, 4479.
14. S. Takano, Y. Imamura, and K. Ogasawara, Chem. Lett., 1981, 1385.
15. P. W. Jeffs, N. A. Cortese, and J. Wolfram, J. Org. Chem., 1982, 47, 3881.
16. M. Bonin, J. R. Romero, D. S. Grierson, and H.-P. Husson, Tetrahedron Lett., 1982, 23, 3369.
17. D. S. Grierson, M. Harris, and H.-P. Husson, J. Am. Chem. Soc., 1980, 102, 1064.
18. Y. Matsumura, K. Maruoka, and H. Yamamoto, Tetrahedron Lett., 1982, 23, 1929.
19. A. Al-Shamma, S. D. Drake, L. E. Guagliardi, L. A. Mitscher, and J. K. Swayze, Phytochemistry, 1982, 21, 485.
20. E. Brown and A. Bonté, Bull. Soc. Chim. Fr., Part 2, 1981, 281.
21. W. V. Brown and B. P. Moore, Aust. J. Chem., 1982, 35, 1255.
22. A. Nahrstedt, J.-D. Kant, and V. Wray, Phytochemistry, 1982, 21, 101.
23. A. Schulze and H. Oediger, Liebigs Ann. Chem., 1981, 1725.
24. I. Addae-Mensah, F. B. Torto, B. Torto, and H. Achenbach, Planta Medica, 1981, 41, 200.
25. S. Linke, J. Kurz, and H.-J. Zeiler, Tetrahedron, 1978, 34, 1979.
26. J. W. Daly, Fortschr. Chem. Org. Naturst., 1982, 41, 205.
27. K. Hattori, Y. Matsumura, T. Miyazaki, K. Maruoka, and H. Yamamoto, J. Am. Chem. Soc., 1981, 103, 7368.
28. J. W. Daly, B. Witkop, T. Tokuyama, T. Nishikawa, and I. L. Karle, Helv. Chim. Acta, 1977, 60, 1128.
29. R. Fujimoto and Y. Kishi, Tetrahedron Lett., 1981, 22, 4197.
30. Y. Inubushi and T. Ibuka, Heterocycles, 1982, 17, 507.
31. T. Ibuka, Y. Mitsui, K. Hayashi, H. Minakata, and Y. Inubushi, Tetrahedron Lett., 1981, 22, 4425.
32. T. Ibuka, H. Minakata, Y. Mitsui, E. Tabushi, T. Taga, and Y. Inubushi, Chem. Letters, 1981, 1409.
33. D. A. Evans, E. W. Thomas, and R. E. Cherpeck, J. Am. Chem. Soc., 1982, 104, 3695.
34. K. Takahashi, B. Witkop, A. Brossi, M. A. Maleque, and E. X. Albuquerque, Helv. Chim. Acta, 1982, 65, 252.
35. A. A. Ibragimov, Z. Osmanov, B. Tashchodzhaev, N. D. Abudullaev, M. R. Yagudaev, and S. Yu.Yunusov Khim. Prir. Soedin., 1981, 623.
36. Z. Osmanov, A. A. Ibragimov, and S.Yu.Yunusov Khim. Prir. Soedin., 1982, 126.
37. G. Spiteller, Liebigs Ann. Chem., 1981, 2096.
38. G. Voss and H. Gerlach, Liebigs Ann. Chem., 1982, 1466.
39. B. R. Sharma and P. Sharma, Planta Medica, 1981, 43, 102.
40. G. W. K. Cavill, P. L. Robertson, J. J. Brophy, D. V. Clark, R. Duke, C. J. Orton, and W. D. Plant, Tetrahedron, 1982, 38, 1931.
41. M. Nitta, A. Sekiguchi, and H. Koba, Chem. Letters, 1981, 933.
42. T. Hiyama, M. Shinoda, H. Saimoto, and H. Nozaki, Bull. Chem. Soc. Jpn., 1981, 54, 2747.
43. K. Utimoto, S. Kato, M. Tanaka, Y. Hoshino, S. Fujikura, and H. Nozaki, Heterocycles, 1982, 18, 149.
44. H. L. Holland, in 'The Alkaloids, Chemistry and Physiology,' ed. R.G.A. Rodrigo, Vol. XVIII, Chapter 5, Academic Press, New York, 1981.
45. O. Luanratana and W. J. Griffin, Phytochemistry, 1982, 21, 449.
46. W. C. Vincek, B. R. Martin, M. D. Aceto, H. L. Tripathi, E. L. May, and L. S. Harris, J. Pharm. Sci, 1981, 70, 1292.
47. M. Shibagaki, H. Matsushita, S. Shibata, A. Saito, Y. Tsujino, and H. Kaneko, Heterocycles, 1982, 19, 1641.
48. C. G. Chavdarian and J. I. Seeman, Tetrahedron Lett., 1982, 23, 2519.
49. Y. Tamura, M. Tsunekawa, H. Ikeda, and M. Ikeda, Heterocycles, 1982, 19, 1595.

3
Tropane Alkaloids

BY G. FODOR AND R. DHARANIPRAGADA

1 Occurrence and Structure of New Alkaloids

The chemical structure of baogongteng A (1), a new myotic agent from <u>Erycibe obtusifolia</u>, has been determined.[1] High-resolution mass-spectral data indicated the presence of a 3,6-disubstituted tropane skeleton. Infrared and ^{1}H and ^{13}C n.m.r. data indicated the presence of secondary amine, secondary hydroxyl, and acetoxy functions, and the location of the latter group at C-6 was based on comparison of ^{1}H n.m.r. chemical shifts with those of 6β-acetoxy-3α-tropanol. The hydroxyl group was assigned to the 2β-position from the ^{1}H n.m.r. spectrum of the <u>N</u>-methylated alkaloid.

(1)

(2) a; R^1 = H, R^2 = OH
b; R^1 = OH, R^2 = O-senecioyl
c; R^1 = O-angeloyl, R^2 = OH
d; R^1 = O-tigloyl, R^2 = OH
e; R^1 = R^2 = OH
f; R^1 = H, R^2 = O-senecioyl

Six tropane alkaloids have been isolated[2] from the aerial parts of <u>Schizanthus hookeri</u>. Besides tropan-3α-ol (2a) and its senecioic acid ester (2f), tropane-3α,6β-diol (2d) and its monoesters with angelic (2c), tiglic (2d), or senecioic acid (2b), respectively, were identified.

The extracts of roots of <u>Anisodus tanguticus</u> gave[3] apoatropine (3a) and 3α-(4,4,4-trichloro-2-phenylbutyryloxy)tropane (3b). In the reviewers' opinion, (3b) is most likely an artefact formed from apoatropine (3a) and chloroform by a radical reaction.

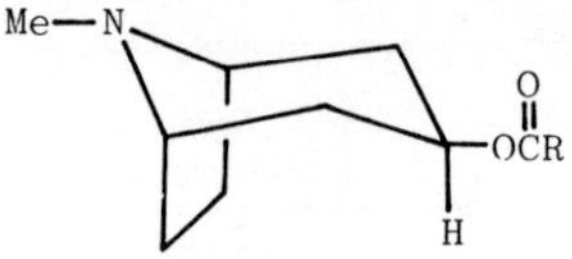

(3) a; R = Ph-C(=CH$_2$)
b; R = Ph-CH(-)-CH$_2$CCl$_3$

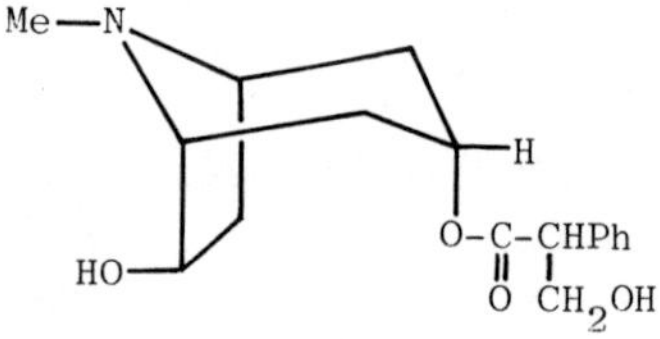

(4)

Tropine, anisodamine (4), and apoatropine were found[4] in the medicinal plant <u>Przewalskia tangutica</u>.

The structure of physoperuvine, the major alkaloid of the roots of <u>Physalis peruviana</u>, previously reported[5] as 3-methylaminocycloheptanone (5), has now been proved to be incorrect.[6,7] Incompatibility of physical data with structure (5) was pointed out[6] and the synthesis of authentic 3-methylaminocycloheptane, different from (±)-physoperuvine, was achieved.[6] Independently a reinvestigation was undertaken.[7]

The i.r. spectrum of the alkaloid has a band at 3200 cm^{-1}, attributed to a bonded hydroxyl or an amino-group, but no band can be ascribed to a carbonyl group, in disagreement with the aminoketone structure (5). The ^{13}C n.m.r. spectrum was indicative of a potential rather than a free ketone carbonyl carbon. All of this is consistent[7] with a tautomeric equilibrium of 4-aminocycloheptanone (6b) with 1-hydroxytropane (6a).[7]

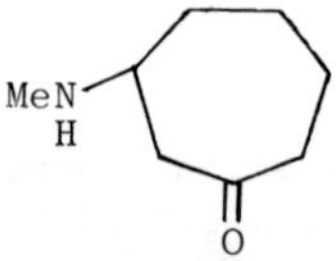

(5) 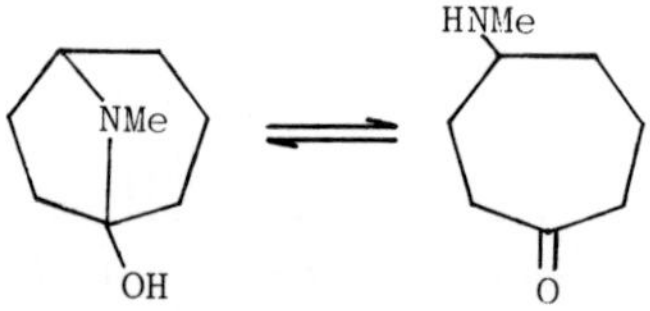   (6a)   (6b) 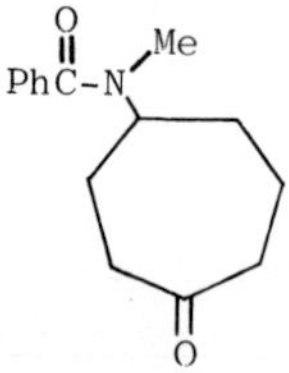   (7)

Final evidence for the tropan-1-ol hydrochloride structure in the crystal lattice was provided by X-ray analysis of physoperuvine hydrochloride. The tautomerism (6a)⇌(6b) was corroborated by the weak positive Cotton effect, $[\theta]_{288} = +174$ (MeOH), compared to that of N-benzoylphysoperuvine, $[\theta]_{288} = +7809$ (MeOH), which can only exist as the benzamido-ketone form (7). A 1:45 ratio of the ketone (6b) to the carbinolamine (6a) was inferred from these data.[7]

2 Synthesis and Chemical Transformations

A new synthetic route to the European Ladybug alkaloid adaline has been elaborated.[8] Conjugate addition of n-amylmagnesium bromide (or methylmagnesium iodide) to cyclo-octa-2,7-dienone, followed by trapping of the magnesium enolate with phenylselenium bromide, gave adducts (8a) and (8b), which were directly oxidized to the substituted dienones (9a) and (9b). Subsequent addition of benzylamine to dienone (9a), followed by hydrogenolysis of (10a), afforded racemic (±)-adaline (11a).

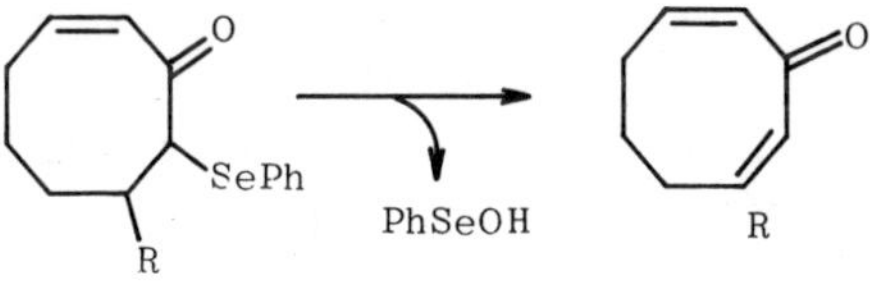

(8) a; R = amyl
    b; R = Me

(9) a; R = amyl
    b; R = Me

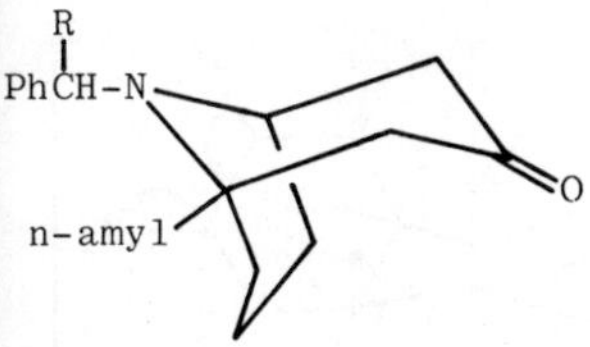

(10) a; R = H
     b; R = Me

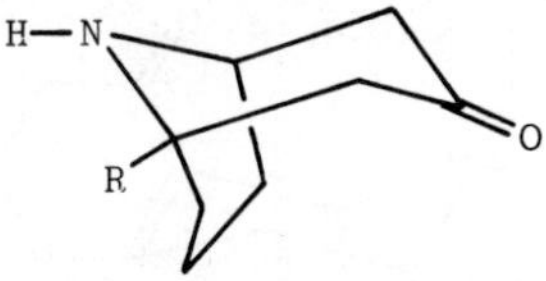

(11) a; R = n-amyl
     b; R = Me

Asymmetric induction was achieved by the addition of (R)-(+)-α-methylbenzylamine to dienone (9a), giving a mixture of diastereomers of (10b) (3:2 ratio) which were subsequently separated into a crystalline and an oily base. Catalytic hydrogenolysis of the solid isomer gave natural (-)-adaline, $[\alpha]_D = -11°$, while hydrogenolysis of the liquid isomer gave the enantiomer, $[\alpha]_D = +12°$.

The enantiomeric forms of the Euphorbia alkaloid (11b) were also synthesized.[8]

Attempts to synthesize the symmetric 1,5-diamyl-3-granatanone failed since the disubstituted dienone 7-amyl-(9a) did not undergo double Michael addition. (±)-Isobellendine (13) has been synthesized[9] from tropinone via the enamine (12) in 30% overall yield (Scheme 1).

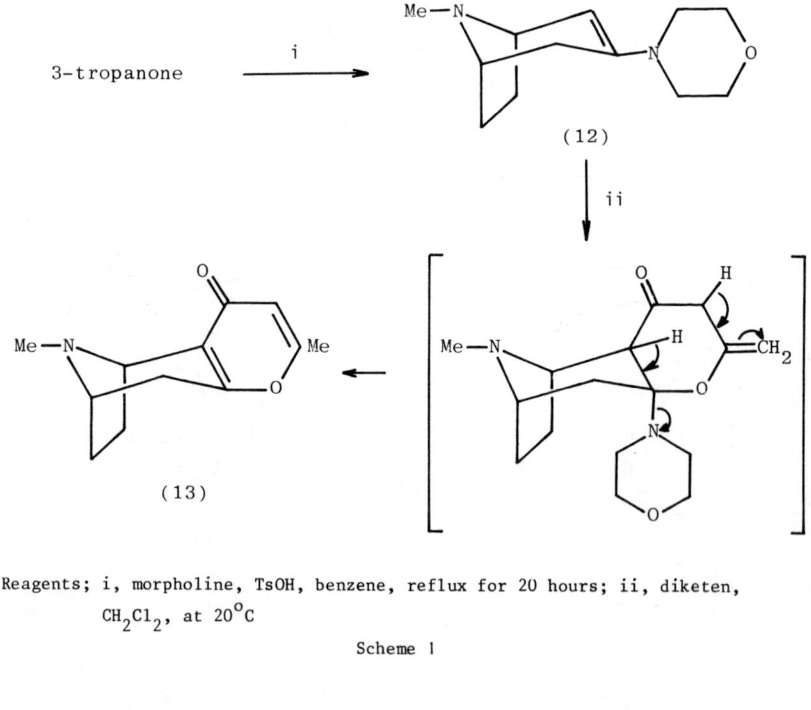

Reagents; i, morpholine, TsOH, benzene, reflux for 20 hours; ii, diketen, CH_2Cl_2, at 20°C

Scheme 1

The chemical and kinetic characteristics[10] of the hydrolysis of cocaine and its major metabolite benzoylecgonine were studied in some detail.

Quaternary derivatives[11] of N-(7-theophyllinyl- and 7-theobrominyl-alkyl)noratropines have been made.

Photolysis[12] of 3α-tropanol in the presence of ketones gave amino-alcohols of type (14) and some of the dimeric product (15).

(14)

(15)

Anisodamine[13] analogs have been synthesized[14] by esterification of 6β-hydroxy-3α-tropanol with mandelic, tropic, cinnamic acids, etc. Also, new esters of 6-tropen-3α-ol (16) are mentioned.

Tropane-3-spiro-4'-imidazol-5'-one (17) has been synthesized and its structure determined by X-ray crystallography;[15] the piperidine ring adopts a distorted chair conformation, and according to n.m.r. data this holds true in solution. Surprisingly, addition of CN⁻ to 3-tropanone imine (18) occurs from the α-side. This was attributed[15] to the larger compression of the 3α-amino-group in (19) than of the linear 3α-cyano-group in (20) by the 6α,7α- (endo) hydrogens; i.e. the addition is product-controlled.

The two stereoisomers of 3-aminotropane-3-carboxylic acids have been prepared[16] by the hydrolysis of the α- and β-tropane-3-spiro-5-hydantoins (21) and (22); thus their configurations were correlated. The N-formyl derivatives of these compounds have been prepared. The 3β-formamidotropane-3α-carboxylic acid is hydrolysed about 5 times as fast as the 3α-formamido isomer. This difference is attributed to steric blockade of the 3-axial aminotropane derivatives by the 6α- and 7α-hydrogens.

(16) (17)

(18) → (19)

(18) → (20)

(21) (22)

The 3α-tropanol β-riboside (23) has been prepared[17] by transglycosylation of the amino-alcohol with 1-β-acetyl-2,3,5-tribenzoylribofuranose, catalysed by boron fluoride etherate, followed by Zemplén deacylation in methanol, with sodium methoxide as a catalyst.

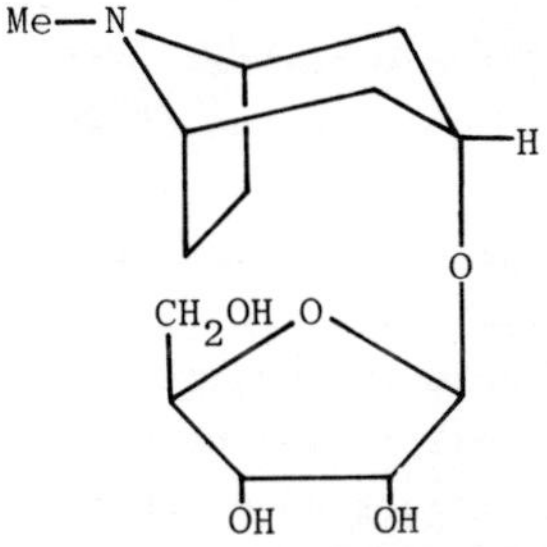

(23)

Conformational analysis of isomeric cocaines has been carried out.[18] The study of ^{1}H and ^{13}C n.m.r. spectra established the chair conformation of the piperidine ring in cocaine, pseudococaine, allococaine, and allopseudococaine. The spectral data also suggest that pseudococaine (24) and allopseudococaine, which are 2α-carbomethoxytropanols, have a larger proportion of axial N-methyl substituents than cocaine (25) and allococaine, where the 2β-placed carbomethoxy-group would interact with an axial N-methyl group. Allopseudococaine was prepared by the reduction of 2-carbomethoxytropinone by sodium borohydride, which proved to be stereoselective.

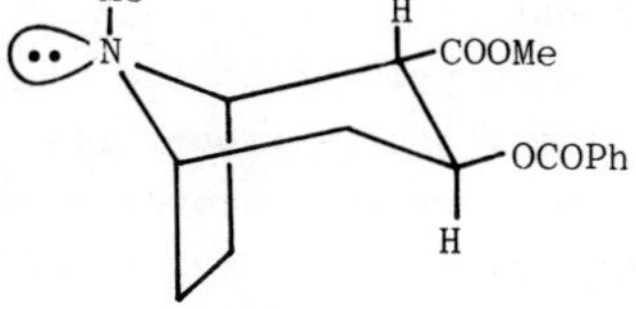

(24) 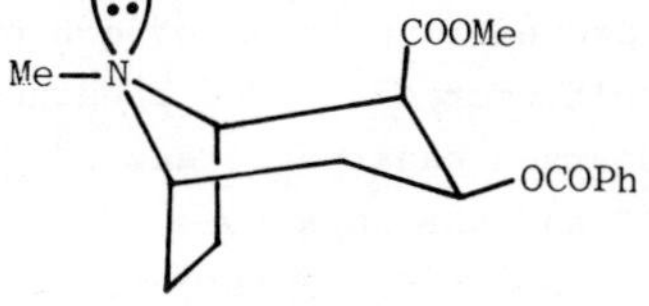   (25)

3 Pharmacology

Atropine was used in a wide variety of studies; among others, of changes in equilibrium performance in monkeys,[19] an interaction with metoclopramide in esophageal sphincter tone,[20] on nicotinic transmission in sympthetic ganglia,[21] on experimentally induced obstruction of an airway in man,[22] on the effect of atropine and of homatropine on the heart,[23] on behavioral effects in man and monkeys,[24] and on the hypnogenic action of stimulation of the basal forebrain.[25]

Cocaine (25) has again received a lot of attention. Stereospecific effects of derivatives of (25) on the uptake of [^{3}H]-dopamine,[26] structural effects on cocaine radioimmunoassay,[27] kinetics of binding of [^{3}H]-(25) in the mouse brain,[28] pituitary adrenocortical response in rats that had been acutely and chronically treated with cocaine,[29] the effect of intranasal (25) on experimental pain in man,[30] effects of acute and daily administration of cocaine on performance under a delayed matching-to-sample procedure,[31] and structural relationships between cocaine-induced or dopaminergic kindling and electrical kindling[32] were studied. Experiments on reinstatement of cocaine-reinforced responding in rats,[33] on the effect of (25) on conditioned taste aversion,[34] and on the effect of pretreatment with lidocaine on cocaine-induced behavior[35] were carried out.

Behavioral and physiological effects of the self-administration of cocaine in humans have been studied.[36] Cocaine and the contractile response of vascular smooth muscle from spontaneously hypertensive rats[37] and the disruption of primate social behavior by d-amphetamine and cocaine were studied.[38] Effects of the self-administration of cocaine by naive rats have been observed.[39]

The pharmacology of scopolamine was extensively investigated. It proved to have an effect on the development of the chick and of rabbit embryo.[40] The binding of ^{3}H-labelled methylscopolamine to dispersed pancreatic acini[41] and the aversive stimulus properties of[42] and the impairing of spatial maze performance by[43] scopolamine were reported. Anisodamine, a 6-hydroxy-atropine (or hyoscyamine) was shown to have an antishock function in experiments.[44,45] Anisodine, a 2,3-dihydroxy-2-phenylpropionic ester of scopine, has a strong depressant effect[46] in rabbits.

4 Analytical Studies

Details of a patented colorimetric method for detection of cocaine have been published.[47] Carbon-13 n.m.r. has been used to detect cocaine in forensic drug analysis.[48] The contents of scopolamine and hyoscyamine in the dried flowers of Datura metel were determined by gas chromatography.[49] Liquid-membrane electrodes[50] that are sensitive to atropinium and novatropinium cations have been developed. Direct potentiometry and potentiometric titrations have been used to determine atropine and novatropine in pharmaceutical preparations, with satisfactory results. A sensitive gas chromatography-mass spectrometry assay has been developed for atropine.[51] High-performance liquid chromatographic analysis of cocaine in human plasma was carried out.[52]

References

1 T.Yao, Z.Chen, D.Yi, and G.Xu, Yaoxue Xuebao, 1981, 16, 582 (Chem. Abstr., 1982, 96, 48972c).
2 V.Gambaro, C.Labbe, and M.Castillo, Bol. Soc. Chil. Quim., 1982, 27, 296 (Chem. Abstr., 1982, 97, 3537u).
3 J.-S.Yang, Y.-W.Chen, and H.-C.Feng, Chung Tsao Yao, 1981, 12, 4 (Chem. Abstr., 1981, 95, 138446t).
4 P.K.Hsiao and L.Ho, Yao Hsueh T'ung Pao, 1980, 15, 41 (Chem. Abstr., 1981, 95, 138447u).
5 M.Sahai and A.B.Ray, J. Org. Chem., 1980, 45, 3265. See these Reports 1981, 11, 36.
6 A.R.Pinder, J. Org. Chem., 1982, 47, 3607; also personal communication to one of the Reviewers (GF).
7 A.B.Ray, Y.Oshima, H.Hikino, and C.Kabuto, Heterocycles, 1982, 19, 1233.
8 R.K.Hill and L.A.Renbaum, Tetrahedron, 1982, 38, 1959.
9 M.Lounasmaa, T.Langenskiöld, and C.Holmberg, Tetrahedron Lett., 1981, 22, 5179.
10 R.Erasmo, Diss. Abstr. Int. B, 1981, 42, 151 (Chem. Abstr., 1981, 95, 138538z).
11 K.H.Klingler, R.Aurich, and S.Habersang, USSR P. 858566 (Chem. Abstr., 1982, 96, 85832c).
12 L.M.Kostochka and A.P.Skoldinov, Zh. Org. Khim., 1981, 17, 2459 (Chem. Abstr., 1982, 96, 85820x).
13 Anisodamine appears to be racemic,6β-hydroxyatropine or 6β-hydroxyhyoscyamine; see these Reports, 1981, 12, 49.
14 J.Xie, J.Zhou, C.Zhang, and J.Yang, Yaoxue Xuebao, 1981, 16, 767 (Chem. Abstr., 1982, 96, 143127q).
15 E.Galvez, M.Martinez, G.G.Trigo, F.Florencio, J.Vilches, S.Garcia-Blanco, and J.Bellanato, J. Mol. Struct., 1981, 75, 241.
16 G.G.Trigo, C.Avendano, E.Santos, H.N.Christensen, and M.E.Handlogten, Can. J. Chem., 1980, 58, 2296.
17 C.Chavis, E.Dumont, F.Grodenic, and J.L.Imbach, J. Carbohydr., Nucleosides, Nucleotides, 1981, 8, 507.
18 F.I.Carroll, M.L.Coleman, and A.H.Lewin, J. Org. Chem., 1982, 47, 13.
19 C.T.Bennett, N.E.Lof, D.N.Farrer, and J.L.Mattsson, Report 1981, SAM-TR-81-29; Gov. Rep. Announce. Index (U.S.), 1982, 82, 1316 (Chem. Abstr., 1982, 96, 210871t).

20 J.G.Brock-Utne, G.E.Dimopoulos, J.W.Downing, and M.G.Moshal, S. Afr. Med. J., 1982, 61, 465.
21 M.Yamada, Kurume. Med. J., 1981, 28, 119 (Chem. Abstr., 1982, 96, 210764k).
22 N.M.Eiser and A.Guz, Clin. Respir. Physiol., 1982, 18, 449.
23 Y.Tamou, Toho Igakkai Zasshi, 1981, 28, 246 (Chem. Abstr., 1982, 96, 79628z).
24 J.L.Mattsson, C.T.Bennett, and D.N.Farrer, Report 1981, SAM-TR-81-16; Gov. Rep. Announce. Index (U.S.), 1981, 81, 5506 (Chem. Abstr., 1982, 96, 79824k).
25 G.Benedek, F.Obal,Jr., F.Bari, and F.Obal, Sleep (Basel), 1980, 5, 272.
26 R.E.Heikkila, L.Manzino, and F.S.Cabbat, Subst. Alcohol Action/Misuse, 1981, 2, 115 (Chem. Abstr., 1981, 95, 161776j).
27 R.D.Budd, Clin. Toxicol., 1981, 18, 773.
28 M.E.A.Reith, H.Sershen, and A.Lajtha, J. Recept. Res., 1981, 2, 233.
29 R.Ramos-Aliaga and G.Werner, Res. Commun. Subst. Abuse, 1982, 3, 29 (Chem. Abstr., 1982, 96, 135781h).
30 J.C.Yang, W.C.Clark, J.Dooley, and F.V.Mignogna, Anesth. Analg. (Cleveland), 1982, 61, 358.
31 M.N.Branch and M.E.Dearing, Pharmacol. Biochem. Behav., 1982, 16, 713.
32 M.Sato and M.Okamoto, in 'Kindling', ed. J.A.Wada, Raven, New York, 1981, Vol. 2, p.105.
33 H.DeWitt and J.Stewart, Psychopharmacology (Berlin), 1981, 75, 134.
34 G.D.Mello, D.M.Goldberg, and I.P.Stolerman, J. Pharmacol. Exp. Ther., 1981, 219, 60.
35 K.M.Squillace, R.M.Post, and A.Pert, Neuropsychobiology, 1982, 8, 113.
36 M.W.Fischman and C.R.Schuster, Fed. Proc. Fed. Am. Soc. Exp. Biol., 1982, 41, 241.
37 R.C.Webb and P.M.Vanhoutte, Arch. Int. Pharmacodyn. Ther., 1981, 253, 241.
38 K.A.Miczek and H.Yoshimura, Psychopharmacology (Berlin), 1982, 76, 163.
39 M.Papasava, T.P.S.Oei, and G.Singer, Pharmacol. Biochem. Behav., 1981, 15, 458.
40 W.G.McBride, P.H.Vardy, and J.French, Aust. J. Biol. Sci., 1982, 35, 173.
41 H.E.Appert, T.H.Chiu, G.C.Budd, A.J.Anthony, and J.M.Howard, Cell Tissue Res., 1981, 220, 673.
42 S.W.MacMahon, N.M.Blampied, and R.N.Hughes, Pharmacol. Biochem. Behav., 1981, 15, 389.
43 R.Stevens, Physiol. Behav., 1981, 27, 385.
44 H.Guo and R.Sun, Kexue Tongbao, 1982, 27, 320 (Chem. Abstr., 1982, 97, 454d).
45 C.Tang, L.Wu, and J.Su, Kexue Tongbao, 1981, 26, 1402 (Chem. Abstr., 1982, 96, 97378x).
46 C.Bian and S.Duan, Yaoxue Xuebao, 1981, 16, 801 (Chem. Abstr., 1982, 96, 46211e).
47 A.Reiss, U.S. P. 4320086 (1982) (Chem. Abstr., 1982, 97, 51086t).
48 S.Alm, B.Bomgren, H.B.Boren, H.Karlsson, and A.C.Maehliy, Forensic Sci. Int., 1982, 19, 271.
49 C.Ye and S.Zhang, Zhongcaoyao, 1981, 12, 493 (Chem. Abstr., 1982, 96, 214316p).
50 E.P.Diamandis, E.Athanasiou-Malaki, D.S.Papastathopoulos, and T.P. Hadjiioannou, Anal. Chim. Acta, 1981, 128, 239.
51 M.Eckert and P.H.Hinderling, Agents Actions, 1981, 11, 520.
52 A.N.Masoud and D.M.Krupski, J. Anal. Toxicol., 1980, 4, 305.

4
Pyrrolizidine Alkaloids

BY D. J. ROBINS

A comprehensive review on pyrrolizidine alkaloids has been published.[1] Lists of plant species known to contain pyrrolizidine alkaloids are included, together with the structures of over 200 known alkaloids.

1 Synthesis of Macrocyclic Pyrrolizidine Alkaloids

The outstanding highlights of this year's work are undoubtedly the long-awaited first successful syntheses of macrocyclic pyrrolizidine alkaloids. Devlin and Robins have prepared (+)-dicrotaline (2)[2] and Narasaka et al. have made (±)-integerrimine (23).[3] These are both diesters of retronecine (1) with 11- and 12-membered macrocycles, respectively. In their synthesis of dicrotaline (2), Devlin and Robins treated (+)-retronecine (1) with the trimethylsilyl ether of 3-hydroxy-3-methylglutaric anhydride to give a quantitative mixture of the 7- and 9- monoesters of retronecine. Cyclisation was achieved using the pyridine-2-thiol esters to yield two main products after removal of the trimethylsilyl

		Δδ(H-9)/p.p.m.
(2)	R^1= OH, R^2= Me	1.24
(3)	R^1= Me, R^2= OH	0.98
(4)	$R^1 = R^2$ = Me	1.24
(5)	$R^1—R^2$ = $(CH_2)_4$	1.23
(6)	$R^1 = R^2$ = H	0.62
(7)	$R^1 = R^2$ = Ph	0.00
(8)	R^1= H, R^2= Me	0.51 / 0.99

protecting group. (+)-Dicrotaline (2) was separated by t.l.c. (30% yield) and shown to be identical with natural material isolated from Crotalaria dura seeds. The other product was the epimer (3) of dicrotaline, also obtained in 30% yield. The most interesting feature in the p.m.r. spectrum of dicrotaline (2) is an AB quartet due to the non-equivalent protons at C-9, with a chemical-shift difference [Δδ(H-9)] of 1.24 p.p.m. This is the largest value observed so far for an 11-membered macrocyclic diester of retronecine, but is the same as that recorded for the related unnatural dilactone (4).[4] The absolute configuration at C-13 in dicrotaline (2) was shown to be S by degrading dicrotaline to (R)-mevalonolactone (10) (Scheme 1), by hydrogenolysis of the allylic ester of (2), followed by selective reduction of the ester group in (9) using sodium in liquid ammonia.

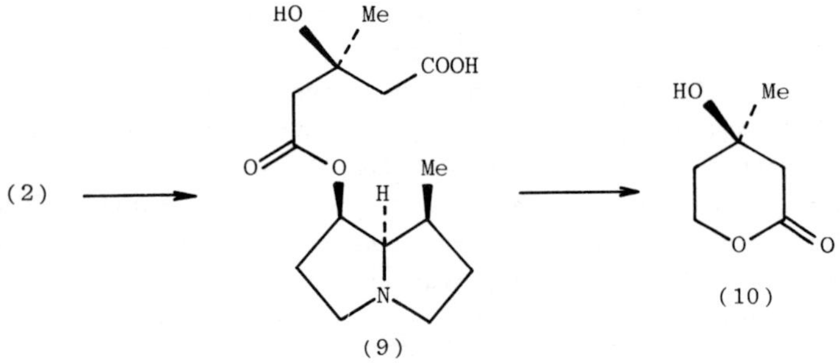

Scheme 1

A series of 11-membered dilactones [(4) - (8)] have been prepared in 50 - 75% yield from (+)-retronecine (1) and the corresponding substituted glutaric anhydride (cf. Vol. 11, p. 49).[5] These pyrrolizidine alkaloid analogues exhibited a range of Δδ (H-9) values shown on the formulae. These values probably reflect the different conformations in the diacid portions. The hydrobromide of (4) is toxic to rats.[6] It is readily metabolised to the corresponding pyrrole derivative and has hepatotoxicity similar to monocrotaline.

The synthesis of (±)-integerrimine (23) required a much longer procedure. Firstly, the acid portion of integerrimine was synthesized by Narasaka and Uchimaru from 2-methyl-2-cyclopentenone as shown in Scheme 2.[7] The carboxylation was stereoselective, giving mainly the ester (11) on methylation, in 80% overall yield. Baeyer-Villiger oxidation of (11) was regioselective, producing the

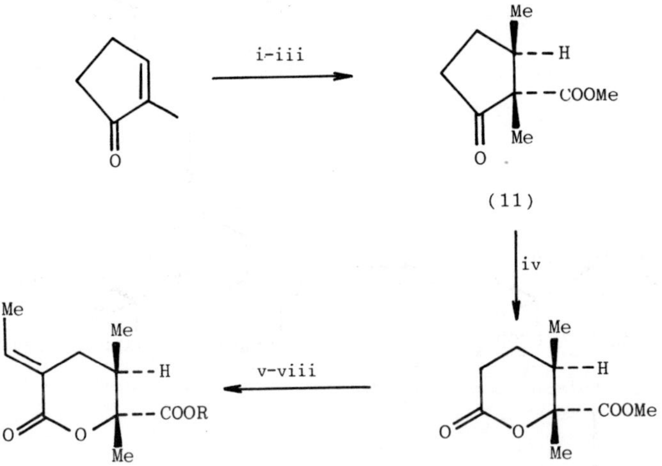

Reagents: i, Me_2CuLi; ii, CO_2; iii, CH_2N_2; iv, $m\text{-}ClC_6H_4CO_3H$; v, $LiNPr^i_2$; vi, MeCHO; vii, 2-fluoro-1-methylpyridinium toluene-p-sulphonate, Et_3N; viii, LiOH

Scheme 2

lactone (12) in 76% yield, together with 13% of the undesired regioisomer. Dehydration of the aldol adduct of (12) was difficult, but eventually gave 60% of the (E)-isomer (13) and 9% of the (Z)-isomer. The desired lactone (14) was obtained by alkaline hydrolysis. The next task was to protect the free carboxyl group in (14), liberate the other carboxyl group to give (15), and convert this into an anhydride (19), ready for coupling to the necine portion.[3] This is outlined in Scheme 3. The free carboxyl group and the t-hydroxyl group in (15) were protected sequentially to give (16). Removal of the silyl group and oxidation of the methoxythiomethyl ether yielded the sulphone (17). After protecting the free acid in (17), the other carboxyl group was liberated by selective hydrolysis of the methanesulphonylmethyl group so that the anhydride (19) could be formed. (+)-Retronecine (1) was prepared by a modification of the method of Geissman and Waiss.[8] The lithium salt (20) of the protected retronecine reacted with

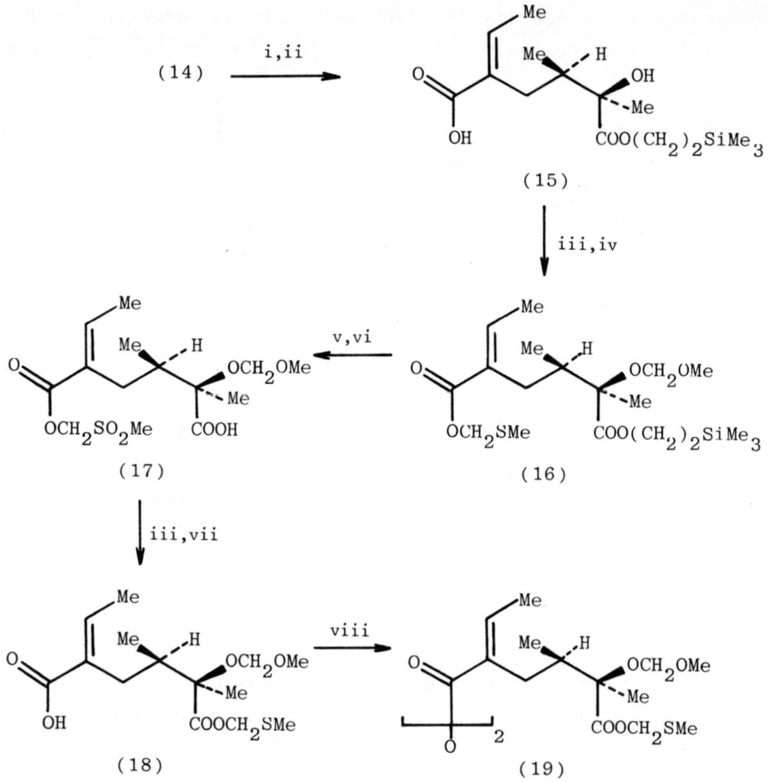

Reagents: i, $HO(CH_2)_2SiMe_3$, 2-chloro-1-methylpyridinium iodide, Et_3N; ii, LiOH, H_2O_2; iii, $MeSCH_2Cl$, NaI, Pr^i_2NEt; iv, $MeOCH_2Cl$, NaI, Pr^i_2NEt; v, Bu_4N^+ F^-; vi, H_2O_2, $(NH_4)_6Mo_7O_{24}$; vii, NaOH; viii, 2-chloro-1-methylpyridinium iodide, Et_3N

Scheme 3

the anhydride (19) to give the monoester (21) (Scheme 4). The silyl group was removed and lactonisation was achieved by intramolecular displacement of the methanesulphonylmethyl group to give (22) and its C-12 diastereoisomer. The N-oxides were reduced, and the mixture of diastereoisomers was separated by t.l.c. Further treatment with acid yielded (±)-integerrimine (23).

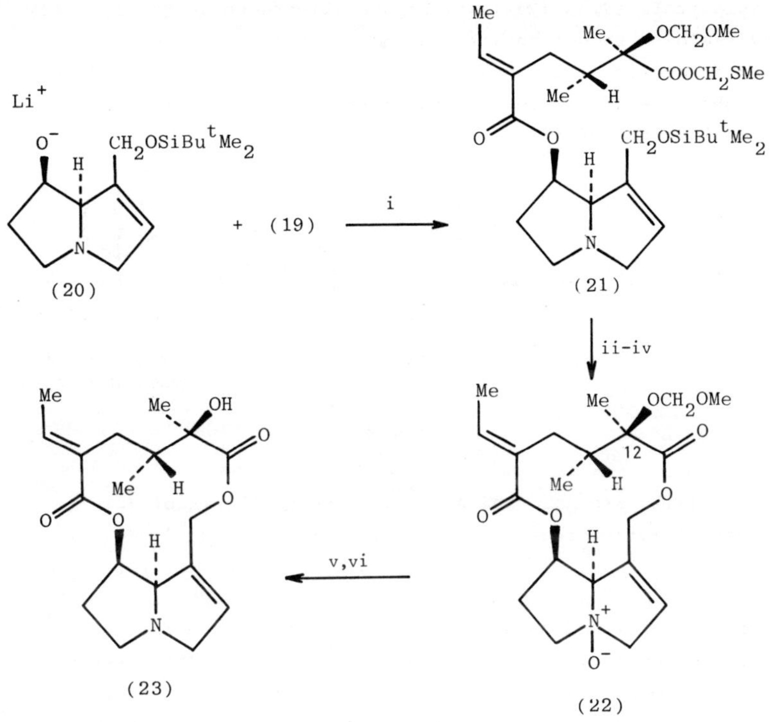

Reagents: i, 4-(dimethylamino)pyridine; ii, $NH_4^+ F^-$; iii, H_2O_2, $(NH_4)_6Mo_7O_{24}$; iv, Bu^nLi; v, Zn, H_2SO_4; vi, H_2SO_4

Scheme 4

2 Synthesis of Necine Bases

Synthesis of (+)-trachelanthamidine (26) was previously achieved using (-)-4-hydroxy-L-proline as starting material.[9] Takano et al. have devised a short route to optically active (26) from an achiral precursor.[10] Deaminative dimerisation of 4-aminobutanal dimethylacetal with Raney nickel in hot benzene gave the symmetrical amine (24) in 71% yield. The corresponding dialdehyde, liberated by acid, underwent a Mannich reaction to afford

1-formylpyrrolizidine (25), which gave (±)-trachelanthamidine (26) in 44% yield on reduction with sodium borohydride (Scheme 5). Asymmetric cyclisation was achieved in the presence of pyridinium (+)-camphor-10-sulphonate, giving (+)-trachelanthamidine (26) with an optical purity of 33%.

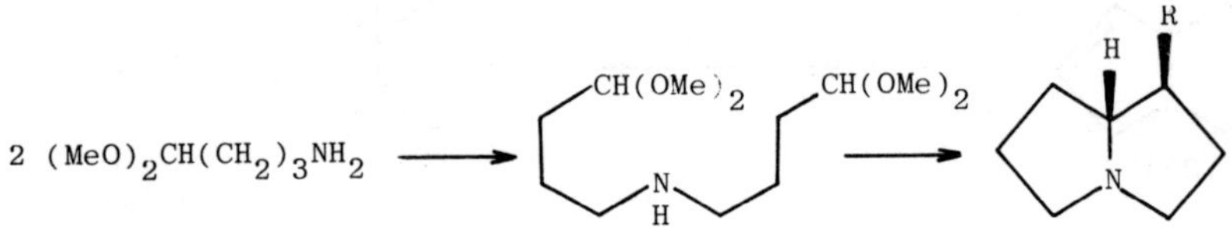

2 $(MeO)_2CH(CH_2)_3NH_2$ → (24) → (25) R = CHO / (26) R = CH_2OH

Scheme 5

Full details are now available[11] of the synthesis of (±)-isoretronecanol [diastereoisomer of (26)] utilising the 1,3-dipolar cycloaddition of a cyclic nitrone to dihydrofuran (cf. Vol. 11, p. 45).

There are still comparatively few synthetic routes to the unsaturated necines. Chamberlin and Chung[12] have used a keten thioacetal group to act as a terminator for a cationic cyclisation and to control the position of the final double bond in their synthesis of (±)-supinidine (30). The key step (Scheme 6) required the generation under basic conditions of the acyliminium ion (27), which cyclised to the pyrrolizidinone (28) in high yield. Migration of the double bond in (28) was achieved by protonation of the lithiated keten thioacetal α to the sulphur atom, leading to the endocyclic compound (29). Finally, hydrolysis of the dithiane and reduction of the amide gave (±)-supinidine (30) in ca 16% overall yield.

The lactone (31) was previously synthesized by Geissman and Waiss[8] from β-alanine as an intermediate in their route to (±)-retronecine (1). Rueger and Benn have now reported[13] a synthesis of the (+)-lactone (31) in 12 steps from (−)-4-hydroxy-L-proline (32) (no yields were given). This therefore constitutes a potential route to (+)-retronecine.

Details[14] of the synthesis of the ring system of the loline alkaloids [as in loline (33)] have now been provided (cf. Vol. 10, p. 50). The authors speculate on the 'close' biogenetic

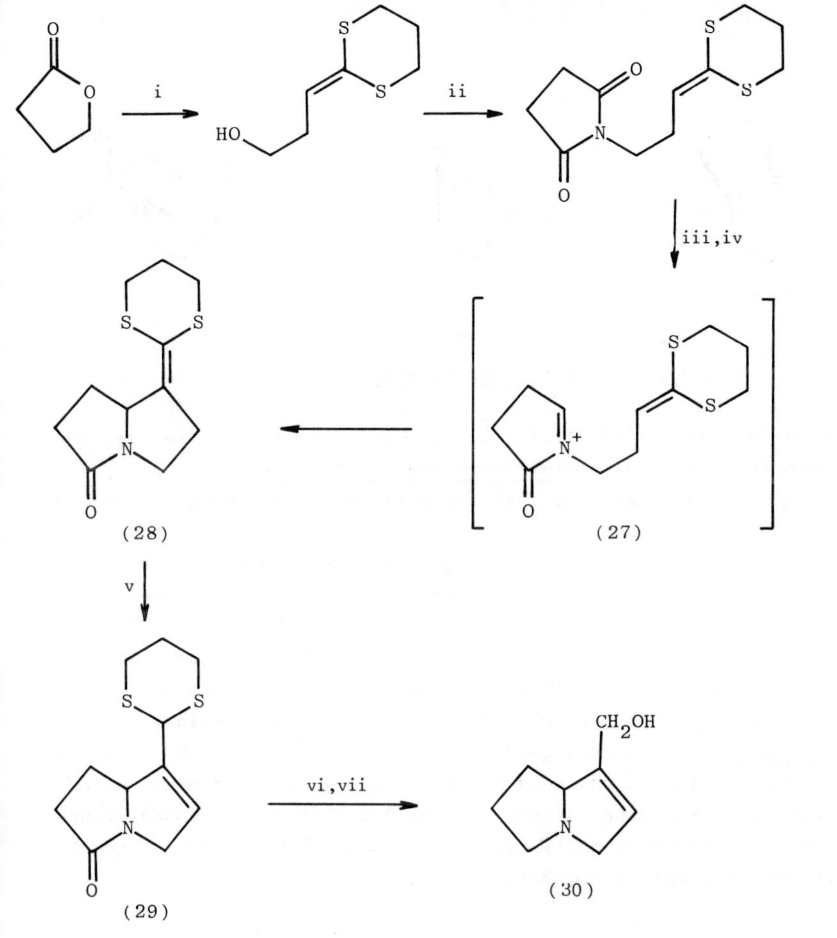

Pyrrolizidine Alkaloids

Reagents: i, $Me_2AlS(CH_2)_2SAlMe_2$; ii, succinimide, PPh_3, EtOOCN=NCOOEt; iii, $NaBH_4$; iv, $MeSO_2Cl$, pyridine, Et_3N; v, $LiNPr^i_2$, hexamethylphosphoramide, then MeOH; vi, $BF_3 \cdot OEt_2$, HgO; vii, $LiAlH_4$

Scheme 6

relationship between loline (33) and retronecine (1), but fail to point out that retronecine has the opposite absolute configuration to loline (33) at C-7 and C-8.

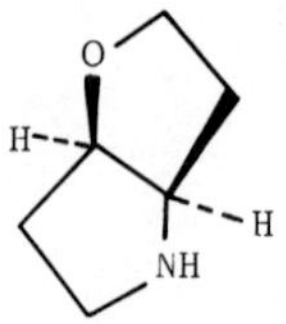

(31) 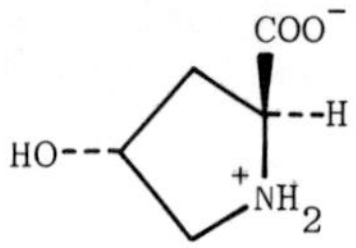   (32) 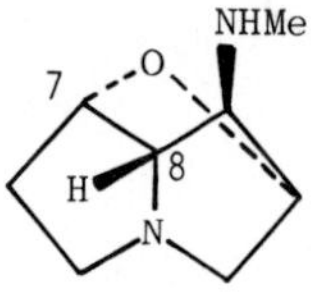   (33)

3 Alkaloids of the Boraginaceae

A new pyrrolizidine alkaloid, together with retronecine (1), has been isolated from <u>Heliotropium</u> <u>ovalifolium</u> Forsk. by Mohanraj et al.[15] The structure (34) for helifoline was deduced from spectral evidence and by hydrolysis of helifoline to a base, believed to be croalbinecine, and presumably angelic acid (not isolated). Croalbinecine has been found only once before as part of the macrocyclic alkaloid croalbidine (<u>cf</u>. Vol. 6, p. 72). Unfortunately, direct comparison with croalbinecine was not possible. However, the bulk of the spectral data for the base from helifoline are in agreement with those of croalbinecine.
H. ovalifolium is reported to be toxic to animals. The source of this poisoning is unlikely to be pyrrolizidine alkaloids, since neither retronecine (1) nor an alkaloid with the proposed structure (34) contains an allylic ester group, which is a structural requirement for hepatotoxicity.

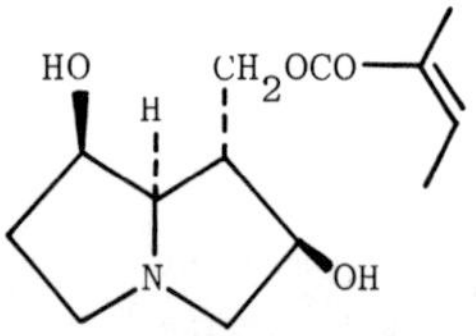

(34)

Extracts from the true forget-me-not, <u>Myosotis</u> <u>scorpioides</u> L., are reported to have a curare-like action on animals. Meinwald and co-workers have extracted a mixture of pyrrolizidine alkaloids from aerial parts of this species, and have shown that this mixture is responsible for the observed biological activity.[16]

Pyrrolizidine Alkaloids

Three new alkaloids [(35)-(37)] were present in this mixture, together with symphytine (38). The structure (35) for 7-acetylscorpioidine was established by spectroscopic data, by hydrogenolysis of (35) to give 7-acetylretronecanol, and by alkaline hydrolysis of (35) to yield retronecine (1), acetic, tiglic, and (2$\underline{S}$,3$\underline{S}$)-viridifloric acids. Scorpioidine (36) was also separated and gave (35) on acetylation. An epimer of scorpioidine, anadoline (39), was previously isolated from two Symphytum species (cf. Vol. 6, p. 75; Vol. 8, p. 50). The remaining two minor alkaloids could not be separated. Alkaline hydrolysis of this mixture gave retronecine (1), tiglic, (2$\underline{S}$,3$\underline{S}$)-viridifloric, and (2$\underline{S}$,3$\underline{R}$)-trachelanthic acids. The p.m.r. spectrum indicated that the mixture consists of symphytine (38) and a new alkaloid, myoscorpine (37), in equal amounts.

(35) R^1 = Ac, R^2 = O-tiglyl, R^3 = H
(36) R^1 = H, R^2 = O-tiglyl, R^3 = Ac
(37) R^1 = tiglyl, R^2 = H, R^3 = OH
(38) R^1 = tiglyl, R^2 = OH, R^3 = H
(39) R^1 = H, R^2 = H, R^3 = O-tiglyl

Symphytine (38) and related pyrrolizidine alkaloids present in various Symphytum preparations have been identified by gas chromatography.[17]

4 Alkaloids of the Compositae

Previous investigations of Tussilago farfara L. (coltsfoot) have yielded two macrocyclic pyrrolizidine alkaloids, senecionine and senkirkine.[1] A new pyrrolizidine ester, tussilagine, has now been isolated from this species by Röder et al.[18] The structure (40) for tussilagine is in accord with the spectral data. The relative configuration of the hydrogens at C-1 and C-8 is deduced to be cis from the p.m.r. spectral vicinal coupling constant, $J_{1,8}$ = 10 Hz. The hydroxyl and ester groups are also believed to be cis, since hydrogen bonding is observed in the i.r. and p.m.r. spectra.

(40)

(41)

A new macrocyclic alkaloid, senecicannabine (41), has been isolated from Senecio cannabifolius by Furuya and co-workers, together with the known alkaloids jacozine and seneciphylline.[19] The mass spectrum of (41) was typical for a macrocyclic diester of retronecine (1), and the p.m.r. chemical-shift difference of 1.38 p.p.m. for the C-9 protons is ascribed to a 12-membered ring. The presence of the two epoxide groups was indicated by the ^{13}C n.m.r. spectrum. Hydrolysis of senecicannabine yielded (+)-retronecine (1), and the proposed structure (41) was confirmed by X-ray analysis.

Two known dihydropyrrolizinone alkaloids, pterophorine and an isomer of senaetnine, were isolated from S. kleinia Less.[20] The latter compound is also present in Notonia petraea.[21] Neosenkirkine and senkirkine have been isolated from S. pierotii.[22] Senecionine, neoplatyphylline, and platyphylline (46) are present in S. congestus.[23] Senecionine and seneciphylline were found in S. pancicii Degen.[24] Otosenine is a minor constituent of S. aegyptius L. and senecionine is present in S. desfontainei Druce.[25] A retronecine ester is reported to be present in S. otites Kunze ex DC.[26]

5 Alkaloids of the Leguminosae

The N-oxides of senecionine and 1-methylenepyrrolizidine were isolated from the perennial shrub Crotalaria anagyroides H.B. & K.[27] Isocromadurine and crispatine (macrocyclic diesters of retronecine) and turneforcidine (1α-hydroxymethyl-7β-hydroxy-8α-pyrrolizidine) were obtained from the pericarps of C. candicans.[28] Usaramine is present in Chinese C. mucronata.[29]

6 Alkaloids in Micro-organisms

A metabolite isolated from Streptomyces olivaceus by Box and Corbett[30] has been assigned the dihydropyrrolizinone structure (42) on the basis of spectroscopic data. This structure and the absolute configuration were established by synthesis of (42) from L-aspartic acid monomethyl ester in ca. 14% overall yield (Scheme 7).

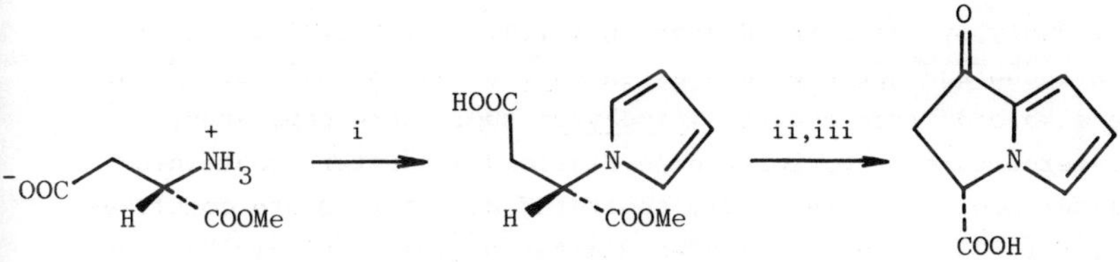

Reagents: i, 2,5-diethoxytetrahydrofuran, AcOH, at 80°C; ii, P_2O_5; iii, NaOH in aqueous MeOH

Scheme 7

7 Alkaloids in Insects

The major volatile component of the scent organ in male moths of two Creatonotos species is hydroxydanaidal (43).[31] The formation of this presumed pheromone depends upon the ingestion of plants containing pyrrolizidine alkaloids by larvae of these moths. It is suggested[31] that the presence of pyrrolizidine alkaloids controls the development of these scent organs.

Males of the flea beetle Gabonia gabriela feed on withered Heliotropium plants. This may provide a source of pyrrolizidine alkaloids but the significance of this observation is not yet clear.[32]

(43)

8 General Studies

Separation of pyrrolizidine alkaloid mixtures has been carried out previously by ion-pair adsorption t.l.c. (cf. Vol. 12, p. 66). This technique has now been extended for use on h.p.l.c.,[33] and the alkaloids from comfrey (Symphytum spp.) were separated. Diastereoisomeric pyrrolizidine alkaloid monoesters containing a vicinal diol were previously separated as their borate complexes (cf. Vol. 12, p. 62). Related diastereoisomeric alkaloids have now been separated by column chromatography on silica gel pretreated with sodium hydroxide.[34]

The high-resolution p.m.r. and ^{13}C n.m.r. spectra of some monoester alkaloids have been recorded.[35] Trachelanthates and viridiflorates [cf. (37) and (38)] can be differentiated in these spectra. Further ^{13}C n.m.r. spectra of macrocyclic[36] and mono- and diester[37] pyrrolizidine alkaloids have been reported. The mass-spectral fragmentation patterns of pyrrolizidine alkaloids have been discussed (in Chinese).[38]

The X-ray structure of junceine (44) has been determined by Stoeckli-Evans.[39] Like all 11-membered macrocyclic diesters of retronecine except trichodesmine (45), the ester groups are syn-parallel and directed below the plane of the macrocycle. X-Ray structures have also been reported for platyphylline (46),[40] lasiocarpine (47),[41] 7-angelylheliotridine (48),[42] and (again) monocrotaline N-oxide.[43] The quantitative determination of platyphylline (46) by a complexometric back titration has been discussed.[44]

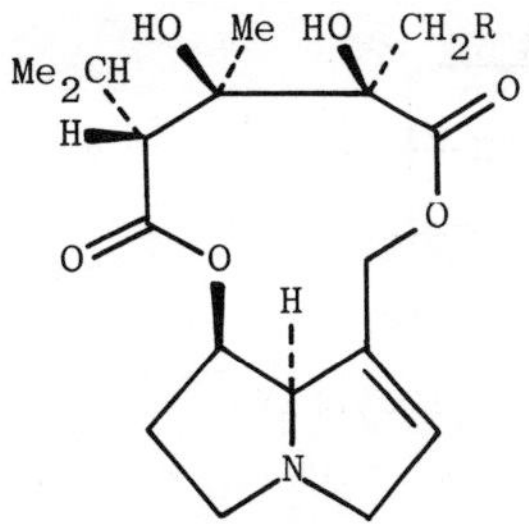

(44) R = OH
(45) R = H

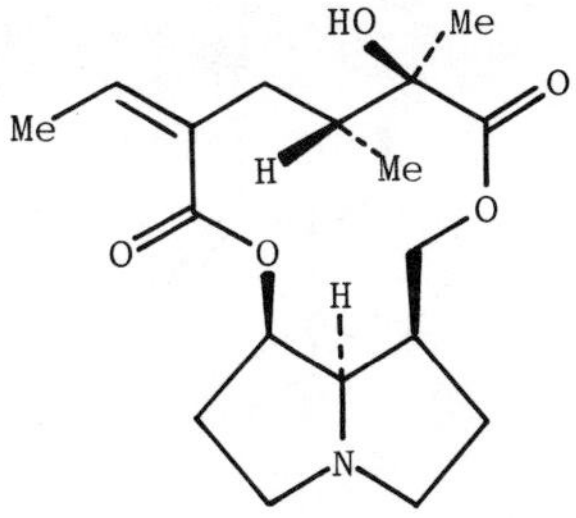

(46)

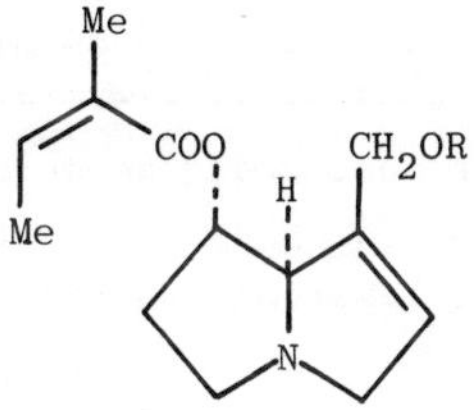

(47) R = —C(=O)—CH(OH)—C(OH)Me$_2$ / CH(OMe)—Me

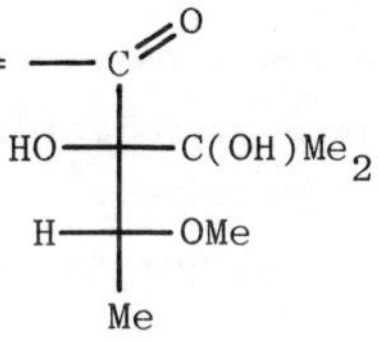

(48) R = H

9 Pharmacological and Biological Studies

The health hazards of pyrrolizidine alkaloids have been reviewed.[45] Indicine N-oxide (49) is an antitumour agent which is being used in clinical trials. It has now been synthesized in radioactive form to aid studies of its biological activity.[46] [9-^{3}H]Retronecine was prepared as described previously (cf. Vol. 7, p. 62). Esterification with (-)-trachelanthic acid (isopropylidene derivative) was carried out using N,N'-dicyclohexylcarbodiimide and a catalytic amount of 4-(dimethylamino)pyridine. Acid treatment and oxidation with m-chloroperoxybenzoic acid then yielded indicine N-oxide (49).

The diastereoisomeric N-oxides (50) have been prepared from (+)-retronecine (1) and (±)-2-hydroxy-2-phenylbutanoic acid, using N,N'-carbonyldiimidazole as the coupling reagent (cf. Vol. 8, p. 48).[47] This diastereoisomeric mixture is a more active anti-

(49) R = —C(=O)—C(Me₂CH)(OH)—C(HO)(H)—Me

(50) R = COC(OH)(Ph)CH₂Me

cancer agent than indicine N-oxide. The conformations of the separated diastereoisomers were studied in solution by p.m.r. spectroscopy.

A series of quaternary ammonium derivatives of 9-thioretronecine have been made, and (51) was the most effective on blocking neuromuscular activity in mice.[48] The preparation of a wide range of 8-substituted pyrrolizidines and their quaternary ammonium salts has been patented.[49] These compounds display a variety of biological activities, including muscle relaxation and dilation of coronary arteries.

Pyrrolizidine alkaloids may be present in some traditional herbs used in Sri Lanka, and this may account for the high incidence of chronic liver disease in this country.[50] Chinese workers are becoming interested in the structure-activity relationships of native pyrrolizidine alkaloids.[51] Veno-occlusive disease in some Israeli Arabs may be due to pyrrolizidine alkaloids.[52] The carcinogenic activity of an alkaloidal extract from the phytotherapeutic drug Senecio nemorensis ssp. fuchsii has been examined.[53]

The effects of pyrrolizidine alkaloids on animals continue to receive wide attention. Pulmonary hypertension is induced in rats fed monocrotaline[54-56] or Crotalaria spectabilis.[57] Senecionine and senecphylline covalently bind to hepatic macromolecules, and can be transferred into the milk of lactating mice.[58,59] The effects on the hepatic drug-metabolising enzymes in rats fed S. jacobaea[60] or milk from goats fed on the same species[61] have been studied. The hepatotoxic effects of S. jacobaea were apparently enhanced by dietary copper in rats,[62] and mineral-vitamin supplements failed to prevent this poisoning in calves.[63] Dried C. juncea did not appear to be toxic to cattle.[64] Rainbow trout (Salmo gairdneri) suffered severe hepatic lesions and veno-

(51)

(52) X = Cl or F

occlusive disease when dosed with pyrrolizidine alkaloids.[65]

The protective effects of zinc,[66] butylated hydroxyanisole,[67,68] ethoxyquin,[68,69] and disulphiram [68] against pyrrolizidine alkaloid poisoning in rats and mice have been studied. Disulphiram was the least effective. A cytological analysis of the regenerating liver in rats treated with integerrimine has been made.[70]

The toxic metabolites of pyrrolizidine alkaloids are dihydropyrrolizines, produced in the liver by oxidases. These derivatives are potent alkylating agents. Dehydroretronecine has been shown to alkylate the $\underline{N}^2$ position of deoxyguanosine.[71] The anti-tumour activity of a series of synthetic dihydropyrrolizine [e.g. (52)] and hydroxymethylpyrrole derivatives has been determined.[72] Liver cell enlargement has been observed in rats given hydroxymethylpyrrole derivatives followed by hepatotoxins such as dimethylnitrosamine.[73]

References

1. D.J. Robins, Fortschr. Chem. Org. Naturst., 1982, **41**, 115.
2. J.A. Devlin and D.J. Robins, J. Chem. Soc., Chem. Commun., 1981, 1272.
3. K. Narasaka, T. Sakakura, T. Uchimaru, K. Morimoto, and T. Mukaiyama, Chem. Lett., 1982, 445.
4. D.J. Robins and S. Sakdarat, J. Chem. Soc., Chem. Commun., 1980, 282.
5. J.A. Devlin, D.J. Robins, and S. Sakdarat, J. Chem. Soc., Perkin Trans. 1, 1982, 1117.
6. A.R. Mattocks, Chem.-Biol. Interact., 1981, **35**, 301.
7. K. Narasaka and T. Uchimaru, Chem. Lett., 1982, 57.
8. T.A. Geissman and A.C. Waiss, J. Org. Chem., 1962, **27**, 139.

9. D.J. Robins and S. Sakdarat, J. Chem. Soc., Chem. Commun., 1979, 1181; J. Chem. Soc., Perkin Trans. 1, 1981, 909.
10. S. Takano, N. Ogawa, and K. Ogasawara, Heterocycles, 1981, 16, 915.
11. T. Iwashita, T. Kusumi, and H. Kakisawa, J. Org. Chem., 1982, 47, 230.
12. A.R. Chamberlin and J.Y.L. Chung, Tetrahedron Lett., 1982, 23, 2619.
13. H. Rueger and M. Benn, Heterocycles, 1982, 19, 23.
14. S.R. Wilson, R.A. Sawicki, and J.C. Huffman, J. Org. Chem., 1981, 46, 3887.
15. S. Mohanraj, P. Kulanthaivel, P.S. Subramanian, and W. Herz, Phytochemistry, 1981, 20, 1991.
16. J.F. Resch, D.F. Rosberger, J. Meinwald, and J.W. Appling, J. Nat. Prod., 1982, 45, 358.
17. P. Stengl, H. Wiedenfeld, and E. Röder, Dtsch. Apoth.-Ztg., 1982, 122, 851.
18. E. Röder, H. Wiedenfeld, and E.-J. Jost, Planta Med., 1981, 43, 99.
19. Y. Asada, T. Furuya, M. Shiro, and H. Nakai, Tetrahedron Lett., 1982, 23, 189.
20. F. Bohlmann, C. Zdero, and R.K. Gupta, Phytochemistry, 1982, 20, 2024.
21. F. Bohlmann and C. Zdero, Phytochemistry, 1979, 18, 1063.
22. Y. Asada and T. Furuya, Planta Med., 1982, 44, 182.
23. E. Röder, H. Wiedenfeld, and E.-J. Jost, Planta Med., 1982, 44, 182.
24. J. Jizba, M. Budesinsky, T. Vanek, A. Boeva, K. Dimitrova, F. Santavy, and L. Novotny, Collect. Czech. Chem. Commun., 1982, 47, 664.
25. A.-A.M. Habib, Planta Med., 1981, 43, 290.
26. A.Q. Reyes, M.O. Rodriguez, and O.M. Silva, Bol. Soc. Chil. Quim., 1982, 27, 305 (Chem. Abs., 1982, 96, 196575).
27. R. Rastogi and T.R. Rajagopalan, Pharmazie, 1982, 37, 75.
28. O.P. Suri, R.S. Jamwal, R.K. Khajuria, C.K. Atal, and C.N. Haksar, Planta Med., 1982, 44, 181.
29. G.-Q. Han and C. Tieu, Pei-ching I Hsueh Yuan Hsueh Pao, 1981, 13, 81 (Chem. Abs., 1981, 95, 138474).
30. S.J. Box and D.F. Corbett, Tetrahedron Lett., 1981, 22, 3293.
31. D. Schneider, M. Boppré, J. Zweig, S.B. Horsley, T.W. Bell, J. Meinwald, K. Hansen, and E.W. Diehl. Science, 1982, 214, 1264.
32. M. Boppré and G. Scherer, Syst. Entomol., 1981, 6, 347.
33. H.J. Huizing and T.M. Malingré, J. Chromatogr., 1981, 214, 257.
34. S. Mohanraj, W. Herz, and P.S. Subramanian, J. Chromatogr., 1982, 238, 530.
35. S. Mohanraj and W. Herz, J. Nat. Prod., 1982, 45, 328.
36. R.J. Molyneux, J.N. Roitman, M. Benson, and R.E. Lundin, Phytochemistry, 1982, 21, 439.
37. E. Röder, H. Wiedenfeld, and P. Stengl, Arch. Pharm. (Weinheim, Ger.), 1982, 315, 87.
38. Y.-W. Fang, Z.-L. Bian, and G.-H. Wang, Hua Hsueh Hsueh Pao, 1981, 39, 139 (Chem. Abs., 1981, 95, 133207); J. Zhao and Y.-W. Fang, Fenxi Huaxue, 1981, 9, 418 (Chem. Abs., 1982, 96, 218077).
39. H. Stoeckli-Evans, Acta Crystallogr., Sect. B, 1982, 38, 1614.
40. H. Wiedenfeld, E. Röder, A. Kirfel, and G. Will, Arch. Pharm. (Weinheim, Ger.), 1982, 315, 165.
41. D.G. Hay, M.F. Mackay, and C.C.J. Culvenor, Acta Crystallogr., Sect. B, 1982, 38, 155.
42. H. Wiedenfeld, E. Röder, A. Kirfel, and G. Will, Arch. Pharm. (Weinheim, Ger.), 1981, 314, 737.
43. S. Wang and N. Hu, Sci. Sin. (Engl. Ed.), 1981, 24, 1536.
44. G.A. Pashaeva and A.F. Alekperov, Azerb. Med. Zh., 1981, 58, 13 (Chem. Abs., 1981, 95, 156640).
45. R. Schoental, Toxicol. Lett., 1982, 10, 323.
46. J.R. Piper, P. Kari, and Y.F. Shealy, J. Labelled Compd. Radiopharm., 1981, 18, 1579.
47. L.T. Gelbaum, M.M. Gordon, M. Miles, and L.H. Zalkow, J. Org. Chem., 1982, 47, 2501.
48. R. Manavalan, K.A. Suri, O.P. Suri, and C.K. Atal, Indian Drugs, 1981, 18, 258.

49. S. Miyano, K. Sumoto, M. Morita, and F. Sato, Eur. Pat. Appl. EP 39903 (Chem. Abs., 1982, 96, 85419).
50. S.N. Arseculeratne, A.A.L. Gunatilaka, and R.G. Panabokke, J. Ethnopharmacol., 1981, 4, 159.
51. X. Wang, G. Han, J. Cui, J. Pan, C. Li, and G. Zheng, Beijing Yixueyuan Xuebao, 1981, 13, 310 (Chem. Abs., 1982, 97, 33110).
52. J. Ghanem and C. Hershko, Isr. J. Med. Sci., 1981, 17, 339.
53. H. Habs, Dtsch. Apoth.-Ztg., 1982, 122, 799; H. Habs, M. Habs, H. Marquardt, E. Röder, D. Schmael, and H. Wiedenfeld, Arzneim.-Forsch., 1982, 32, 144.
54. M. Kido, T. Hirose, K. Tanaka, T. Kurozumi, and Y. Shoyama, Jpn. J. Med., 1981, 20, 170.
55. R.A. Roth, L.A. Dotzlaf, B. Baranyi, C.H. Kuo, and J.B. Hook, Toxicol. Appl. Pharmacol., 1981, 60, 193.
56. K.S. Hilliker, T.G. Bell, and R.A. Roth, Am. J. Physiol., 1982, 242, H537.
57. B. Meyrick and L. Reid, Am. J. Pathol., 1979, 94, 37.
58. C.E. Green, H.J. Segall, and J.L. Byard, Toxicol. Appl. Pharmacol., 1981, 60, 176.
59. D.F. Eastman, G.P. Dimenna, and H.J. Segall, Drug Metab. Dispos., 1982, 10, 236.
60. B.J. Garrett, P.R. Cheeke, C.L. Miranda, D.E. Goeger, and D.R. Buhler, Toxicol. Lett., 1982, 10, 183.
61. C.L. Miranda, P.R. Cheeke, D.E. Goeger, and D.A. Buhler, Toxicol. Lett., 1981, 8, 343.
62. C.L. Miranda, M.C. Henderson, and D.R. Buhler, Toxicol. Appl. Pharmacol., 1981, 60, 418.
63. A.E. Johnson, Am. J. Vet. Res., 1982, 43, 718.
64. S. Srungbooamee and C. Maskasame, Sattawaphaet San, 1981, 32, 91 (Chem. Abs., 1982, 96, 137560).
65. J.D. Henricks, R.O. Sinnhuter, M.C. Henderson, and D.R. Buhler, Exp. Mol. Pathol., 1981, 35, 170.
66. C.L. Miranda, M.C. Henderson, R.L. Reed, J.A. Schmitz, and D.R. Buhler, J. Toxicol. Environ. Health, 1982, 9, 359.
67. C.L. Miranda, D.R. Buhler, P.R. Cheeke, and J.A. Schmidtz, Toxicol. Lett., 1982, 10, 177.
68. H.L. Kim and L.P. Jones, Res. Commun. Chem. Pathol. Pharmacol., 1982, 36, 341.
69. C.L. Miranda, H.M. Carpenter, P.R. Cheeke, and D.R. Buhler, Chem.-Biol. Interact., 1981, 37, 95.
70. M.C. Gimmler and N.B. Nardi, Rev. Bras. Genet., 1982, 5, 111.
71. K.A. Robertson, Cancer Res., 1982, 42, 8.
72. W.K. Anderson, Cancer Res., 1982, 42, 2168; W.K. Anderson, C.P. Chang, P.F. Corey, J.M. Halat, A.N. Jones, H.L. McPherson, J.S. New, and A.C. Rick, Cancer Treat. Rep., 1982, 66, 91; W.K. Anderson and H.L. McPherson, J. Med. Chem., 1982, 25, 84.
73. A.R. Mattocks, Toxicol. Lett., 1981, 8, 201.

5
Indolizidine Alkaloids

BY J. A. LAMBERTON

1 Swainsonine

Swainsonine, previously shown to be 8aβ-octahydroindolizine-1α,2α,8β-triol (cf. vol.11, p.59) and swainsonine N-oxide have been isolated from Utah spotted locoweed (Astragalus lentiginosus).[1] Inhibition by these alkaloids of the lysosomal enzyme α-mannosidase is suggested as the cause of locoism, a chronic neurological disease of range animals.[1] Swainsonine and swainsonine N-oxide are equally effective in inhibiting hydrolysis of mannosides by either mouse liver homogenate or jack-bean α-mannosidase.[1] Inhibition by swainsonine of the processing of asparagine-linked glycoproteins in both cell-free extracts and in animal cells in culture is considered to occur through inhibition of the mannosidase that releases mannose from high-mannose glycopeptide.[2] Swainsonine also prevents the processing of oligosaccharide chains of influenza virus haemagglutinin.[3]

2 Prosopis Alkaloids

Juliprosine, isolated from Prosopis juliflora A.DC., has the structure (1). Reduction of the dihydroindolizinium system gives a hexahydroderivative identical with the product from catalytic hydrogenation of the closely related alkaloid juliprosopine (cf.

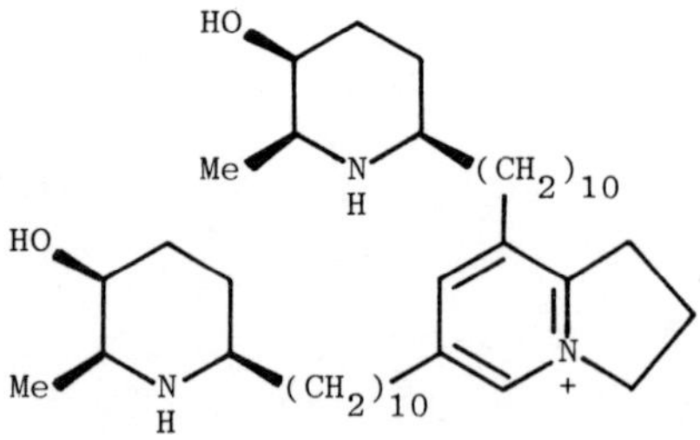

(1)

vol.12, p.69). Comparison of $[M]_D$ values of simpler *Prosopis* alkaloids of known absolute configuration with that of juliprosine shows juliprosine to have the same absolute configuration as spectaline.[4]

3 *Elaeocarpus* Alkaloids

In an alternative approach[5] to the nitrone method for the synthesis of *Elaeocarpus* alkaloids (*cf.* vol.11, p.61), a 1,3 dipolar cycloaddition of 1-pyrroline 1-oxide to a substituted enone gives the isoxazolidine (2), which was converted in a number of steps into the compound (3), the C-7 epimer of elaeokanine C. Oxidation of the epimer (3) afforded the diketone (4), an intermediate in a previously published synthesis of (±)-elaeokanine C.

Acid-catalyzed cyclization of the hydroxylactam (5) to give the chloro compound (6) is the basis for a new synthesis of (±)-elaeokanine B (7).[6]

The synthesis of indolizidine alkaloids, including tylophorine and δ-coniceine as well as *Elaeocarpus* alkaloids, by the intramolecular imino Diels-Alder method (*cf.* vol.12, p.71) has been published with more detail.[7] Synthesis of (±)-elaeokanine B (7) by this method gave the two diastereoisomeric alcohols of this structure, and it has been suggested that natural (±)-elaeokanine B may be a mixture of these two diastereoisomers.[7] Oxidation of the mixture of diastereoisomeric alcohols afforded (±)-elaeokanine A.[7]

4 *Dendrobates* Alkaloids

The absolute configuration assigned to natural gephyrotoxin by X-ray analysis (*cf*. vol.8, p.63) has been questioned.[8] A synthesis of gephyrotoxin from L-pyroglutamic acid (8) through the vinylogous amide (9) has given dextrorotatory gephyrotoxin, whereas natural gephyrotoxin is levorotatory, and it has been suggested that the absolute configuration of natural gephyrotoxin should be revised to that depicted in (10).[8]

New syntheses of (±)-gephyrotoxin[9] and of depentylperhydrogephyrotoxin[10] have been described.

In a further study of the pumiliotoxins from the Panamanian frog *Dendrobates pumilio* the double bond in the side-chain of pumiliotoxin B has been shown from nuclear Overhauser effects to have the *E* configuration, and comparison with model compounds showed that the diol group is *threo*.[11] The *threo* configuration for the allylic diol system of pumiliotoxin B has also been established in another study, and a method for the synthesis of this functionality with excellent stereocontrol has been developed.[12]

The structure and relative stereochemistry of gephyrotoxin (GTX) 223 AB, a minor alkaloid in skin extracts from a number of tropical frogs of the genus *Dendrobates*, have been established as 3α-butyl-5β-propyl-8aα-octahydroindolizine (12).[13] Natural gephyrotoxin 223 was shown to be identical with one of four diastereoisomers obtained by catalytic hydrogenation of (13), and in the same study it was found to be identical with the synthetic 3-butyl-5-propyloctahydroindolizine that was earlier considered[14] to be a stereoisomer of natural gephyrotoxin 223 (*cf*. vol.11, p.62). One of the diastereoisomers of gephyrotoxin 223 has been prepared from the cyanoamine (14) in a reaction with n-propylmagnesium bromide which gives stereospecifically 3α-butyl-5β-propyl-8aβ-octahydroindolizine (15).[15]

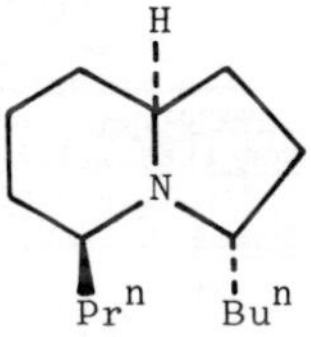

(12)

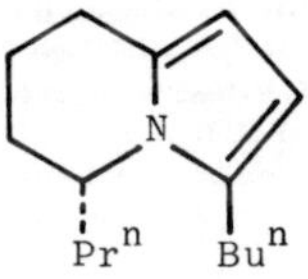

(13)

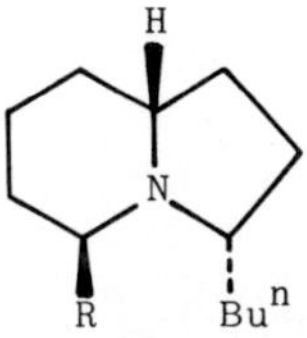

(14) R = CN
(15) R = Pr^n

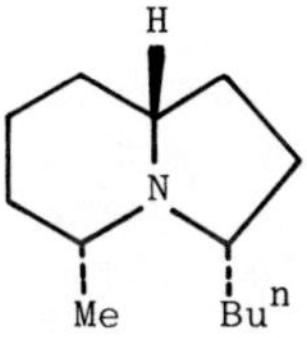

(16)

5 Other Syntheses

An improvement has been effected in the synthesis of ipalbidine.[16] δ-Coniceine has been prepared in several steps that include the hydroboration-carbonylation of a bisolefinic amine.[17] A stereospecific total synthesis of (±)-monomorine (16), a trail pheromone of the Pharaoh ant, has been achieved.[18]

References

1 R.J.Molyneux and L.F.James, Science, 1982, 216, 190.
2 A.D.Elbein, R.Solf, P.R.Dorling, and K.Vosbeck, Proc. Natl. Acad. Sci. U.S.A. 1981, 78, 7393.
3 A.D.Elbein, P.R.Dorling, K.Vosbeck, and M.Horisberger, J. Biol. Chem., 1982, 257, 1573.
4 P.Dätwyler, R.Ott-Longoni, E.Schöpp, and M.Hesse, Helv. Chim. Acta, 1981, 64, 1959
5 H.Otomasu, N.Takatsu, T.Honda, and T.Kametani, Heterocycles, 1982, 19, 511.
6 B.P.Wijnberg and W.N.Speckamp, Tetrahedron Lett., 1981, 22, 5079.
7 N.A.Khatri, H.F.Schmitthenner, J.Shringarpure, and S.M.Weinreb, J. Am. Chem. Soc., 1981, 103, 6387.
8 R.Fujimoto and Y.Kishi, Tetrahedron Lett., 1981, 22, 4197.
9 D.J.Hart, J. Org. Chem., 1981, 46, 3576.
10 D.J.Hart, J. Org. Chem., 1981, 46, 367.
11 T.Tokuyama, K.Shimada, M.Uemura, and J.W.Daly, Tetrahedron Lett., 1981, 23, 2121.
12 L.E.Overman and R.J.McCready, Tetrahedron Lett., 1982, 23, 2355.
13 T.F.Spande, J.W.Daly, D.J.Hart, Y.-M.Tsai, and T.L.Macdonald, Experientia, 1981, 37, 1242.
14 T.L.Macdonald, J. Org. Chem., 1980, 45, 193.
15 R.V.Stevens and A.W.M.Lee, J. Chem. Soc., Chem. Commun., 1982, 103.
16 J.E.Cragg, S.H.Hedges, and R.B.Herbert, Tetrahedron Lett., 1981, 22, 2127.
17 M.E.Garst and J.N.Bonfiglio, Tetrahedron Lett., 1981, 22, 2075.
18 R.V.Stevens and A.W.M.Lee, J. Chem. Soc., Chem. Commun., 1982, 102.

6
Quinolizidine Alkaloids

BY M. F. GRUNDON

A number of new alkaloids have been isolated and stereochemical studies by the Polish group have been extended, but the greatest advances this year have been in the synthesis of Lythraceae alkaloids.

1 The Lupinine-Cytisine-Sparteine-Matrine-*Ormosia* Group

1.1 Occurrence.- Isolation studies are summarised in Table 1,[1-18] indicating that ten new alkaloids have been obtained. Hartmann and co-workers have continued their investigation of the constituents of <u>Lupinus polyphyllus</u> plants and cell suspension cultures by g.l.c.-m.s. (cf. Vol. 11, p.64) and have identified a number of 13-hydroxylupanine esters;[19] the same group has applied the technique to <u>Sarothamnus scoparius</u> and to its root parasite, <u>Orobanche rapum-genistae</u>.[8]

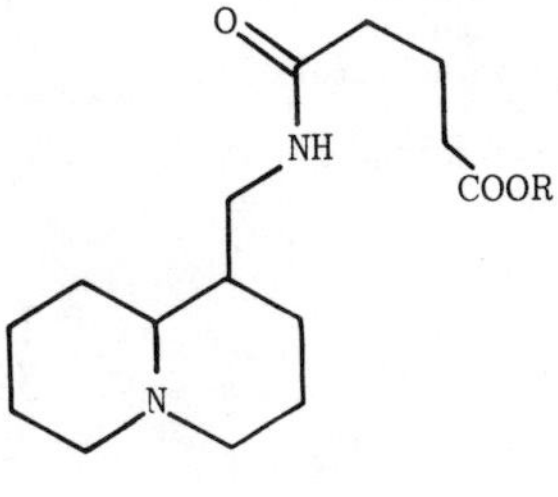

Sophorine (1) R = Bu
(2) R = Me

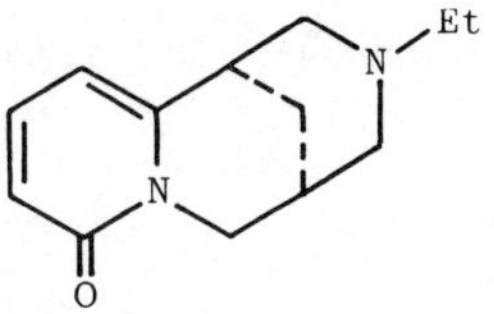

N-Ethylcytisine (3)

1.2 Structural and Stereochemical Studies.- Structure (1) for sophorine (isolated from <u>Sophora alopecuroides</u>[9]) is reminiscent of that assigned to the <u>N</u>-oxide of compound (2) (cf. Vol. 12, p.75); the latter is regarded as an artefact formed by solvolysis of epilamprolobine <u>N</u>-oxide and the butyl ester (1) may have a similar origin. A new alkaloid of <u>Echinosophora koreensis</u> was shown to be <u>N</u>-ethylcytisine (3) on the basis of its formation by ethylation of (-)-cytisine.[2]

Table 1 Isolation of alkaloids of the lupinine-cytisine-sparteine-matrine-Ormosia group

Species	Alkaloid (Structure)	Ref.
Camoensia brevicalyx	Camoensidine (7) Camoensine (5) *12-Hydroxycamoensidine (8) *12-α-Hydroxycamoensine (6) *Alkaloid (9)	1
Echinosophora koreensis	*(-)-N-Ethylcytisine (3)	2
Genista subcapitata	Cytisine N-Methylcytisine Sparteine	3
Laburnum anagyroides	Cytisine N-Methylcytisine	4
	(-)-N-(3-Oxobutyl)cytisine	5
Lupinus holosericeus (cf. Vol. 12, p.74)	5,6-Dehydrolupanine β-Isosparteine Sparteine	6
L. oscar-haughtii (cf. Vol. 12, p.73)	(+)-Tetrahydrorhombifoline	7
L. truncatus (cf. Vol. 12, p.73)	Tetrahydrorhombifoline	7
Sarothamnus scoparius	Ammodendrine Anagyrine Angustifoline 5,6-Dehydrolupanine *11,12-Dehydrosparteine (10) *4,13-Dihydroxylupanine (11) 4-Hydroxylupanine 13-Hydroxylupanine esters α-Isolupanine N-Methylangustifoline 17-Oxolupanine	8
Sophora alopecuroides	*Sophorine (1)	9
S. franchetiana	*Tsukushinamine-B (13) *Tsukushinamine-C (14)	10
S. macrocarpa	Cytisine 3-Hydroxymatrine 5-Hydroxymatrine (sophoranol) Matrine N-oxide	11 12 11,12 11
S. mollis	Ammodendrine (-)-Baptifoline 5,6-Dehydrolupanine (-)-N-Formylcytisine Lupinine Rhombifoline (+)-Sparteine	13
S. viciifolia	Sophocarpine Sophocarpine N-oxide	14

Table 1 - continued

Species	Alkaloid (Structure	Ref.
Spartinum junceum	Anagyrine	15,16
	α-Isolupanine	15
	(±)-Lupanine	
	N-Methylcytisine	16
Sweetia elegans	(±)-6-epi-Podopetaline (Sweetinine) (23)	17
Virgilia divaricata	Lupinine	18
V. oroboides	*Virgiboidine (4)	

* New alkaloids

Virgiboidine, a tricyclic alkaloid of <u>Virgilia oroboides</u> and V. divaricata, was assigned structure (4); the presence of the butenyl group is apparent from the mass spectrum, in which the base peak at m/z 207 arises by allylic cleavage, and from the ^{1}H n.m.r. spectrum, but evidence for placing the amide carbonyl group at C-10 was not given.[18]

The quinolizidine-indolizidines isolated from <u>Camoensia brevicalyx</u> include the known alkaloids camoensine (5) and camoensidine (7) (cf. Vol. 7, p.72). Spectroscopic data, particularly the characteristic cleavage under electron impact into A/B and D ring fragments, suggested that two new alkaloids are 12-α-hydroxycamoensine (6) and 12-hydroxycamoensidine (8). It is apparent from the ^{1}H n.m.r. and mass spectra that another new alkaloid does not contain a normal D ring; structure (9) has been proposed tentatively.[1]

A new alkaloid of <u>Sarothamnus scoparius</u>[8] seems to be the same dehydrosparteine obtained in transformation of lupin alkaloids by plant cell cultures and regarded as 11,12-dehydrosparteine (10) from its mass spectrum and its non-identity with 5,6-dehydrosparteine.[20] Another new constituent of S. scoparius is thought to be 4,13-dihydroxylupanine (11) on the basis of the mass spectrum, which is characteristic of a 13-hydroxylupanine but differs from that of 10,13- and 12,13-dihydroxylupanine.[8] The structure and absolute configuration of tsukushinamine-A (12) was reported recently (cf. Vol. 10, p.69; Vol. 12, p.76) and now two isomers, tsukushinamine-B (13) and -C (14) have been isolated from <u>Sophora franchetiana</u>; tsukushinamine-A is converted into the C isomer by heating at 200°C.[10]

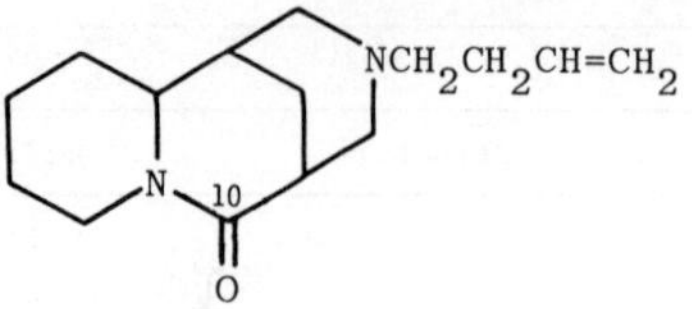

(4)

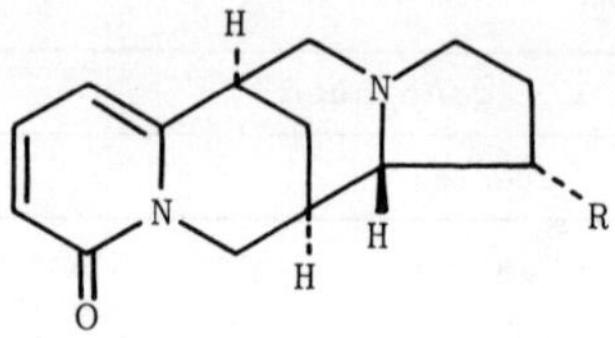

(5) R = H

12-α-Hydroxycamoensine (6) R = OH

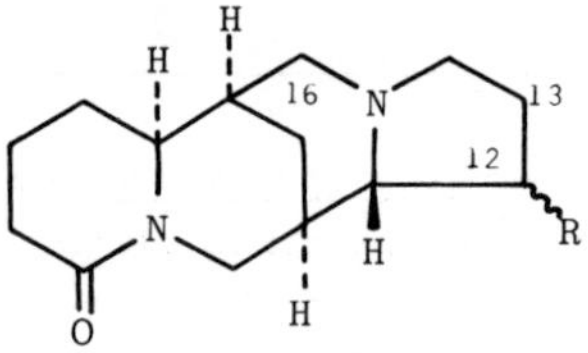

(7) R = H

12-α-Hydroxycamoensidine (8) R = OH

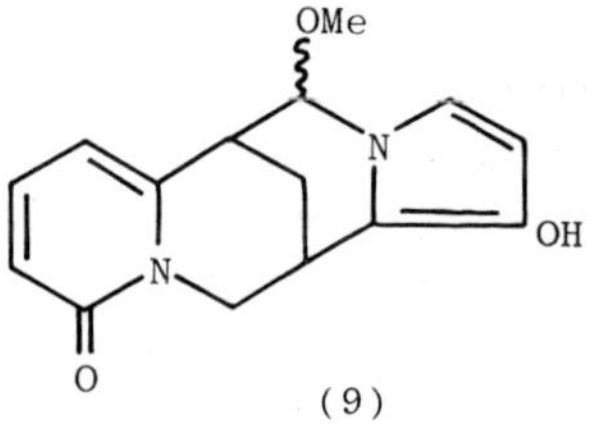

(9)

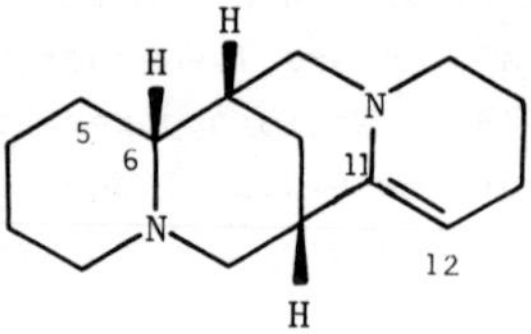

11,12-Dehydrosparteine (10)

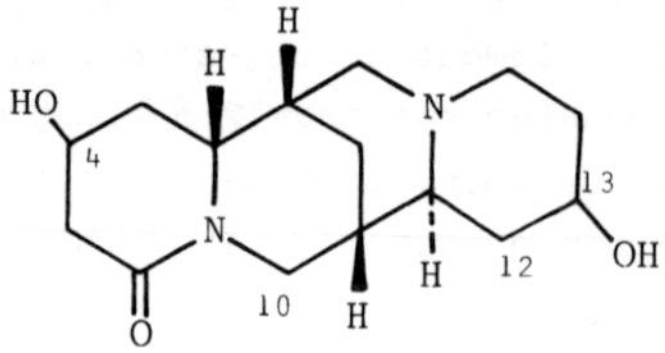

4,13-Dihydroxylupanine (11)

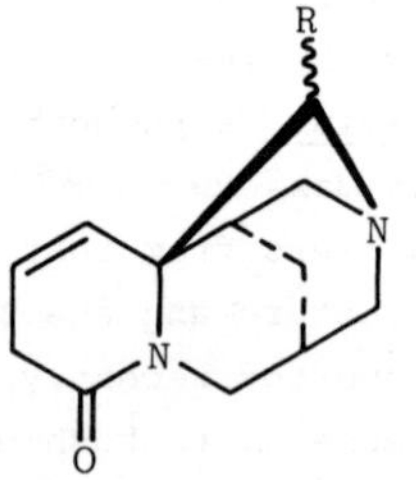

Tsukushinamine-A (12) R = β-CH₂CH=CH₂
Tsukushinamine-B (13) R = α-CH₂CH=CH₂

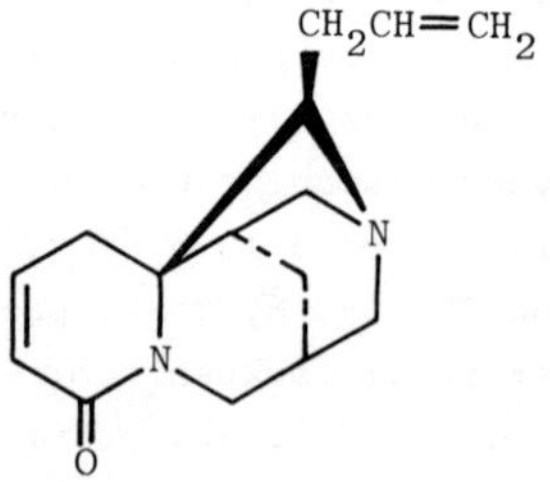

Tsukushinamine-C (14)

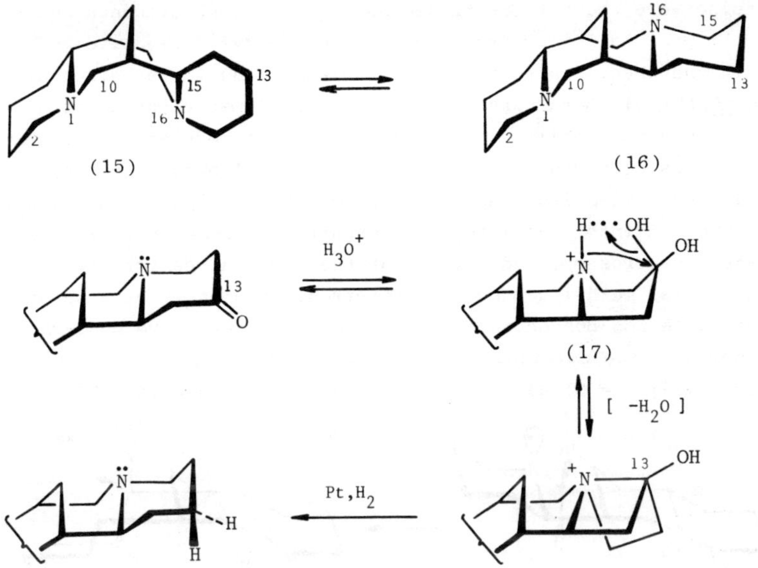

Quinolizidine Alkaloids

Scheme 1

Polish chemists have continued to study the stereochemistry of sparteine and its derivatives in relation to the all-chair conformation (15) ('cisoidal') and the chair,chair,boat,chair conformation (16) ('transoidal'). Infrared studies indicated that the hydrate of 10-oxosparteine hydrochloride is in the transoidal conformation, cf. (16), whereas the corresponding acid salts of 15-oxosparteine have a distorted cisoidal arrangement in the solid state.[21]

It was also shown by i.r. spectroscopy that the mono- and di-perchlorates of 2,13-dioxosparteine (13-oxo-lupanine) can also exist in cisoidal and transoidal conformations.[22] These data are relevant to the mechanism of the catalytic hydrogenation of γ-aminoketones to amines in the presence of ≥ 2.5 moles HCl observed with 1-methyl-4-piperidone, 13-oxo-lupanine, 13-oxo-α-lupanine and 13-oxo-sparteine (cf. Vol. 8, p.69). A new investigation of the reaction involving ^{13}C n.m.r. spectroscopy in ^{2}HCl indicated that the protonated hydrate (17) was formed readily; a new mechanism (Scheme 1) was proposed.[23] The configurations of α-2-methyl- and α-2-phenylsparteine were shown to be transoidal, cf. (16)(substituents equatorial); protonation occurs at N-1 in the methyl derivative and at N-16 in the phenylsparteine to give bridged ions in the cisoidal form, cf.(15), in both cases.[24] The equatorial methyl group of 17-β-methyl-lupanine stabilises the transoidal conformations (18). Reaction of this compound with mercuric acetate followed by borohydride gave 17-β-methyl-α-isolupanine (19; R = O) as principal product, which was converted into 17-β-methyl-α-isosparteine (19; R = H_2) (Scheme 2). Infrared spectroscopy indicated that the cisoidal conformations containing axial methyl groups are preferred, apparently because of unfavourable non-bonded interactions in the alternative boat-chair arrangement; the cisoidal conformation of the perchlorate salt of compound (19; R = O) has been confirmed by X-ray analysis.[25]

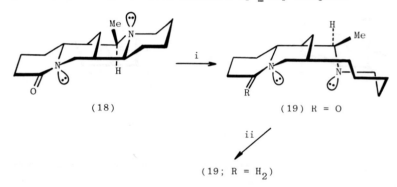

Reagents: i, Hg(OAc)$_2$, 5% AcOH, at 65°C; ii, LiAlH$_4$ or Pt, H$_2$
Scheme 2

Russian studies of the stereochemistry of alkaloids of the matrine group by X-ray analysis have been extended to cis-matrine (20),[26] sophoridine N-oxide (21),[27] and lehmannine N-oxide (22).[28]

The stereochemical complexity of some Ormosia alkaloids was discussed last year (cf. Vol. 12, p.76-77). Now the alkaloid sweetinine, first obtained from Sweetia panamensis, has been isolated from S. elegans and identified as (±)-6-epi-podopetaline (23) by conversion into its formaldehyde adduct, (±)-homo-6-epi-podopetaline.[17]

1.3 Synthesis.- The nitrone (24), used previously in a [2 + 3]-cycloaddition reaction to prepare lupinine (25) (cf. Vol. 8, p.68), has now been applied to a new synthesis of the alkaloid (Scheme 3).[29]

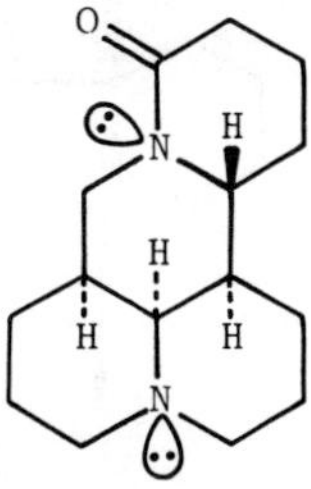

(20)

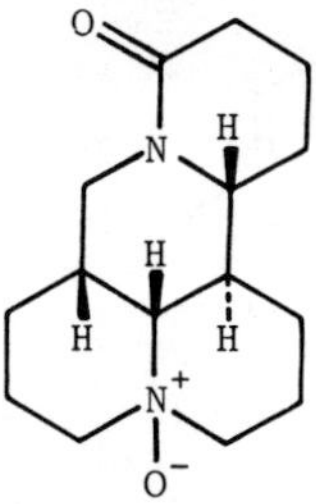

(21)

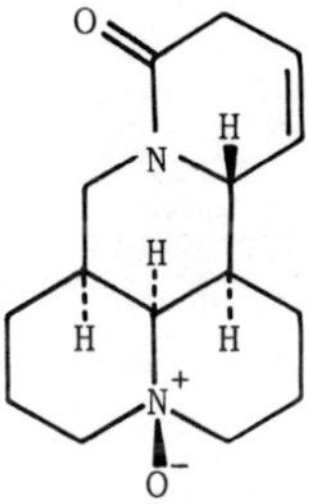

(22)

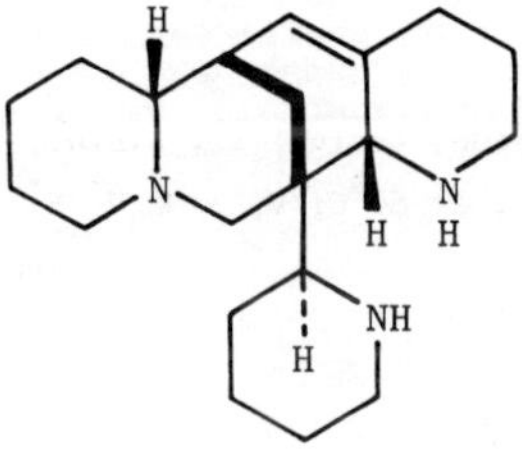

(23)

A full account[30] of the structure determination and synthesis of the alkaloids myrtine and epi-myrtine is available (cf. Vol. 9, p.71; Vol. 11, p.68).

2 Sesquiterpenoid Alkaloids

α-Thiohemiaminal derivatives of deoxynupharidines (27) and (28) have been prepared from 6-dehydrodeoxynupharidine (26), sulphensuccinimides proving to be useful reagents for the conversion.[31]

In a projected synthesis of Nuphar quinolizidine alkaloids, it was shown by X-ray analysis that the reaction of the pyridine derivative (29) with bromine gave the indolizinium salt (30) rather than the expected quinolizinium derivative (31).[32]

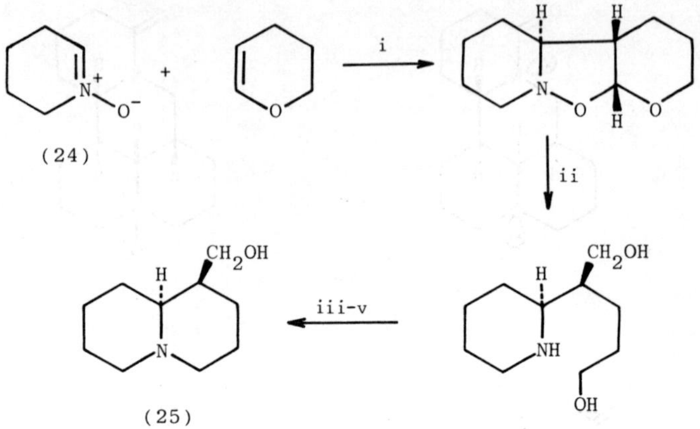

(24) (25)

Reagents: i, EtOH, reflux; ii, $LiAlH_4$, THF, reflux; iii, Et_2NSiMe_3; iv, Me_3SiI, CCl_4, at $80^{o}C$; v, $BzMe_3N^+ F^-$, THF

Scheme 3

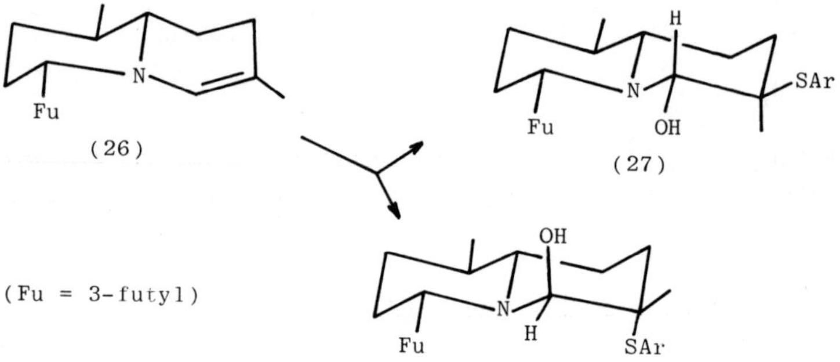

(26) (27) (28)

(Fu = 3-futyl)

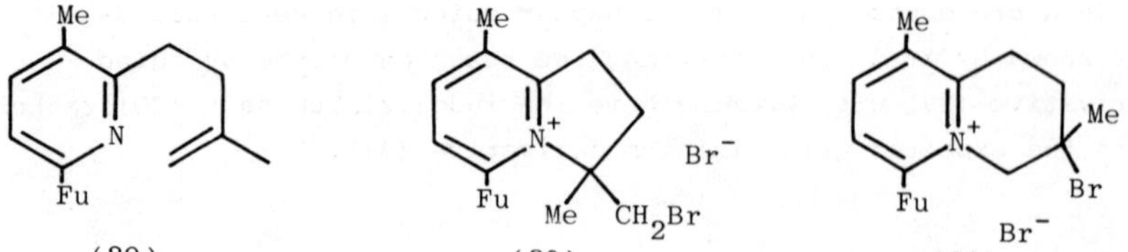

(29) (30) (31)

3 Alkaloids of the Lythraceae

A new synthetic route to 4-arylquinolizidine alkaloids again involves nitrone (24) (Scheme 4).[33] A mixture of diastereoisomeric adducts (32) was converted into the alcohols (33) and (34), separated as their acetates; hydrogenolysis of (34) yielded the alkaloid (36) and the stereoisomeric alkaloid (35) was obtained from alcohol (33) after inversion of the C-2 centre using the Mitsunobu reaction.

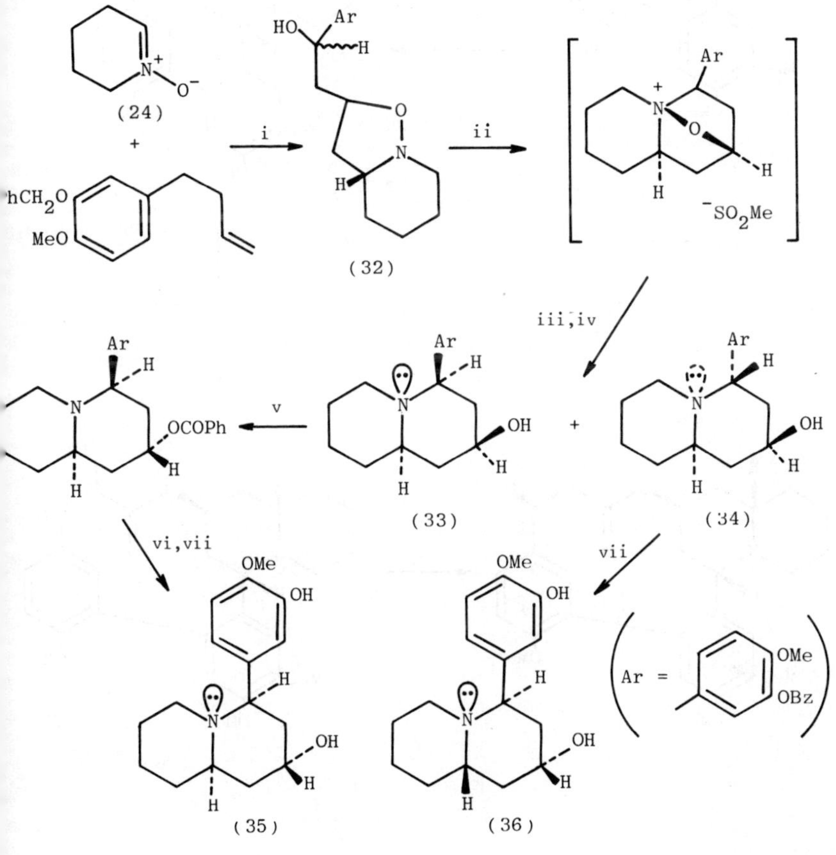

Reagents: i, PhMe, reflux; ii, MeSO$_2$Cl, pyridine; iii, Zn, 50% AcOH, then Ac$_2$O, pyridine; iv, NaOH, aq. MeOH; v, (EtOOC-N$\Rightarrow$)$_2$, PPh$_3$, PhCOOH; vi, NaOMe, MeOH; vii, 10% Pd/C, H$_2$

Scheme 4

A key feature of a new synthesis of the macrocyclic alkaloid O-methyl-lythridine (37) is initiation of ring closure by fluorodesilylation of a trimethylsilyl acetate (38) (Scheme 5).[34]

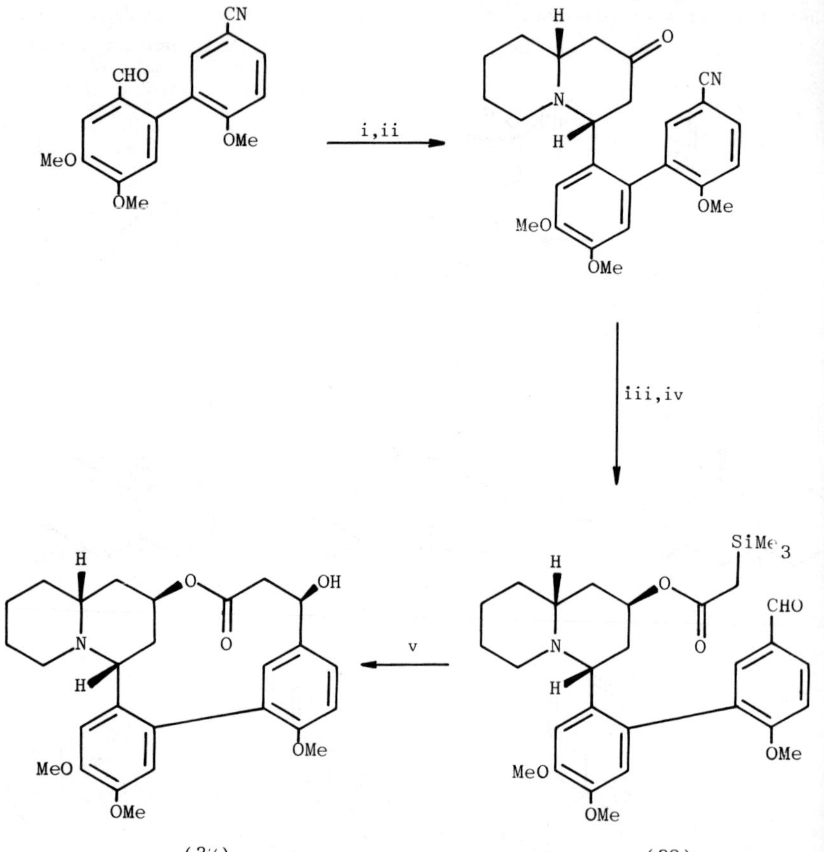

Reagents: i, pelletierine, aq. NaOH, THF, MeOH; ii, MeOH, reflux; iii, L-Selectride, THF, at -78°C; iv, trimethylsilylketen, THF, at -20°C; v, $Bu^n_4N^+ F^-$, THF at -78°C, then water

Scheme 5

Quinolizidine Alkaloids

Earlier syntheses of quinolizidines involved Mannich reactions of pelletierine. In a new approach, applied to the synthesis of (±)-vertaline (Scheme 6), allylic strain controls the stereochemical course of cyclisation of the N-acyliminium ion (40); the quinolizidine (39) was converted into vertaline by a procedure similar to that used previously (Vol. 8, p.74).[35]

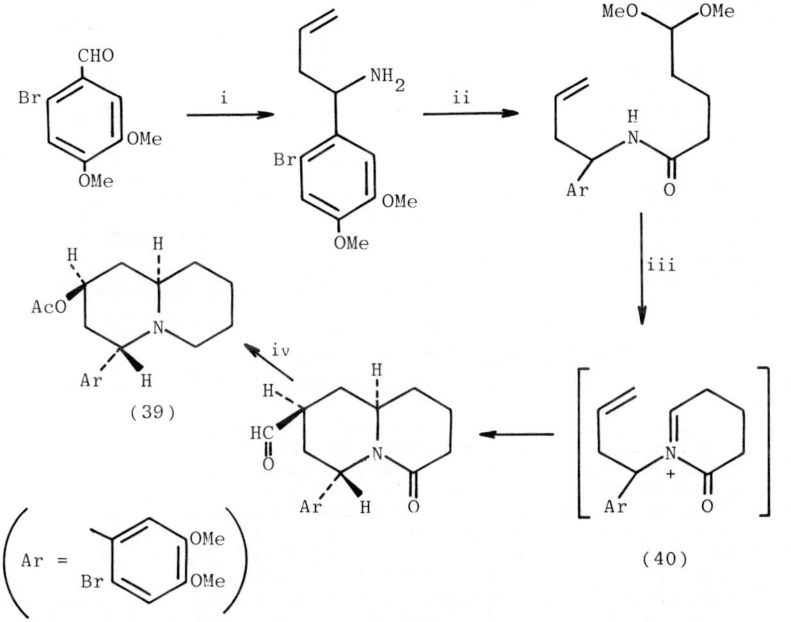

Reagents: i, $LiN(SiMe_3)_2$, THF, then $H_2C=CHCH_2MgBr$, Et_2O; ii, $AlMe_3$, $MeOOC(CH_2)_3-CH(OMe)_2$; iii, HCOOH, CH_2Cl_2, at 25°C; iv, BF_3, THF, then Ac_2O

Scheme 6

References

1 P.G.Waterman and D.F.Faulkner, Phytochemistry, 1982, 21, 215.
2 I.Murakoshi, M.Watanabe, J.Haginiwa, S.Ohmiya, and H.Otomasu, Phytochemistry, 1982, 21, 1470.
3 N.Nakov, D.Obreshkova, and Ch.Akhtardzhiev, Farmatsiya (Sofia), 1981, 31, 41 (Chem. Abstr., 1981, 95, 183 912).
4 J.Jurenitsch, M.Poehm, and G.Weilguny, Pharmazie, 1981, 36, 370 (Chem. Abstr., 1981, 95, 111 805).
5 A.I.Gray, M.C.Henman, and C.J.Meegan, J. Pharm. Pharmacol., 1981, 33, 95P.

6   W.J.Keller, J. Nat. Prod., 1981, 44, 357.
7   M.F.Balandrin and A.D.Kinghorn, J. Nat. Prod., 1981, 44, 495.
8   M.Wink, L.Witte, and T.Hartmann, Planta Med., 1981, 43, 342.
9   F.G.Kamaev, S.Kuchkarov, F.K.Kushmuradov, and Kh.A.Aslanov, Khim. Prir. Soedin., 1981, 604 (Chem. Abstr., 1982, 96, 123 059).
10 S.Ohmiya, K.Higashiyama, H.Otomasu, J.Haginiwa, and I.Murakoshi, Phytochemistry, 1981, 20, 1997.
11 R.Negrete, N.Backhouse, and B.K.Cassels, Contrib. Cient. Tecnol.(Univ. Tec Estado, Santiago), 1981, 11, 31 (Chem. Abstr., 1982, 97, 36 120).
12 R.E.Negrete, N.C.Backhouse, and B.K.Cassels, Bol. Soc. Chil. Quim., 1982, 27, 263 (Chem. Abstr., 1982, 97, 3 534).
13 I.Murakoshi, E.Kidoguchi, M.Ikram, M.Israr, N.Shafi, J.Haginiwa, S.Ohmiya, and H.Otomasu, Phytochemistry, 1982, 21, 1313.
14 S.Li and H.-J.Chang, Yao Hsueh Tung Pao, 1981, 16, 52 (Chem. Abstr., 1981, 95, 138 464).
15 P.Martinod, M.Arteaga, and I.Alcivar, Politecnica, 1981, 6, 7 (Chem. Abstr., 1982, 97, 69 323).
16 Yu.D.Sadykov and M.Khodzhimatov, Dokl. Akad. Nauk Tadzh. SSR, 1981, 24, 736 (Chem. Abstr., 1982, 96, 139 690).
17 M.F.Balandrin and A.D.Kinghorn, J. Nat. Prod., 1981, 44, 619.
18 J.L.van Eijk and M.H.Radema, Planta Med., 1982, 44, 224.
19 M.Wink, H.M.Schiebel, L.Witte, and T.Hartmann, Planta Med., 1982, 44, 15.
20 M.Wink, T.Hartmann, and L.Witte, Planta Med., 1980, 40, 31.
21 A.Perkowska and M.Wiewiórowski, Bull. Acad. Pol. Sci., Ser. Sci. Chim., 1980, 28, 249.
22 W.Wysocka, Bull. Acad. Pol. Sci., Ser. Sci. Chim., 1980, 28, 19.
23 W.Wysocka, Bull. Acad. Pol. Sci., Ser. Sci. Chim., 1980, 28, 263.
24 W.Boczon, Pol. J. Chem., 1981, 55, 339.
25 M.Wiewiórowski and A.Perkowska, Bull. Acad. Pol. Sci., Ser. Sci. Chim., 1980, 28, 499.
26 B.T.Ibragimov, S.A.Talipov, Yu.K.Kushmuradov, and T.F.Aripov, Khim. Prir. Soedin., 1981, 597 (Chem. Abstr., 1982, 96, 143 145).
27 B.T.Ibragimov, G.N.Tishchenko, S.A.Talipov, Yu.K.Kushmuradov, T.F.Aripov, and S.Kuchkarov, Khim. Prir. Soedin., 1981, 588 (Chem. Abstr., 1982, 96, 123 063)
28 B.T.Ibragimov, S.A.Talipov, Yu.K.Kushmuradov, T.F.Aripov, and S.Kuchkarov, Khim. Prir. Soedin., 1981, 757 (Chem. Abstr., 1982, 97, 56 094).
29 T.Iwashita, T.Kusumi, and H.Kakisawa, J. Org. Chem., 1982, 47, 230.
30 P.Slosse and C.Hootele, Tetrahedron, 1981, 37, 4287.
31 R.T.LaLonde and T.S.Eckert, Can. J. Chem., 1981, 59, 2298.
32 J.Szychowski, J.Wróbel, and A.Leniewski, Bull. Acad. Pol. Sci., Ser. Sci. Chim., 1980, 28, 9.
33 S.Takano and K.Shishido, J. Chem. Soc., Chem. Commun., 1981, 940.
34 D.E.Seitz, R.A.Milius, and J.Quick, Tetrahedron Lett., 1982, 23, 1439.
35 D.J.Hart and K.Kanai, J. Org. Chem., 1982, 47, 1555.

7
Quinoline and Acridone Alkaloids

BY M. F. GRUNDON

A considerable number of new quinoline and acridone alkaloids has been identified and the recognition of two new groups of dimeric quinolinone alkaloids is of particular interest. There has been little activity in the quinazoline alkaloid area and this section is omitted this year.

1 Quinoline Alkaloids

1.1 Occurrence.- New alkaloids and known alkaloids that have been obtained from new sources are listed in Table 1;[1-29,47] the ubiquitous furoquinoline alkaloid skimmianine (1; R^1 = H, R^2 = R^3 = OMe) has been identified in twelve more rutaceous species. The rather rare 2,2-dimethylpyranoquinolone flindersine (39) has been obtained from a Zanthoxylum species for the first time[25] and the identification of the wide-spread N-methylflindersine (40) and its known and new derivatives (Table 1) is a particular feature of isolation studies this year.

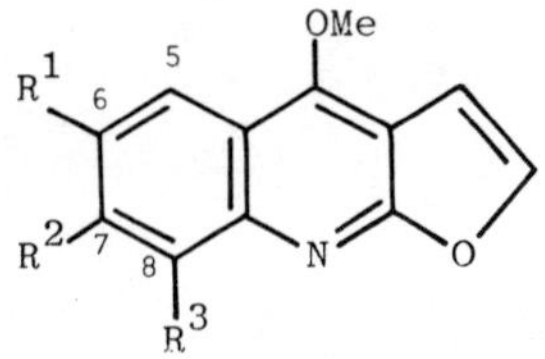

(1)

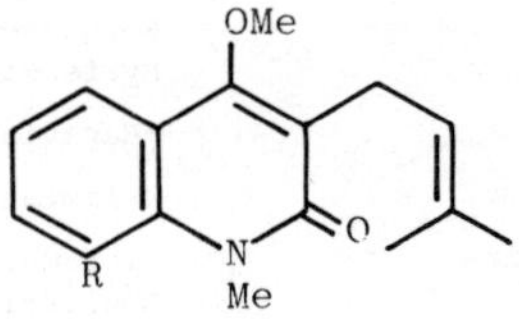

(2) R = H
(3) R = OH

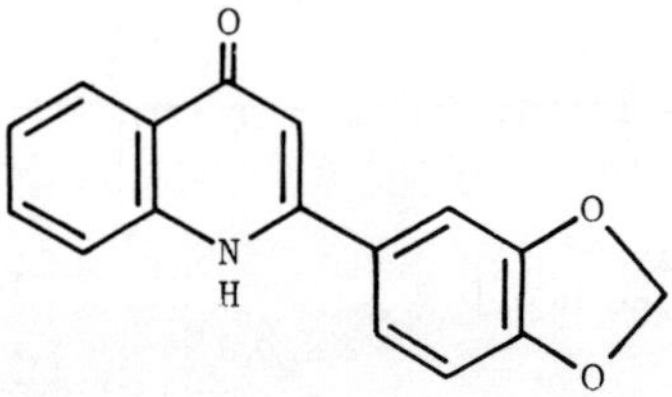

(4)

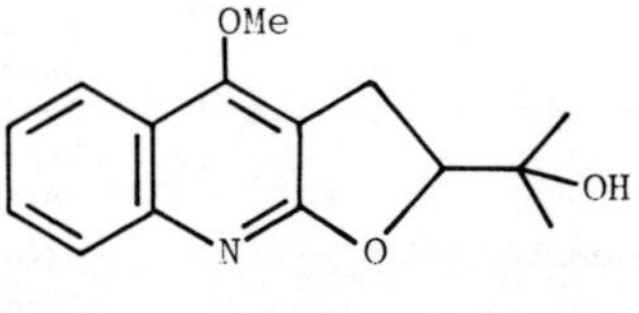

(5)

Table 1 Isolation of quinoline alkaloids

Species	Alkaloid (Structure)	Ref.
Afraegle paniculata	Atanine (15) Dictamnine (1; $R^1 = R^2 = R^3 = H$) Haplopine (1; $R^1 = H$, $R^2 = OH$, $R^3 = OMe$)	1
Bauerella simplicifolia subsp. *neo-scotica*	Acronycidine (5,7,8-trimethoxy dictamnine)	47
Boronella aff. *verticillata*	Dictamnine	2
Dutaillyea drupacea	Dictamnine *Dutadrupine (65) (-)-Edulinine (32) Evolitrine (1; $R^1 = R^3 = H$, $R^2 = OMe$) Kokusaginine (1; $R^1 = R^2 = OMe$, $R^3 = H$) Pteleine (1; $R^1 = OMe$, $R^2 = R^3 = H$)	3
D. oreophila	Dictamnine Evolitrine Kokusaginine Pteleine	3
Euxylophora paraensis	(-)-Lemobiline (38) N-Methylflindersine (40) *Paraensidimerin D (68) Skimmianine (1; $R^1 = H$, $R^2 = R^3 = OMe$) *Alkaloid (35) *Alkaloid (49)	4
Flustra foliacea	*Alkaloid (8)	5
Haplophyllum alberti	Evoxine (1; $R^1 = H$, $R^2 = OCH_2CH(OH)C(OH)Me_2$, $R^3 = OMe$) Skimmianine	6
H. dahuricum	γ-Fagarine (1; $R^1 = R^2 = H$, $R^3 = OMe$) Skimmianine	7
H. foliosum	Norgraveoline (4) Myrtopsine (34)	8
H. popovii	*Hapovine (6)	9
H. tuberculatum	Skimmianine	10
Melicope indica	Dictamnine Evolitrine N-Methylatanine (2)	11
Monnieria trifolia	Evoxine Haplopine	12
Oricia gabonensis	Skimmianine	13
O. renieri	Kokusaginine Maculosidine (1; $R^1 = R^3 = OMe$, $R^2 = H$) Oricine (43) Skimmianine *Vepridimerines B (73), C (70), and D (71) Veprisine (44) *Alkaloid (42)	13
Ruta chalepensis	*Isotaifine (62) *8-Methoxytaifine (63)	14
	*Taifine (61)	15

Table 1 - continued

Species	Alkaloid (Structure)	Ref.
Teclea verdoorniana	*Tecleaverdine (60) *Tecleine (58)	16
Toddalia asiatica var. *gracilis*	Dictamnine γ-Fagarine Skimmianine	17
Vepris louisii	*Vepridimerines A (72), B (73) and C (70) *Veprisilone (30) *Veprisinium salt (31)	18 19 20
Vepris stolzii	γ-Fagarine Skimmianine Veprisine *Alkaloid (45) *Alkaloid (46) *Alkaloid (47)	21
Zanthoxylum arborescens	*8-Hydroxy-7-methoxydictamnine (1; R^1 = H, R^2 = OMe, R^3 = OH) *7-Methoxy-8-(3-methylbut-2-enyloxy)-dictamnine (1; R^1 = H, R^2 = OMe, R^3 = $OCH_2CH=CMe_2$)	22
Z. budrunga	Dictamnine	23
Z. bungeaum	Skimmianine *Zanthobungeanine (41)	24
Z. coca	Flindersine (39)	25
Z. inerma	Dictamnine	26
Z. integrifoliolum	Dictamnine Haplopine *Integriquinolone (7; R = OH) 4-Methoxy-1-methyl-2-quinolone (7; R = H) Myrtopsine (+)-Platydesmine (5) Robustine (1; R^1 = R^2 = H, R^3 = OH) Skimmianine	27
Z. nitidum	Skimmianine	28
Z. simulans	(-)-Araliopsine (9) (-)-Edulinine (±)-Ribalinine (10) Skimmianine	29

* New alkaloids

1.2 Non-terpenoid Quinolines.- Hapovine, a new alkaloid from Haplophyllum popovii, was shown by spectroscopic studies to be 2-(n-decapenta-6',9',12'-trienyl)-4-quinolinone (6).[9] A new preparation of 2-alkylquinolinones, 2-alkyl-1-methylquinolinones and 2-alkyl-4-methoxyquinolines (Scheme 1)[30] resulted in the first synthesis of the alkaloids (11; *n* = 8) (cf. Vol. 10, p.76), (12; *n* = 11), and (12; n = 13) (cf. Vol. 6, p.105). Integriquinolone (7; R = OH) was obtained from Zanthoxylum integrifoliolum[27] and the bromoquinoline (8) was isolated from the marine bryozoan Flustra foliacea.[5]

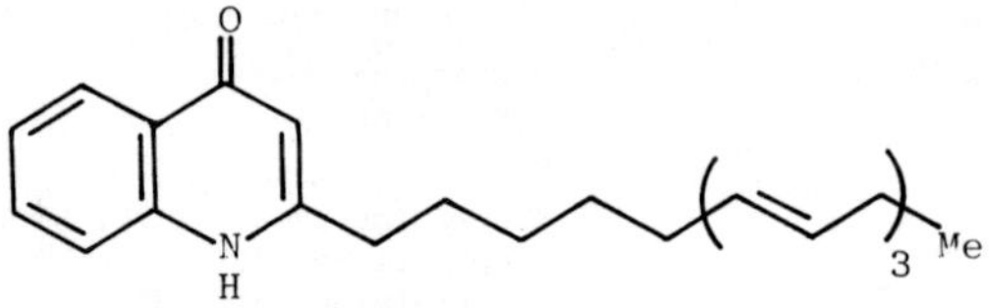

Hapovine (6)

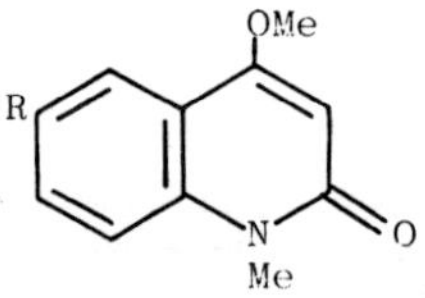

(7)

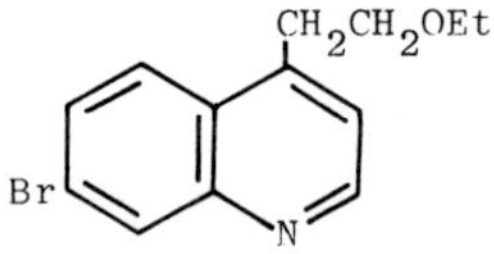

(8)

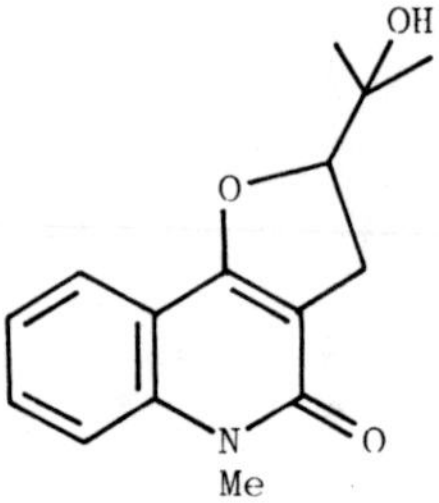

(9)

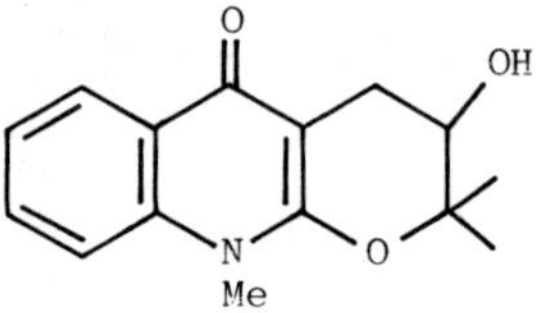

(10)

The co-enzyme methoxatin (13), a constituent of bacterial methanol dehydrogenases, has excited considerable interest since its isolation in 1979. A total synthesis of methoxatin has now been described (Scheme 2).[31]

1.3 Prenylquinolinones and Hemiterpenoid Tricyclic Alkaloids.-
3-Prenylquinolinones are well-known constituents of rutaceous plants and are biosynthetic and synthetic precursors of hemiterpenoid quinoline alkaloids; all aspects of their chemistry feature

in the literature this year. Thus, new sources of atanine (15), N-methylatanine (2), and edulinine (32) were discovered (Table 1), a full account of the isolation and properties of glyosolone (3) has appeared[32] (cf. Vol. 9, p.81), and further studies of the synthesis of atanine and related compounds have been reported.

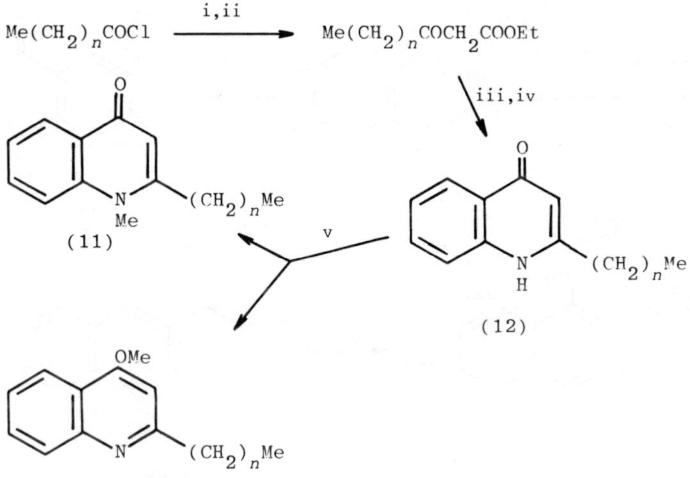

Reagents: i, MeCOCH$_2$COOEt, NaH, Et$_2$O; ii, NaOEt, EtOH; iii, PhNH$_2$, TsOH, PhH, reflux; iv, Ph$_2$O, reflux; v, MeI, K$_2$CO$_3$, DMF

Scheme 1

A new approach to the synthesis of 3-prenyl-2-quinolinones begins with a 3-carboxymethyl-2-quinolinone and gives a mixture of atanine (15) and its 3-methylbut-1-enyl isomer (14) in high yield; atanine was converted into khaplofoline (16)[33] (Scheme 3). The reaction of 4-hydroxy-1-methyl-2-quinolinone (21) was shown previously to give compounds (26) and (27), with the bisprenyl derivative (19) as principal product.[34] A systematic study of the corresponding reaction of 4-hydroxy-2-quinolinone (20), using a phase-transfer catalyst, is now reported to give the bisprenyl derivative (18) (50%) and five other products (22), (23), (24), (26), and (28) (Scheme 4); khaplofoline (16) and its angular isomer (17) were obtained from compound (18), and methylation of the 4-hydroxy-quinolinone (26) gave atanine (15). N-Prenylquinolinones were obtained by prenylation of 4-methoxy-2-quinolinone (Scheme 4).[35]

Preparation of the bisprenylquinolinone (19) from (26) is also recorded.[36]

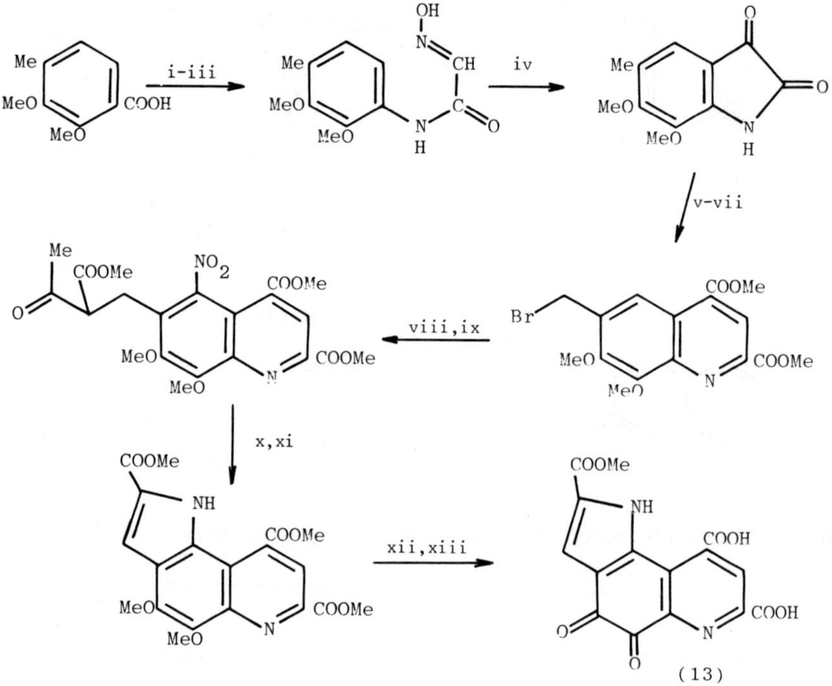

Reagents i, ClCOOEt, NEt$_3$, NaN$_3$, PhMe, reflux; ii, KOH, H$_2$O; iii, Cl$_3$CCH(OH)$_2$, NH$_2$OH·HCl, aq. Na$_2$SO$_4$; iv, polyphosphoric acid, at 100°C; v, MeCOCOOH, 30% KOH, at 95°C; vi, H$_2$SO$_4$, MeOH; vii, N-bromosuccinimide, CCl$_4$, reflux; viii, HNO$_3$, H$_2$SO$_4$, at -20°C; ix, MeCOCH$_2$COOMe, NaH, THF; x, PhN$_2^+$ BF$_4^-$, aq. pyridine, then NaBH$_4$; xi, H$_2$, Pd/C, HCl, MeOH; xii, AgO, HNO$_3$, THF; xiii, LiOH, aq. THF

Scheme 2

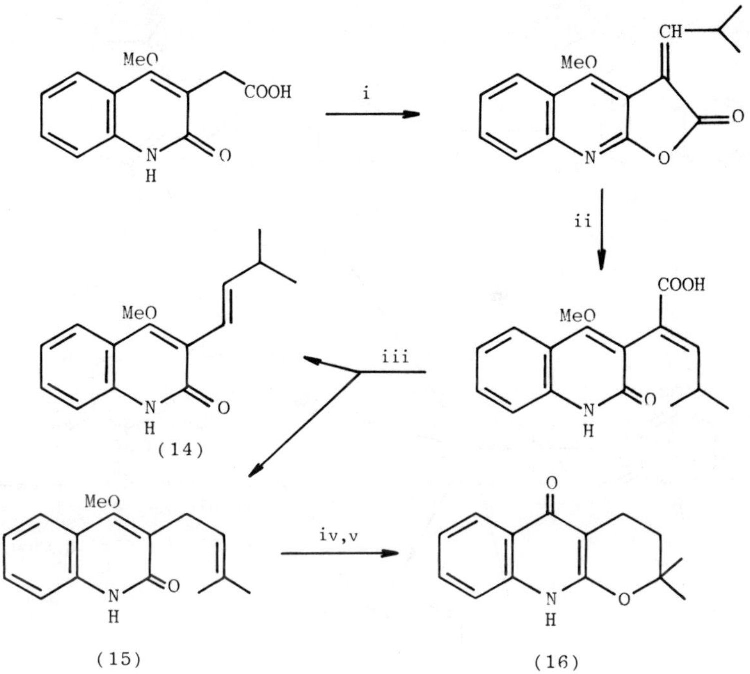

Reagents: i, Me$_2$CHCHO, Ac$_2$O, NaOAc, AcOH; ii, aq. OH$^-$, then acidify; iii; Cu, Ph$_2$O; iv; polyphosphoric acid; v, HCl, EtOH, reflux

Scheme 3

New alkaloids obtained from the stem bark of Vepris louisii include the hydroxyketone veprisilone (30)[19] and the quinolinium derivative veprisinium salt (31).[20] The structure of veprisilone was determined by its spectroscopic properties and by its conversion into the same diol (33) obtained from preskimmianine (29), which is also a constituent of V. louisii (cf. Vol. 11, p.73). Veprisinium salt is responsible for the significant antibacterial activity of the bark of V. louisii, and its structure was apparent from spectral and chemical data and from its formation from preskimmianine (Scheme 5). The (S) configuration was assigned to veprisinium salt on the basis of its c.d. spectrum.

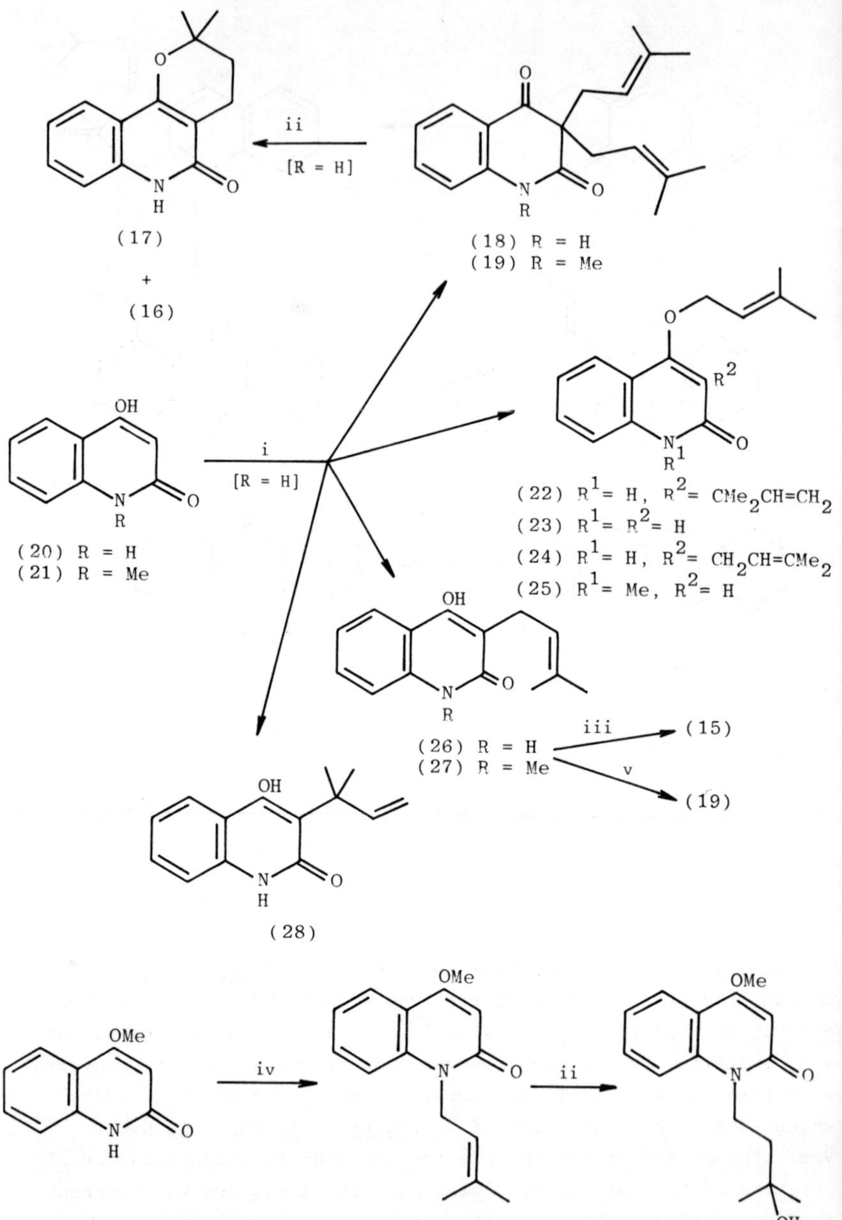

Reagents: i, Me$_2$C=CHCH$_2$Br, Bu$_4$N$^+$ Cl$^-$, PhMe, NaOH; ii, HCl, EtOH; iii, CH$_2$N$_2$, MeOH; iv, Me$_2$C=CHCH$_2$Cl, Bu$_4$N$^+$ Cl$^-$, PhMe, NaOH; v, Me$_2$C=CHCH$_2$Br, then MeI, K$_2$CO$_3$, Me$_2$CO

Scheme 4

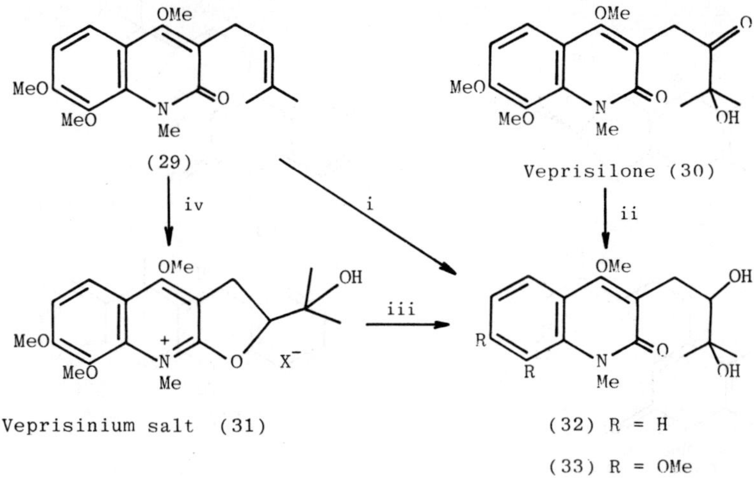

Reagents: i, m-ClC$_6$H$_4$CO$_3$H, then OH$^-$; ii, NaBH$_4$; iii, NaOH, aq. MeOH; iv, m-ClC$_6$H$_4$CO$_3$H, H$_2$SO$_4$, Bu^{t}OH

Scheme 5

Myrtopsine (34), first obtained from Myrtopsis sellingii (cf. Vol. 8, p.82), has been isolated from Zanthoxylum integrifoliolum[27] and from Haplophyllum foliosum;[8] the trans-arrangement of substituents in the dihydrofuran ring, assigned on the basis of ^{1}H n.m.r. spectroscopy,[8] is in accord with the proposal that myrtopsine or one of its derivatives is a biosynthetic precursor of furoquinoline alkaloids.[37]

The structure of an O-prenylquinolinone derivative (35) isolated from Euxylophora paraensis was established by spectroscopic studies.[4] This alkaloid is the diol corresponding to ravenine (36), which was shown to be the biosynthetic precursor of its abnormal Claisen rearrangement product ravenoline (37) and probably the cyclised product, lemobiline (38);[38] the co-occurrence of the diol (35) and lemobiline in E. paraensis[4] is thus of considerable interest.

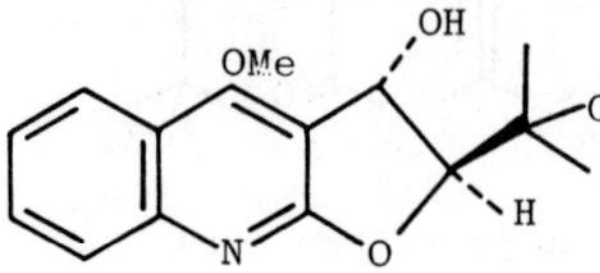

(34)

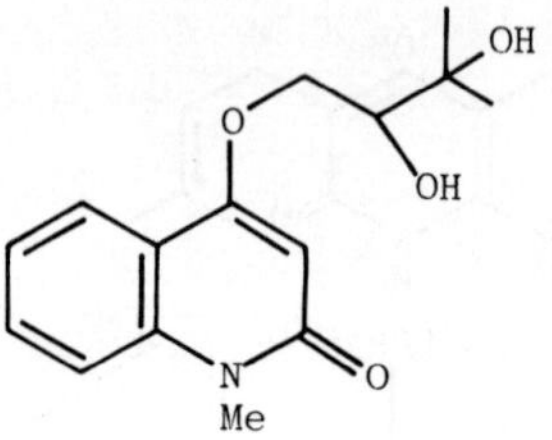

(35)

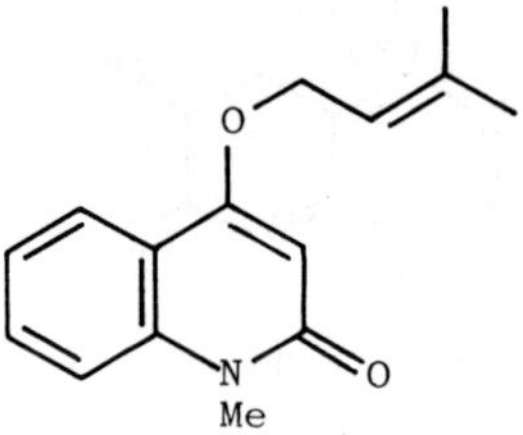

(36)

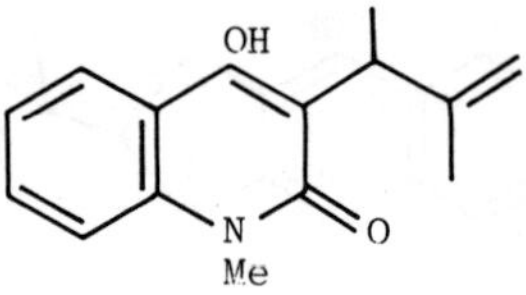

(37)

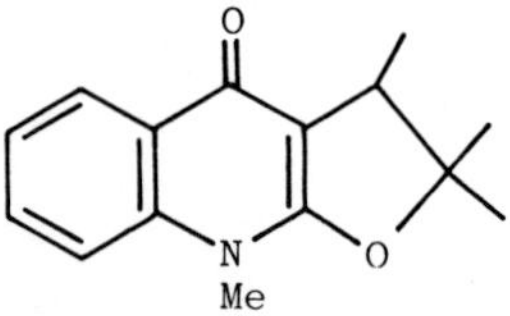

(38)

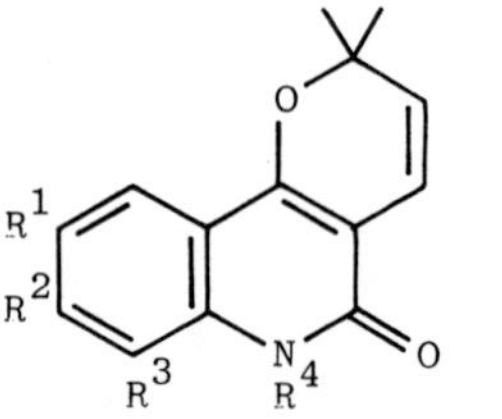

(39) $R^1 = R^2 = R^3 = R^4 = H$

(40) $R^1 = R^2 = R^3 = H$, $R^4 = Me$

Zanthobungeanine (41) $R^1 = R^2 = H$, $R^3 = OMe$, $R^4 = Me$

(42) $R^1 = R^3 = H$, $R^2 = OMe$, $R^4 = Me$

(43) $R^1 = R^2 = OMe$, $R^3 = H$, $R^4 = Me$

(44) $R^1 = H$, $R^2 = R^3 = OMe$, $R^4 = Me$

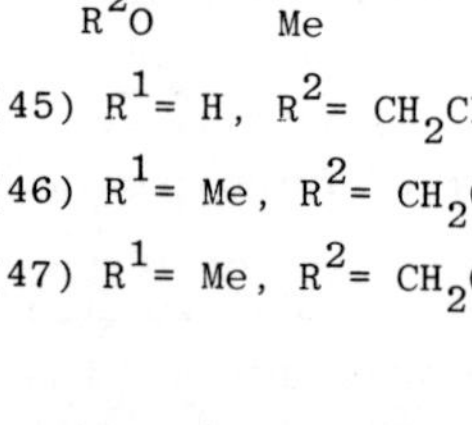

(45) $R^1 = H$, $R^2 = CH_2CH=CMe_2$

(46) $R^1 = Me$, $R^2 = CH_2CH=CMe_2$

(47) $R^1 = Me$, $R^2 = CH_2CH-CMe_2$ with O bridge

2,2-Dimethylpyranoquinolinones related to flindersine (39) continue to attract attention. Five new alkaloids of this group were isolated; the methoxy-N-methylflindersines (42) from <u>Oricia renieri</u>[13] and (41) (Zanthobungeanine) from <u>Zanthoxylum bungeaum</u>[24] and the three derivatives (45), (46), and (47) of N-methylflindersine from <u>Vepris stolzii</u>.[21] The structures of the alkaloids of the latter species were determined by UV, ^{1}H n.m.r., and mass spectroscopy and by acid hydrolysis to products shown by positive Gibbs tests to contain aromatic protons para to phenolic OH groups. Zanthophylline (48) (cf. Vol. 9, p.82) has been synthesised from a 4-hydroxy-3-prenyl-2-quinolinone via 8-methoxyflindersine (Scheme 6).[36] An optically active alkaloid (49) isolated from E. paraensis[4] can be regarded as a hydrate of N-methylflindersine (40), which occurs in the same plant.

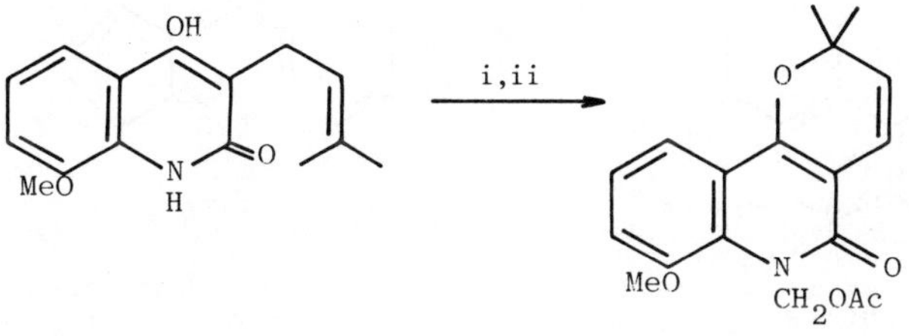

Reagents: i, 2,3-dichloro-5,6-dicyanobenzoquinone, PhH, reflux; ii, AcOCH$_2$Cl, NaH, THF

Scheme 6

1.4 Furoquinoline Alkaloids.- A new alkaloid of Zanthoxylum arborescens was shown to be 8-hydroxy-7-methoxydictamnine (50) by methylation to skimmianine (1; R^1 = H, R^2 = R^3 = OMe) and by its non-identity with haplopine (1; R^1 = H, R^2 = OH, R^3 = OMe). The structure (51) for a second new furoquinoline from Z. arborescens was established by spectroscopic studies and by its formation from reaction of alkaloid (50) with dimethylallyl bromide.[22] The constitution (53) for perfamine was proposed partly on the basis of its conversion into a hydroxymethoxydictamnine believed to be compound (50) (cf. Vol. 9, p.81), but since this product is not identical with the alkaloid from Z. arborescens, a new structure (52) has been assigned to perfamine; ^{1}H n.m.r. resonances at 8.80 (H$_A$) and 6.2 p.p.m. (H$_B$) are in accord with this proposal.[22]

Bhattacharyya and Serur established the structures of delbine (54), m.p. 229-231°C, and montrifoline (56), m.p. 191-193°C, the leaf alkaloids of Monnieria trifolia; direct comparison of delbine with heliparvifoline (55), m.p. 245-247°C, showed that the compounds were not identical (cf. Vol. 12, p.86). Clearly unaware of the

Brazilian work, Moulis et al.[12] isolated alkaloids, m.p. 229°C and m.p. 188-190°C, to which were assigned structures (55) (heliparvifoline; cf. Vol. 7, p.82) and (57) (evolatine; m.p. 201-202°C), respectively; comparison of melting points and spectral data apparently indicate that these compounds are the 7-methoxyquinoline delbine (54) and montrifoline (56), and not the isomeric 5-methoxy derivatives as supposed.

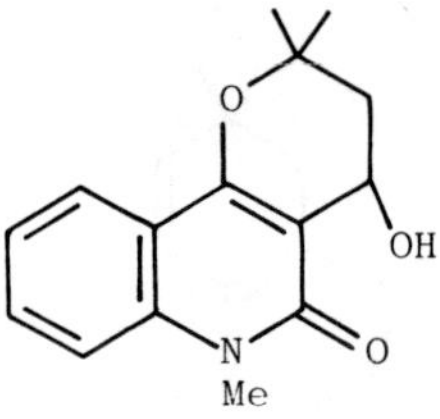

(49)

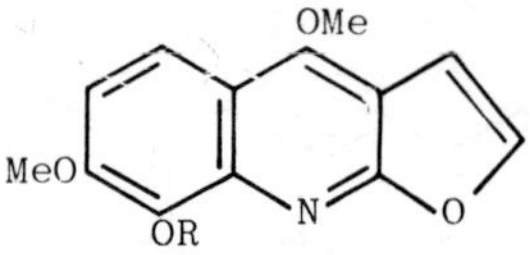

(50) R = H
(51) R = $CH_2CH=CMe_2$

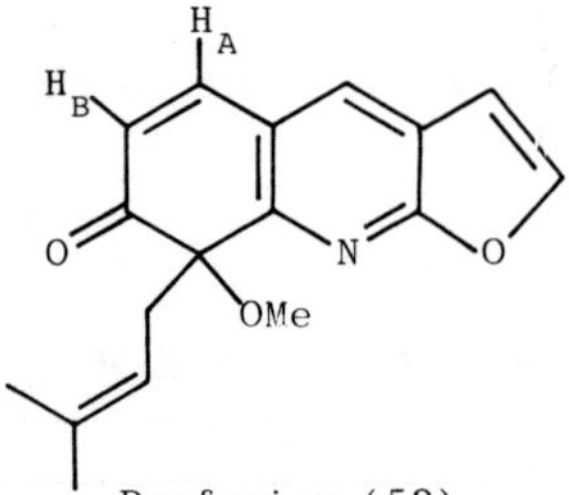

Perfamine (52)

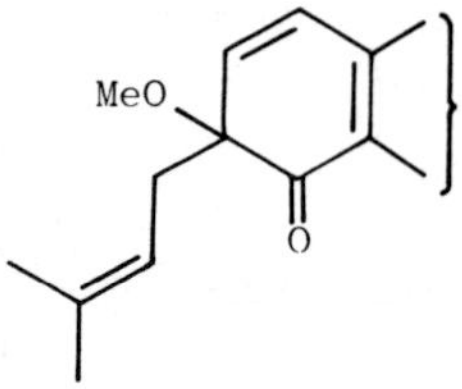

(53)

A hydroxymethylenedioxydictamnine was obtained from Teclea sudanica but in insufficient quantity for complete structural determination.[39] The same alkaloid, now named tecleine, has been isolated from T. verdoorniana and structure (58) has been established by spectroscopic studies and by methylation to flindersiamine (1; R^1R^2 = OCH_2O, R^3 = OMe).[16] A full account of the isolation and properties of tecleaverdoornine (59) (cf. Vol. 9, p.81) suggests that the failure of the alkaloid to undergo the iso rearrangement with methyl iodide is due to intramolecular hydrogen-bonding of the OH group to the heterocyclic nitrogen atom.[16] The same publication describes the conversion of tecleaverdoornine into a new alkaloid, tecleaverdine (60).

Tecleine (58) R = H
(59) R = CH$_2$CH=CMe$_2$
Tecleaverdine (60) R = CH$_2$CH(OH)C(OH)Me$_2$

Taifine (61) R^1 = OMe, R^2 = H
Isotaifine (62) R^1 = H, R^2 = OMe
8-Methoxy-taifine (63) R^1 = R^2 = OMe

N-Ethyl-4-quinolinones, named taifine (61),[15] isotaifine (62), and 8-methoxytaifine (63),[14] were isolated from <u>Ruta chalepensis</u>. The distinction between an O-ethyl group and an N-ethyl group in taifine was made by comparing the ^{1}H n.m.r. spectra in CDCl$_3$ and CF$_3$CO$_2$D and observing a significant downfield shift of the methylene resonance in the acidic solvent due to protonation of the nitrogen atom.[15] The method of extraction (refluxing ethanol and heating the alkaloid mixture with ethanolic KOH) may be responsible for the presence of N-ethyl groups in these alkaloids.

Dutadrupine (65), the new alkaloid of <u>Dutaillyea drupacea</u>,[3] is an addition to the small group of furoquinoline alkaloids containing a 2,2-dimethylpyran ring. The structure of the alkaloid was apparent from the ^{1}H n.m.r. spectrum in comparison with that of isodutadrupine (64) obtained in the usual way. Dutadrupine was synthesised from the appropriate hydroxydictamnine (Scheme 7).[40]

Dictamnine, isodictamnine, haplopine, maculosidine (1; R^1 = R^3 =

OMe, R^2 = H), and maculine (1; R^1R^2 = OCH_2O, R^3 = H) are phototoxic to certain bacteria and yeasts in long-wave u.v. light, but are less active than the furocoumarin, 8-methoxypsoralen.[41]

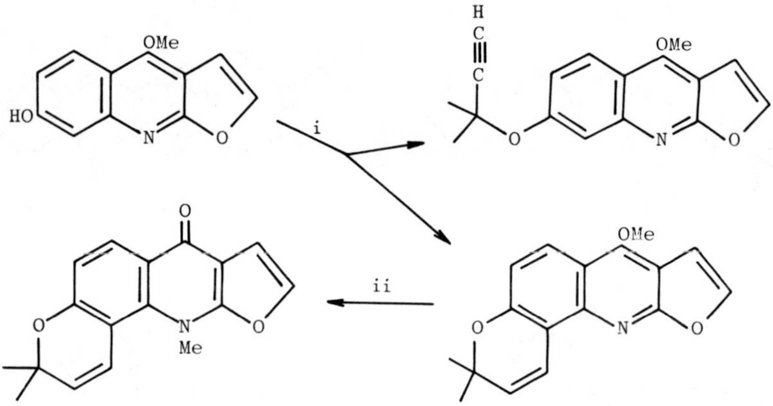

(64)   Dutadrupine (65)

Reagents: i, $Me_2C(Cl)C{\equiv}CH$, K_2CO_3, KI, Me_2CO, reflux; ii, MeI, at 80°C

Scheme 7

1.5 Dimeric Quinolinone Alkaloids.- The first dimeric quinolinone alkaloids, pteledimerine and the closely related pteledimeridine, are constituents of Ptelea trifoliata (cf. Vol. 10, p.80 and Vol. 11, p.75). Two new types of dimer have now been identified, represented by paraensidimerin D and by the vepridimerines.

Several dimeric alkaloids were isolated from the heartwood of Euxylophora paraensis and the structure of one of them, paraensidimerin D (68), was established by X-ray analysis.[4] The cis fusion of the central rings suggests that the alkaloid may originate by Diels-Alder addition of a quinolinone quinone methide to N-methylflindersine (40); this route was supported by an independent study in which the endione (67), generated in situ from a 2-quinolinone (66), was shown to react readily and specifically with N-methylflindersine to give the Diels-Alder adduct (69), containing the same ring system as paraensidimerin D (Scheme 8).[42]

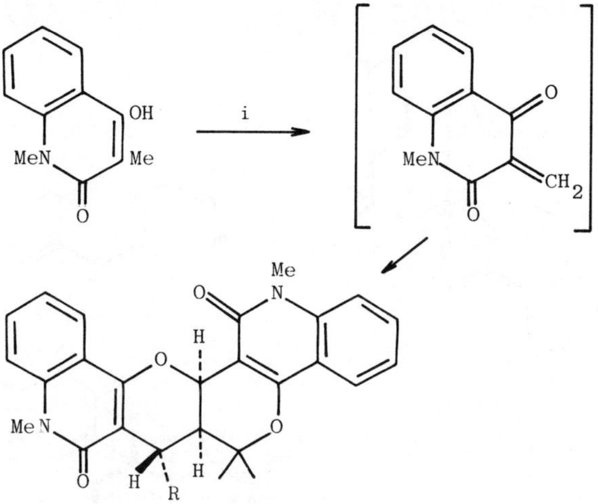

Paraensidimerine D (68) R = CH=CMe$_2$
(69) R = H

Reagents: i, N-methylflindersine (40), 2,3-dichloro-5,6-dicyanobenzoquinone, PhH, reflux

Scheme 8

The barks of <u>Vepris louisii</u> and <u>Oricia renieri</u> contain isomeric racemic quinolinone dimers, vepridimerines A-C from <u>V. louisii</u> and vepridimerines B-D from <u>O. renieri</u>. Structures (70)-(73) (relative stereochemistry shown) were determined mainly by ^{1}H- and ^{13}C-n.m.r. spectroscopy. Vepridimerines A and B have two 2-quinolinone functions and differ in the configuration of H$_d$ and H$_e$; 2- and 4-quinolinone groups are present in each of the other alkaloids, with H$_d$ and H$_e$ <u>cis</u>- in vepridimerine C and <u>trans</u>- in the D isomer. The C$_{10}$ alicyclic moiety of the vepridimerines is similar to that present in isoalfileramine, the cyclisation product of the alkaloid alfileramine from <u>Zanthoxylum punctatum</u> (<u>cf</u>. Vol. 11, p.244), and is also found in a synthetic 2,2-dimethylbenzopyran dimer.[43] Paraensidimerins isolated from <u>Euxylophora paraensis</u> have been shown to be the four racemates, <u>cf</u>. (72) and (73), related to tetra-demethoxy-vepridimerines A and B.[44] The dimeric quinolinone alkaloids of this group may arise by Diels-Alder addition of one molecule of <u>o</u>-hydroxy-dienes (74; R = H or OMe) to the internal double bond of a second molecule, followed by cyclisation.

Vepridimerine C (70) α-H$_e$
Vepridimerine D (71) β-H$_e$

(74)

Vepridimerine A (72) α-H$_d$, α-H$_e$
Vepridimerine B (73) α-H$_d$, β-H$_e$

2 Acridone Alkaloids

2.1 *Occurrence*.- Table 2[13,45-55] lists known acridone alkaloids isolated from fresh sources as well as new alkaloids; the latter group includes fourteen compounds obtained by Furukawa and co-workers from *Citrus*, *Glycosmis*, and *Severinia* species. *Oricia renieri* contains the o-aminobenzophenone tecleonone (81)[13] (cf. Vol. 7, p.90). In an independent study, the presence of rutacridone epoxide (103) in roots of *Ruta graveolens* has been confirmed[53] (cf. Vol. 12, p.91). One source of the Chinese drug "Chen-pi" is the peel of *Citrus depressa*, and the first acridones from *Citrus* have been isolated from the root bark of this species.[49]

(75) R^1= OH, R^2= R^3= OMe, R^4= H
(76) R^1= OMe, R^2R^3= -OCH$_2$O-, R^4= H
(77) R^1= OH, R^2= R^4= H, R^3= OMe
(78) R^1= R^2= R^3= R^4= OMe
(79) R^1= R^2= R^3= OMe, R^4= H

(80)

(81)

Table 2 Isolation of acridone alkaloids

Species	Alkaloid (Structure)	Ref.
Atalantia monophylla [cf. Vol. 2, p.95 & Vol. 7, p.91 for (87) & (92)]	N-Methylatalaphyllinine (102) *1,5-Dihydroxy-2,3-dimethoxy-N-methyl-acridone (85)	45
A. wightii	N-Methylatalaphylline (88) N-Methylatalaphyllinine	46
Bauerella simplicifolia subsp. *neo-scotica* (formerly classified as *B. baueri*, cf. Vol. 5, p.110; cf. Vol. 12, p.91)	Melicopicine (78) 1,2,3-Trimethoxy-N-methylacridone (79)	47
Boenninghausenia albiflora (cf. Vol. 9, p.87)	1-Hydroxy-7-methoxyacridone (80) 1-Hydroxy-3-methoxy-N-methylacridone (77)	48
Citrus depressa	*Citracridone-I (93) *Citracridone-II (94) *Citpressine-I (82) *Citpressine-II (83) *Prenylcitpressine (84) Alkaloid (95)	49
Glycosmis citrifolia	*Glyfoline (86)	51
	Furacridone (106) *Furofoline-II (107)	52
	*Glycocitrine-I (89) *Glycocitrine-II (90) *O-Methylglycocitrine-II (91)	51
	*Pyranofoline (96)	52
	*Alkaloid (97)	50
Oricia renieri	Arborinine (75) Evoxanthine (76)	13
Ruta graveolens	*Gravacridonol (105)	53
	*Hydroxyrutacridone epoxide (104)	54

*New alkaloids

Species	Alkaloid (Structure)	Ref.
Severinia buxifolia	Atalaphyllinine (101) N-Methylatalaphyllinine *N-Methylseverifoline (100) *Severifoline (99)	55

* New Alkaloids

2.2 New Alkaloids.- The N-methylacridone alkaloids isolated from Citrus depressa contain oxygen atoms substituted at C-2, C-3, and C-5, the five new alkaloids citpressine-I (82), citpressine-II (83), prenylcitpressine (84), and the 2,2-dimethylpyrano derivatives citracridone-I (93) and citracridone-II (94) possessing an additional oxygen function at C-6. Structures were determined by spectroscopic methods and by interconversions. For example, in the ^{1}H n.m.r. spectrum of citracridone-I (93), long-range coupling of C-1' and C-2 indicated that the pyrano ring was attached to ring C of the acridone structure and n.o.e. enhancement of the signal at δ 6.63 (H-1') on irradiation at 3.75 (N-CH$_3$) confirmed the angular annelation in the alkaloid; the n.o.e. technique was also used to good effect to establish the substitution pattern in ring A. Methylation of citracridone-I (93) and citpressine-I (82) with diazomethane gave citracridone-II (94) and citpressine-II (83), respectively; cyclisation of prenylcitpressine (84) with formic acid gave the dihydro derivative of citracridone-I.[49]

Constituents of Glycosmis citrifolia include the 4-prenyl-N-methylacridone alkaloids glycocitrine-I (89), glycocitrine-II (90) and its O-methyl ether (91) as well as the dihydroxytetramethoxyacridone glyfoline (86); the structures of these compounds were determined by methods similar to those used for the Citrus alkaloids.[51] Pyranofoline (96) is an interesting new alkaloid of G. citrifolia; the linear arrangement of the pyrano ring was shown by the u.v., ^{1}H n.m.r., and ^{13}C n.m.r. spectra and the presence of an OH group at C-5 was established by the n.o.e. observed in the n.m.r. spectrum of the mono-methoxymethyl derivative.[52] Another unusual constituent of G. citrifolia is alkaloid (97);[50] this is the first monoterpenoid acridone obtained from natural sources although the synthesis of compound (98), lacking an OH group at C-5, has been reported.[56]

Citpressine-I (82) $R^1 = R^2 = H$
Citpressine-II (83) $R^1 = Me$, $R^2 = H$
renyl-
  citpressine (84) $R^1 = H$, $R^2 = CH_2CH=CMe_2$

(85) $R^1 = R^3 = H$, $R^2 = OH$
Glyfoline (86) $R^1 = OH$, $R^2 = R^3 = OMe$

(87) R = H
(88) R = Me

Glycocitrine-I (89) $R^1 = OH$, $R^2 = Me$
Glycocitrine-II (90) $R^1 = R^2 = H$
O-Methylglycocitrine-II (91) $R^1 = H$, $R^2 = Me$

The acridone alkaloids of <u>Severinia buxifolia</u> are a closely related group of prenylpyrano derivatives consisting of the new alkaloids severifoline (99) and N-methylseverifoline (100) and their known 5-hydroxy analogues, atalaphyllinine (101) and N-methylatalaphyllinine (102).[55]

Two new dihydrofuroacridones have been isolated from <u>Ruta graveolens</u>, hydroxyrutacridone epoxide (104) from the roots and callus tissue culture[54] and gravacridonol (105) from the roots;[53] the structures were established by spectroscopy. Rutacridone epoxide was converted into the alkaloids gravacridondiol and gravacridondiol monomethyl ether (Scheme 9)[54] (<u>cf</u>. Vol. 4, p.126).

Furacridone (106), previously obtained from <u>Ruta graveolens</u> as an inseparable mixture containing 20% of 1-hydroxy-3-methoxy-N-methylacridone (<u>cf</u>. Vol. 8, p.84) and synthesised last year (Vol. 12, p.90), has now been isolated from <u>Glycosmis citrifolia</u>.[52] Furofoline-II (107), the first hydroxyisopropylfuroacridone, has also been found in <u>G. citrifolia</u>.[52]

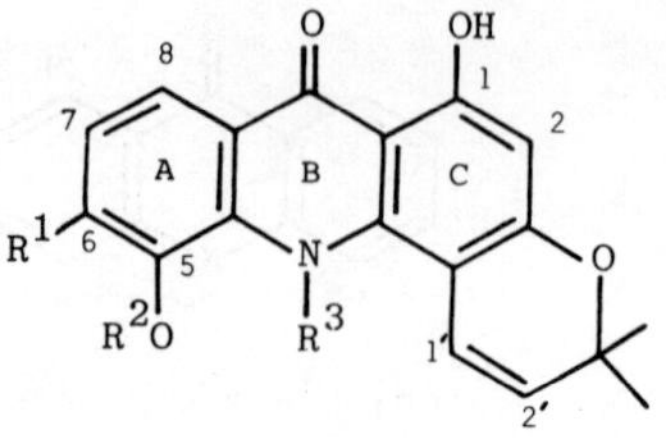

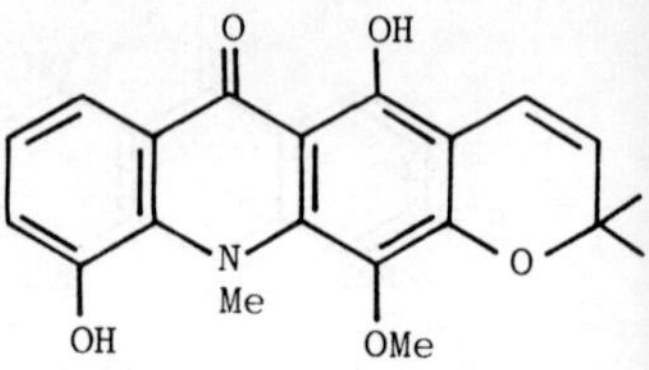

(92) $R^1 = R^2 = R^3 = H$

Citracridone-I (93) $R^1 = OH$, $R^2 = R^3 = Me$

Citracridone-II (94) $R^1 = OMe$, $R^2 = R^3 = Me$

(95) $R^1 = H$, $R^2 = R^3 = Me$

Pyranofoline (96)

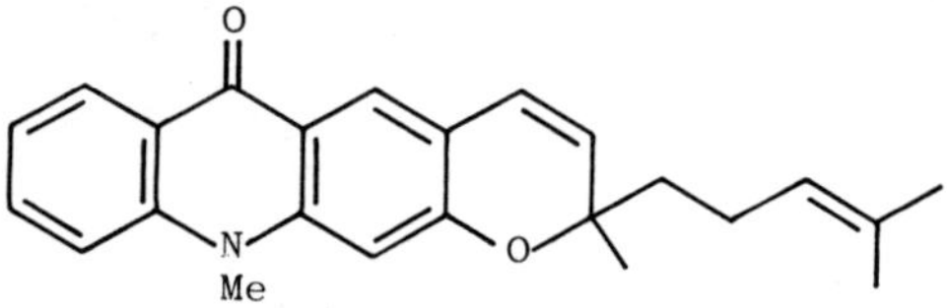

(97) R = OH

(98) R = H

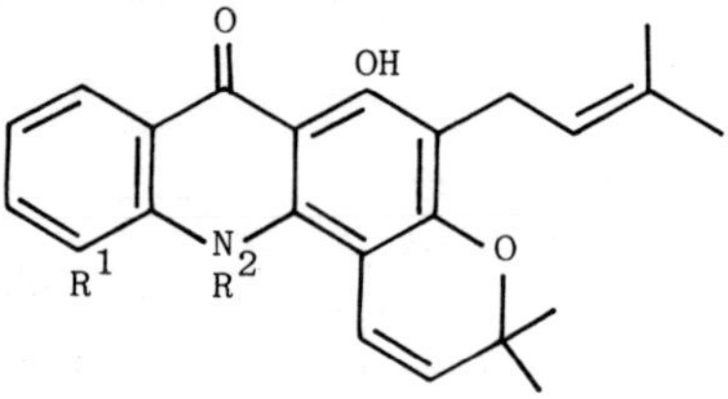

Severifoline (99) $R^1 = R^2 = H$

N-Methylseverifoline (100) $R^1 = H$, $R^2 = Me$

(101) $R^1 = OH$, $R^2 = H$

(102) $R^1 = OH$, $R^2 = Me$

Structure (85) for a new alkaloid of <u>Atalantia monophylla</u> was based on the ^{1}H n.m.r. spectrum.[45] An alkaloid obtained from <u>Boenninghausenia albiflora</u>, formerly regarded as 1,7-dihydroxy-N-methylacridone, has now been shown from the ^{13}C n.m.r. spectrum to be the isomeric 1-hydroxy-7-methoxyacridone (80).[48]

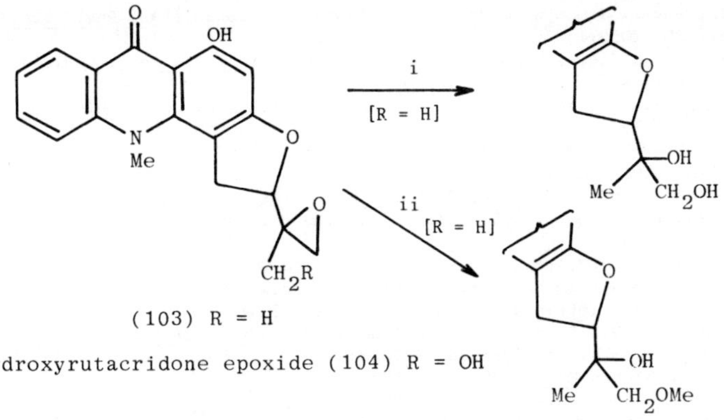

(103) R = H
Hydroxyrutacridone epoxide (104) R = OH

Reagents: i, (COOH)$_2$, aq. Me$_2$CO or KOH, aq. Me$_2$CO, reflux; ii, KOH, MeOH

Scheme 9

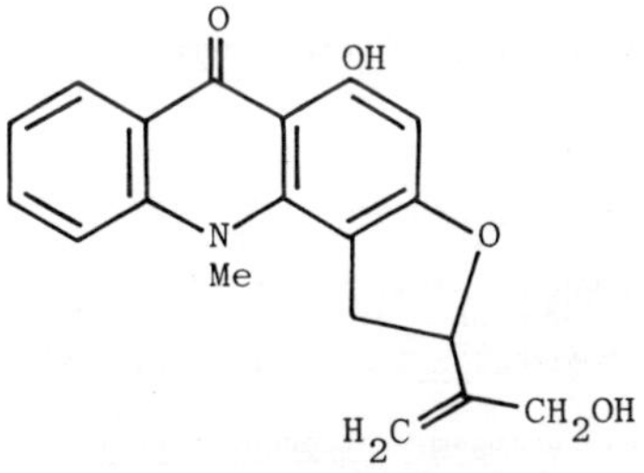

Gravacridonol (105)

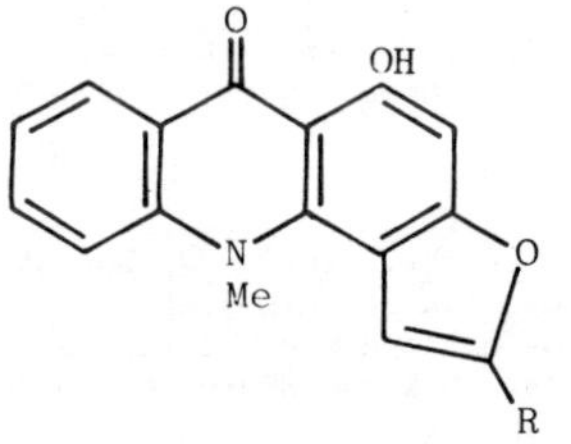

Furofoline-II (106) R = H
(107) R = C(OH)Me$_2$

References

1. J.Reisch, M.Mueller, and I.Mester, Planta Med., 1981, 43, 285.
2. F.Bevalot, J.Vaquette, and P.Cabalion, Plant. Med. Phytother., 1980, 14, 218.
3. G.Baudouin, F.Tillequin, M.Koch, J.Pusset, and T.Sevenet, J. Nat. Prod., 1981, 44, 546.
4. L.Jurd and R.Y.Wong, Aust. J. Chem., 1981, 34, 1625.
5. P.Wulff, J.S.Carle, and C.Christophersen, Comp. Biochem. Physiol. B, 1982, 71, 525.
6. D.M.Razakova and I.A.Bessonova, Khim. Prir. Soedin., 1981, 673 (Chem. Abstr., 1982, 96, 31 680).
7. D.Batsuren, E.Kh.Batirov, and V.M.Malikov, Khim. Prir. Soedin., 1981, 659 (Chem. Abstr., 1982, 96, 48 968).
8. V.I.Akhmedzhanova and I.A.Bessonova, Khim. Prir. Soedin., 1981, 613 (Chem. Abstr., 1982, 96, 31 670).
9. D.M.Razakova and I.A.Bessonova, Khim. Prir. Soedin., 1981, 528 (Chem. Abstr., 1982, 96, 100 871).
10. S.A.Khalid and P.G.Waterman, Planta Med., 1981, 43, 148.

11 M.Th.Fauvel, J.Gleye, C.Moulis, F.Blasco, and E.Stanislas, Phytochemistry, 1981, 20, 2059.
12 G.Moulis, J.Gleye, F.Fouraste, and E.Stanislas, Planta Med., 1981, 42, 400.
13 S.A.Khalid and P.G.Waterman, Phytochemistry, 1981, 20, 2761.
14 N.Mohr, H.Budzikiewicz, B.A.H.El-Tawil, and F.K.A.El-Beih, Phytochemistry, 1982, 21, 1838.
15 B.A.H.El-Tawil, F.K.A.El-Beih, H.Budzikiewicz, and N.Mohr, Z. Naturforsch., Teil.B, 1981, 36, 1169.
16 J.F.Ayafor and J.I.Okogun, J. Chem. Soc., Perkin Trans.1, 1982, 909.
17 P.N.Sharma, A.Shoeb, R.S.Kapil, and S.P.Popli, Indian J. Chem., Sect. B, 1981, 20, 936.
18 T.B.Ngadjui, J.F.Ayafor, B.L.Sondengam, J.D.Connolly, D.S.Rycroft, S.A. Khalid, P.G.Waterman, and (in part) N.M.D.Brown, M.F.Grundon, and V.N. Ramachandran, Tetrahedron Lett., 1982, 23, 2041.
19 J.F.Ayafor, B.L.Sondengam, and B.T.Ngadjui, Phytochemistry, 1982, 21, 955.
20 J.F.Ayafor, B.L.Sondengam, and B.T.Ngadjui, Tetrahedron Lett., 1981, 22, 2685 and Planta Med., 1982, 44, 139.
21 S.A.Khalid and P.G.Waterman, J. Nat. Prod., 1982, 45, 343.
22 J.A.Grina, M.R.Ratcliff, and F.R.Stermitz, J. Org. Chem., 1982, 47, 2648.
23 N.Ruanguengsi, P.Tantiratana, R.P.Borris, and G.A.Cordell, J. Sci. Soc. Thailand, 1981, 7, 123 (Chem. Abstr., 1982, 96, 31 659).
24 L.Ren and F.Xie, Yaoxue Xuebao, 1981, 16, 672 (Chem. Abstr., 1982, 96, 48 976).
25 M.R.Torres and B.K.Cassels, Bol. Soc. Chil. Quim., 1982, 27, 260 (Chem. Abstr., 1982, 96, 196 571); M.A.Munoz, M.R.Torres and B.K.Cassels, J. Nat. Prod., 1982, 45, 367.
26 H.Ishii, K.Murakami, K.Takeishi, T.Ishikawa, and J.Haginiwa, Yakugaku Zasshi, 1981, 101, 504 (Chem. Abstr., 1981, 95, 111 726).
27 H.Ishii, I.S.Chen, M.Akaike, T.Ishikawa, and S.T.Lu, Yakugaku Zasshi, 1982, 102, 182 (Chem. Abstr., 1982, 97, 69 240).
28 M.-H.Wang, Yao Hsueh T'ung Pao, 1981, 16, 48 (Chem. Abstr., 1981, 95, 192 260).
29 Z.Chang, F.Liu, S. Wang, T.Zhao, and M.Wang, Yaoxue Xuebao, 1981, 16, 394 (Chem. Abstr., 1982, 97, 20 735).
30 R.Somanthan and K.M.Smith, J. Heterocyclic Chem., 1981, 18, 1077.
31 J.A.Gainor and S.M.Weinreb, J. Org. Chem., 1981, 46, 4317.
32 B.P.Das, D.N.Chowdhury, B.Choudhury, and I.Mester, Indian J. Chem. B, 1982, 21, 176.
33 M.Ramesh, V.Arisvaren, S.P.Rajendran, and P.Shanmugam, Tetrahedron Lett., 1982, 23, 967.
34 T.R.Chamberlain and M.F.Grundon, J. Chem. Soc. (C), 1971, 910.
35 J.Reisch, M.Muellar, and I.Mester, Z. Naturforsch., Teil. B, 1981, 36, 1176.
36 P.Venturella, A.Bellino, and L.M.Marino, Heterocycles, 1981, 16, 1873.
37 M.F.Grundon in 'The Alkaloids', ed. R.H.F.Manske and R.G.A.Rodrigo, Academic Press, 1979, Vol. XVII, p.186.
38 T.R.Chamberlain, J.F.Collins, and M.F.Grundon, Chem. Comm., 1969, 1269.
39 R.R.Paris and A.Stambouli, C.R. Med. Sci., 1959, 248, 3736.
40 F.Tillequin, G.Baudouin, and M.Koch, Heterocycles, 1982, 19, 507.
41 F.H.N.Towers, E.A.Graham, I.D.Spenser, and Z.Abramowski, Planta Med., 1981, 41, 136.
42 M.F.Grundon, V.N.Ramachandran, and B.M.Sloan, Tetrahedron Lett., 1981, 22, 3105.
43 C.S.Barnes, M.I.Strong, and J.L.Occolowitz, Tetrahedron, 1963, 19, 839.
44 L.Jurd, private communication.
45 J.S.Shah and B.K.Sabata, Indian J. Chem., Sect. B, 1982, 21, 16.
46 J.Banarji, N.Ghoshal, S.Sarkar, A.Patra, K.Abraham, and J.N.Shoolery, Indian J. Chem., Sect. B, 1981, 20, 835.
47 B.Couge, F.Tillequin, M.Koch, and T.Sevenet, Plant. Med. Phytother., 1980, 14, 208.
48 Zs.Roza, J.Reisch, I.Mester, and K.Szendrei, Fitoterapia, 1981, 52, 37.
49 T.-S.Wu, H.Furukawa, and C.-S.Kuoh, Heterocycles, 1982, 19, 273.

40 T.-S.Wu and H.Furukawa, Heterocycles, 1982, 19, 825.
41 T.-S.Wu, H.Furukawa, and C.-S.Kuoh, Heterocycles, 1982, 19, 1047.
42 T.-S.Wu, H.Furukawa, and K.-S.Hsu, Heterocycles, 1982, 19, 1227.
43 Zs.Roza, J.Reisch, K.Szendrei, and E.Minker, Fitoterapia, 1981, 52, 93.
44 U.Eilert, B.Wolters, A.Nohrstedt, and V.Wray, Z.Naturforsch, Teil. C, 1982, 37, 132.
45 T.-S.Wu, C.-S.Kuoh, and H.Furukawa, Phytochemistry, 1982, 21, 1771.
46 W.M.Bandaranayaka, M.J.Begley, B.O.Brown, D.G.Clarke, L.Crombie, and D.A. Whiting, J. Chem. Soc., Perkin Trans. 1, 1974, 998.

8
β-Phenylethylamines and the Isoquinoline Alkaloids

BY K. W. BENTLEY

1 β-Phenylethylamines

3-Methoxytyramine and heliamine have been isolated from Backebegia militaris,[1] salicifoline from Magnolia sprengeri,[2] tyramine and N-methyltyramine from Coryphantha missouriensis,[3] and hordenine from C.missouriensis,[3] Papaver litwinowii,[4] Selenicereus pteranthus, and S.grandiflorus;[5] since hordenine was the only phenylethylamine isolated from the two Selenicereus species it is assumed to be identical with cactine previously obtained from this source.[5] All currently known alkaloid-bearing cactus species and the alkaloids that they contain have been tabulated.[6] A synthesis of [α-^{14}C]hordenine has been reported,[7] photo-electron spectra of ephedrine have been studied,[8] analogues of mescaline with the N-dimethylamino group replaced by other tertiary amino groups have been prepared,[9] and attempts have been made to correlate the effects of mescaline on electroencephalograms and behaviour.[10]

2 Isoquinolines

Lemaireocereine has been isolated from Backebegia militaris[1] and a new quaternary alkaloid, pycnarrhine, identified as 7-hydroxy-6-methoxy-2-methyl-3,4-dihydroisoquinolinium chloride, has been isolated from Pycnarrhena longifolia.[11] Corypalline (1; R = H) has been oxidised by lead tetra-acetate to the dienone (2), which reacts with veratrole and with corypalline in the presence of trifluoroacetic acid to give 8-veratrylcorypalline (1; R = 3,4-dimethoxyphenyl) and 8,8'-biscorypalline respectively.[12] The quinone-acetals (3; R = H) and (3; R = Me) undergo Thiele acetylation in glacial acetic acid and concentrated sulphuric acid and the products can be hydrolysed and O-methylated to (±)-tehaunine (4; R = H) and (±)-O-methylgigantine (3; R = Me).[13] 6,7-Dimethoxy-2-methylisoquinolone has been synthesised from dimethoxyhomophthalic anhydride by processes previously reported (see Vol. 11, p.79).[14] The effect of salsolinol on the secretion of prolactin has been studied.[15]

3 Benzylisoquinolines

Alkaloids of this group have been isolated as follows:

Annona cristalensis[16]	coclaurine
Annona muricata[17]	anomurine (5; R = Me), anomuricine (5; R = H), coclaurine, and reticuline
Berberis aristata[18]	taxilamine (6; R = Me, R^1 = OMe, R^2 = OH)
Berberis baluchistanica[19,20]	armepavine, gandharamine (6; R = R^1 = R^2 = H), quettamine chloride (7), secoquettamine (8; R = R^1 = H), and dihydrosecoquettamine (8; RR^1 = bond)
Berberis chilensis[21]	reticuline
Fumaria vaillantii[22]	coclaurine, norjuzicine, and reticuline
Hernandia cordigera[23]	reticuline
Magnolia sprengeri[2]	magnocurarine
Xylopia buxifolia[24]	nor-O-methylarmepavine

The alkaloids whose structures are given in this list are new bases the structures of which were determined by spectroscopic methods.

Racemisation of (−)-norreticuline has been accomplished by aromatisation of the tetrahydroisoquinoline system, N-benzylation, reduction of the isoquinolinium salt, and reductive N-debenzylation.[25] Aryloxy radical intermediates arising from both phenolic rings have been identified by ESR in the oxidation of reticuline with ceric sulphate.[26] The carbanion resulting from the removal of one of the benzylic protons from papaverine has been alkylated with alkyl, amino-alkyl, and hydroxyalkyl halides and ethyl acrylate.[27,28] Solvent effects on the reduction of papaverine to tetrahydropapaverine by sodium borohydride have been studied.[29] The amide from tetrahydropapaverine and 1-methyl-4-phenyl-piperidine-4-carboxylic acid (the parent acid of pethidine) has been prepared and shown to have analgesic properties.[30]

Higenamine has been synthesised by the Bischler-Napieralsky route from 3,4-dibenzyloxy-[31] and 4-benzyloxy-3-methoxyphenylethylamine[32] and norreticuline and macrostomine have been synthesised by essentially similar routes. In the synthesis of norreticuline, 4-hydroxy-3-methoxyphenylethylamine was prepared by partial demethylation of the 3,4-dimethoxy compound[33] and 2-(3,4-dimethoxyphenyl)-2-(1-methyl-2-pyrrolidyl)-ethylamine for the synthesis of macrostomine was prepared from dimethoxyphenylacetonitrile[34] and from dimethoxy-ω-nitrostyrene[35] by conventional methods. (±)-Tetrahydrotakatonine (4; R = p-methoxybenzyl) has been prepared from (3; R = p-methoxybenzyl)[13] and polycarpine has been prepared from 1-(2-benzyloxy-3,4-dimethoxybenzyl)-6,7-dimethoxy-3,4-dihydroisoquinoline by conventional methods.[36] The N-benzyltetrahydroisoquinoline alkaloid sendaverine has been prepared by cyclisation and debenzylation of 2-(3-benzyloxy-4,4'-dimethoxydibenzylamino)-ethyl bromide, obtained by the action of 4-methoxybenzyl bromide on N-(3-benzyloxy-4-methoxybenzyl)aziridine.[37]

Bisquaternary salts of laudanosine, and of laudanosine and simpler dihydroisoquinolines, with long-chain dihalides have been prepared and studied as potential long-term neuromuscular blocking agents.[38-40] The effects of papaverine[41-45] and higenamine[46] on the cardiovascular system, of norlaudanosoline on neurons,[47] of coclaurine and reticuline on the central dopaminergic system[48] and of demethylcoclaurine on levels of cyclic AMP,[49] on β-adrenoreceptors[50] and on asthma[51] have been studied and a method of determining levels of papaverine in blood by HPLC has been described.[52]

4 Bisbenzylisoquinolines

Alkaloids of this group have been isolated from natural sources as follows:

Berberis buxifolia[53]	calafatimine (9)
Berberis empetrifolia[54]	isotetrandrine
Berberis orthobotrys[55]	aromoline, berbamine, and oxyacanthine
Mahonia repens[56]	obaberine, obamegine, oxyacanthine, and thalrugosine
Pycnarrhena longifolia[11]	daphnoline, homoaromoline, krukovine, limacine, and obaberine
Thalictrum minus[57]	obaberine, O-methylthalicberine, thaligosine, thalrugosine
Thalictrum sultanabadense[58]	hernandezine, hernandezine N-oxide, thalabadenzine, and thalidenzine
Tiliacora triandra[59]	tiliacorine, nortiliacorine-A, and tiliacorine-2'-N-oxide.

Calafatimine is a new alkaloid the structure of which was determined by spectroscopic methods. The ^{1}H and ^{13}C nmr spectra of the curine class of alkaloids have been studied in detail.[60] Dihydrodaphnine has been shown to have the structure (10) rather than the isomeric 7'-methoxy-7,6'-ether linked structure by nuclear Overhauser effect difference spectroscopy[61] and magnolamine has been shown to have the structure (11) by fission of the triethyl ether with sodium and liquid ammonia to (+)-O,O-diethylreticuline and (+)-7-O-ethyl-N-methylcoclaurine, the structure being confirmed by synthesis from the isoquinolines (12) and (13), using the Ullmann reaction to form the diphenyl ether linkage, followed by debenzylation.[62]

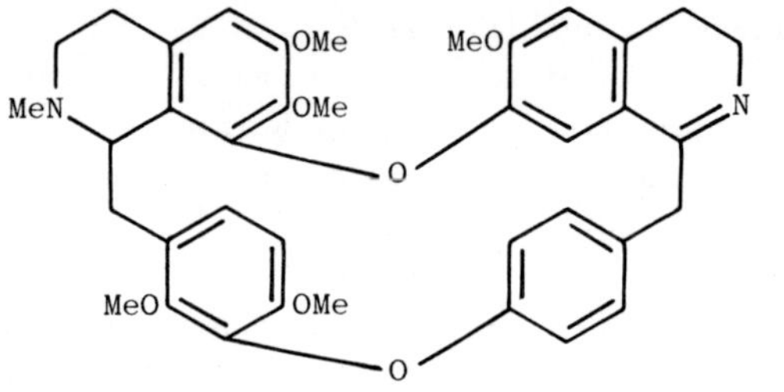

(9)

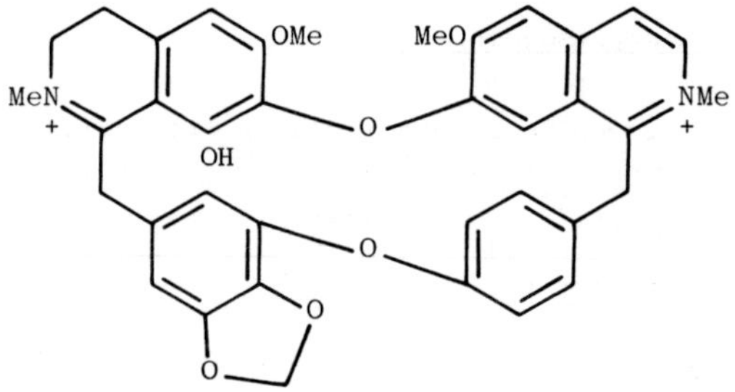

(10)

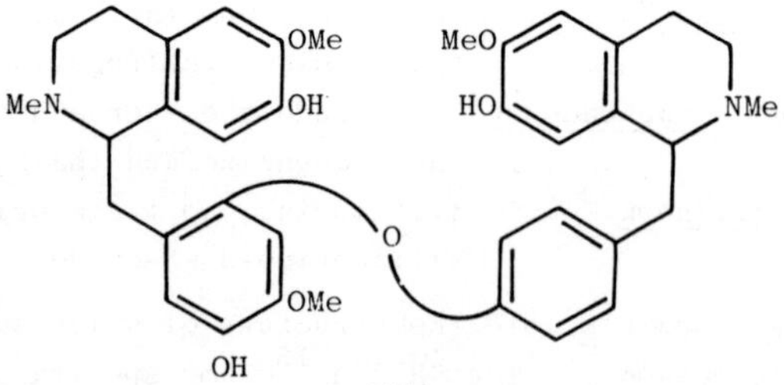

(11)

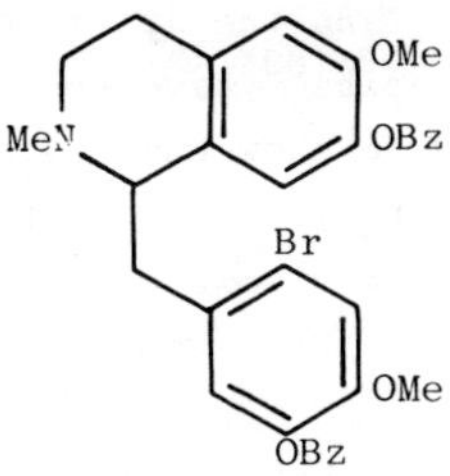

(12)

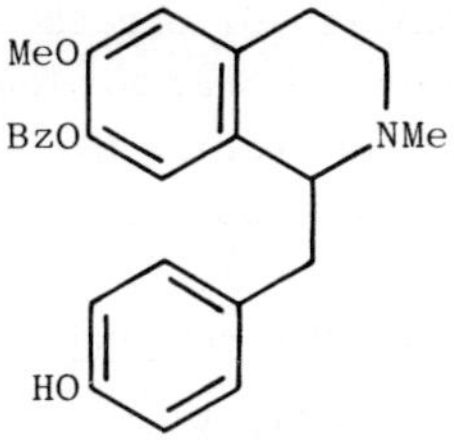

(13)

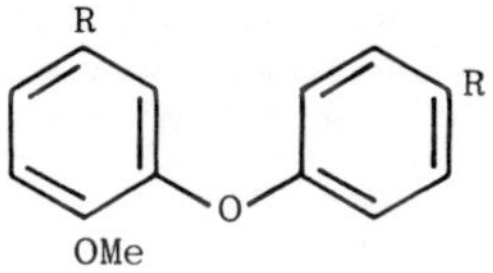

(14)

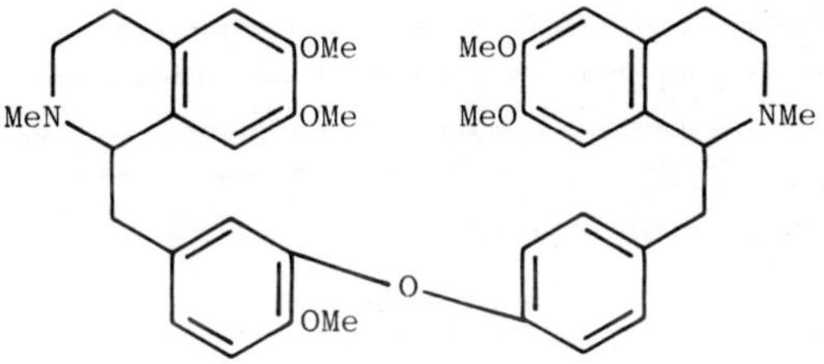

(15)

<u>O</u>-Methyldauricine (15) has been synthesised from the diphenyl ether (14; R = H) by Friedel-Crafts acetylation to the diketone (14; R = COCH$_3$), conversion of this by a Willgerodt reaction with homoveratrylamine and sulphur into a bis-thioamide that was subjected to Bischler-Napieralsky ring-closure, and <u>N</u>-methylation and reduction of the resulting bis-3,4-dihydroisoquinoline.[63,64]

Toxicity studies[65] and analytical procedures[66,67] for (+)-tubocurarine chloride have been described, as have the effects of the alkaloid on the heart,[68] nicotinic receptors,[69,70] placental and foetal weight,[71] and fast excitatory post-synaptic currents.[72] The pharmacodynamics of metocurine,[73] the effects of cepharanthine on mitochondrial function,[74,75] on cellular immunity,[76] and on tumour growth,[77] of cycleanine on nerve ganglia and the neuromuscular junction,[78] of tetrandrine on the heart,[79,80] on cell growth,[81] and

on the toxicity of cardiac glycosides,[82] and the antitumour effect
and preclinical pharmacology of thalidasine[83] have been studied.
(+)-Isochondodendrine has been claimed to show analgesic properties
with a potency approximately one quarter of that of morphine.[84]

5 Pavines and Isopavines

Isopavines have been synthesised from urethane acetals by cyclisation, directly by chlorosulphonic acid at -50°C [85] and, through the related aldehyde, by formic acid.[86] In this way O-methylthalisopavine (17; R = Me) has been prepared from (16; R = R^1 = Me)[85] and the base (17; R = H) from (16; R = CH$_2$Ph, R^1 = CO$_2$Et) with reduction of CO$_2$Et to CH$_3$ and debenzylation;[86] homoisopavines can be prepared in a similar way from homologues of (16).[86] The chloroisopavine (19) has been prepared, together with the 9-chloromorphinandienone, by the cyclisation of the acetoxydienone (18) in trifluoroacetic acid.[87]

6 Analogues of Cularine

Two analogues of cularine with a different arrangement of substituents, sarcocapnine (20) and oxosarcocapnine (21), have been isolated from <u>Sarcocapnos enneaphylla</u>, their structures being determined spectroscopically.[88]

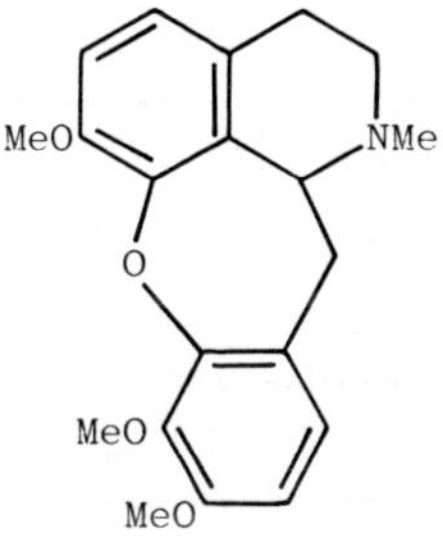

(20)

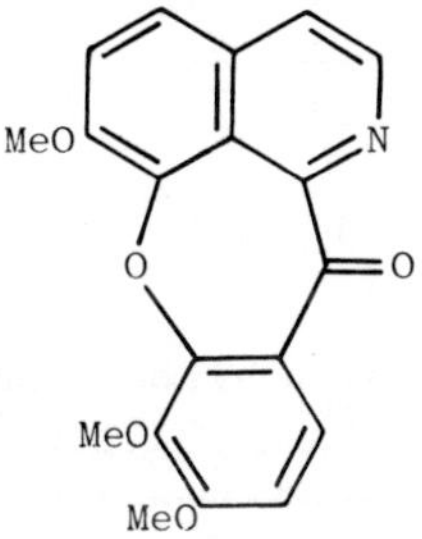

(21)

7 Berberines and Tetrahydroberberines

The following species have been shown to contain the alkaloids stated:

<u>Annona muricata</u>[17]	coreximine
<u>Berberis empetrifolia</u>[54]	berlambine (oxyberberine)
<u>Coptis japonica</u>[89]	berberine, coptisine, jatrorrhizine, and palmatine
<u>Corydalis bulbosa</u>[90]	stylopine
<u>Corydalis ledebouriana</u>[91]	corydaline and tetrahydropalmatine
<u>Corydalis tashiroi</u>[92]	dehydrodiscretamine chloride (22)
<u>Fumaria parviflora</u>[93]	cheilanthifoline
<u>Fumaria schrammii</u>[94]	sinactine and stylopine
<u>Fumaria vaillantii</u>[22]	cheilanthifoline, scoulerine, stylopine, and hydroxymethylstylopine
<u>Mahonia repens</u>[56]	berberine, columbamine, jatrorrhizine, and palmatine

Papaver albiflorum[95]	berberine, canadine, coptisine, corysamine, scoulerine, and stylopine
Papaver lecoquii[95]	berberine, canadine, coptisine, scoulerine, and stylopine
Papaver litwinowii[4]	coptisine, corysamine, and scoulerine
Papaver oreophilum[96]	alborine, berberine, coptisine, and corysamine
Thalictrum dioicum[97]	berberine
Thalictrum minus[57]	berberine, jatrorrhizine, and palmatine
Xylopia buxifolia[24]	discretamine, xylopine, and xylopinine

The absolute configurations of thalidastine and berbastine, the two known 5-hydroxy-berberines, have been determined by inference from the observation that the analogue (23) is dextrorotatory as are these alkaloids. The hydroxy-base (23) was prepared from (+)-tetrahydrojatrorrhizine by oxidation with lead tetra-acetate, followed by hydrolysis to give a 2:1 mixture of the alcohols (24; R = H, R^1 = OH) and (24; R = OH, R^1 = H), the configurations of which were determined by nmr spectroscopy; oxidation of (24; R = OH, R^1 = H) with iodine then gave (23).[98]

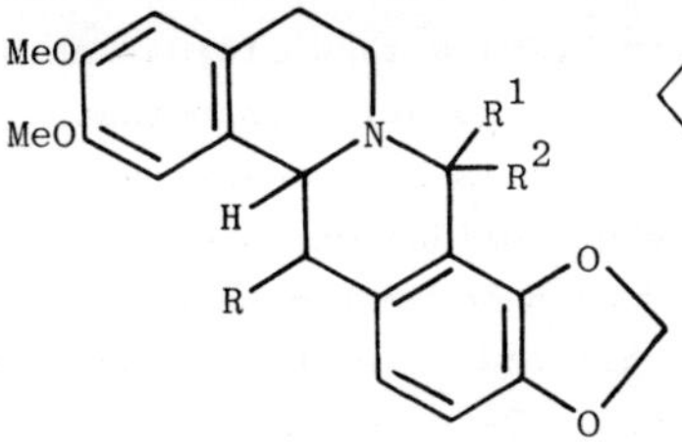

(26)

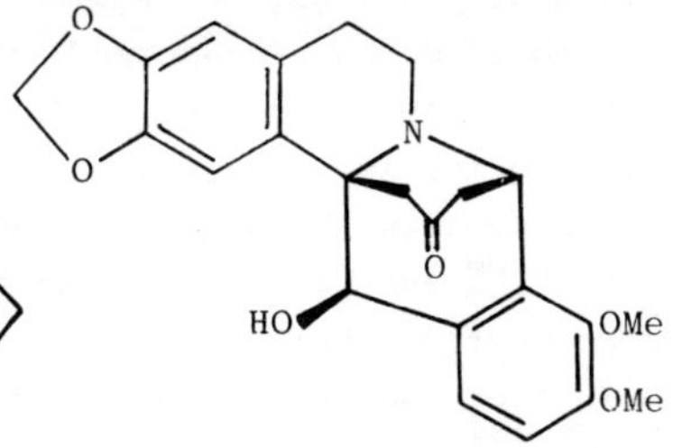

(27)

Reduction of the ester-amides (25; R = CO$_2$Et, R^{1}R^2 = O) and (26; R = CO$_2$Et, R^{1}R^2 = O) to the hydroxymethyl bases (25; R = CH$_2$OH, R^1 = R^2 = H) and (26; R = CH$_2$OH, R^1 = R^2 = H), the O-toluenesulphonyl esters of which, on further reduction, gave (+)-thalictricavine (25; R = Me, R^1 = R^2 = H)[99,100] and (+)-thalictrifoline (26; R = Me, R^1 = R^2 = H),[100,101] confirmed the 13S,14R and 13R,14R absolute configurations for these bases. ^{13}C nmr spectroscopy has been used to determine the stereochemistry of the 10,11-dimethoxy isomer of (25; R = CH$_2$OH, R^1 = R^2 = H).[102]

The structure (27) has been determined for neoxyberberine acetone by X-ray crystallographic methods.[103] Berberine salts have been reacted with acetone, methyl ethyl ketone, acetophenone, and p-aminoacetophenone in sodium hydroxide solution to give ketones of general structure (28).[104]

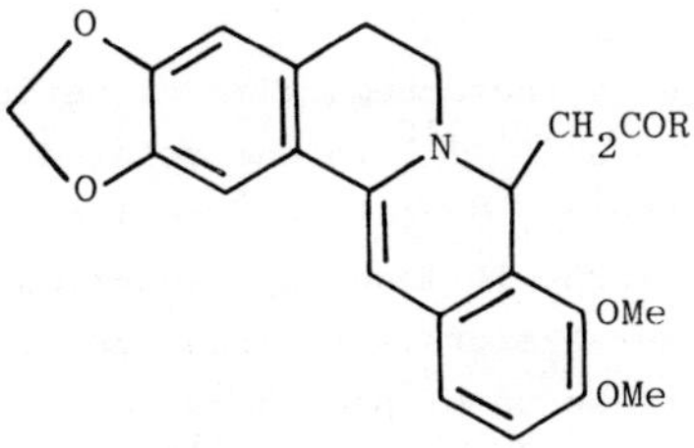

(28)

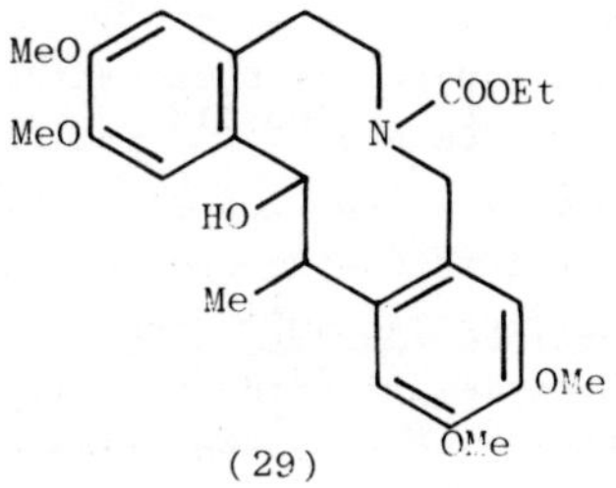

(29)

The reaction of ethyl chloroformate with 13-methylxylopinine has been found to give the ring-opened form (29) with no formation of the isomeric secoberberine system.[105] In solution in methanol, berberine chloride is converted by unsensitised photo-oxidation into the aldehyde (30) and by sensitised photo-oxidation into the methoxy-betaine (31),[106] and this betaine has been converted by processes previously reported (Vol. 11, p.93) into the spiro-base (32), which has been further converted into the 8,11,12-trimethoxy isomer of the betaine (31).[107] A series of salts of tetrahydroberberine and its derivatives, of structures (33; R = R^1 = H), (33; R = H, R^1 = OH), and (33, RR^1 = O), with R^2 in each case a secondary or tertiary amino group, has been prepared as potential antitumour agents.[108]

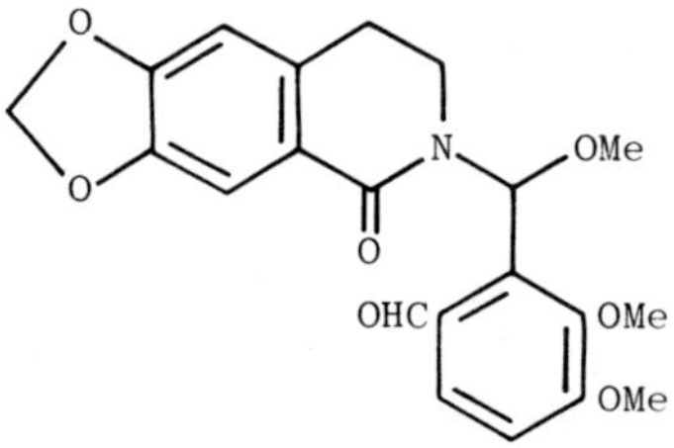

(30)

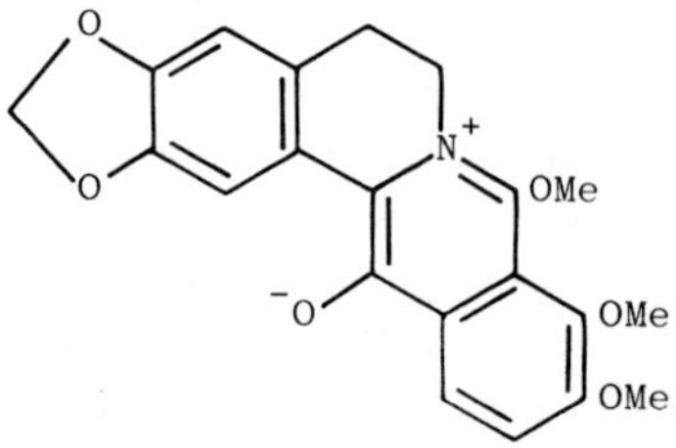

(31)

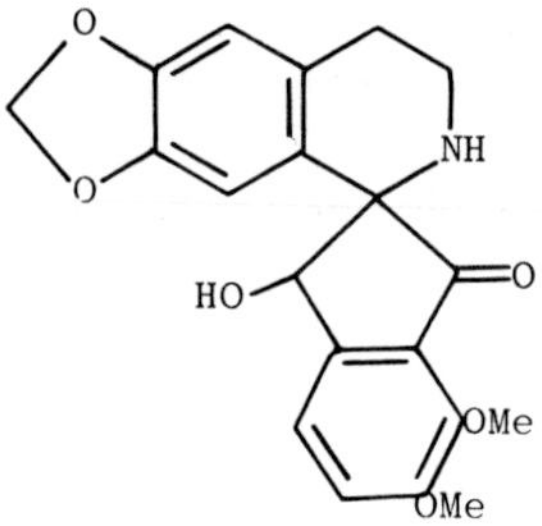

(32)

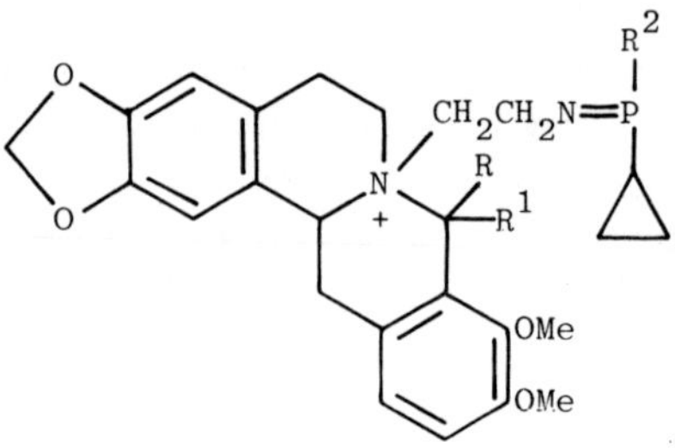

(33)

Xylopinine has been synthesised by photochemical cyclisation of (34; R = CO$_2$Me)[109,110] and (34; R = H)[111,112] to the related lactams followed by conventional processes, by carbon monoxide insertion into 6'-bromonorlaudanosine (35) followed by reduction of the resulting lactam,[113] and by Bischler-Napieralsky ring-closure of the hydroxymethyl amide (36) in the presence of phosphorus oxychloride followed by reduction with sodium borohydride.[114] Tetrahydropalmatine has been synthesised from laudanosine by treatment with butyl-lithium and formaldehyde to introduce a 6'-hydroxy-methyl group, the mesyl ester of which was then cyclised,[115] and mecambridine (37; R = CH$_2$OH) has been synthesised by condensation of formaldehyde and (37; R = H), itself prepared by a conventional synthesis from the appropriate Reissert compound.[116]

The azaberberine alkaloid alamarine (40) has been synthesised from (38) by photochemical or thermal cyclisation to (39) followed by reduction and debenzylation.[117]

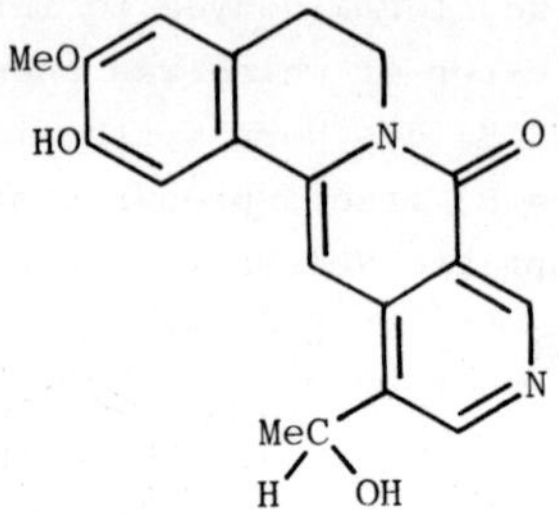

(40)

The effects of berberine chloride on blood pressure,[118] on intestinal secretions[119-121] and on the antibacterial efficacy of sulphonamides,[122] the antibacterial properties of berberine catechuate,[123] and the effects of dehydrocorydaline on the cardiovascular system[124] have been studied.

3 Secoberberines

6'-Hydroxymethyllaudanosine, on treatment with ethyl chloroformate, gives the cyclic ether (41),[125] but the C-methyl analogue gives only the carbonate ester (42).[105] In contrast with these, macrantaline (43) is converted into the stilbene (44).[105]

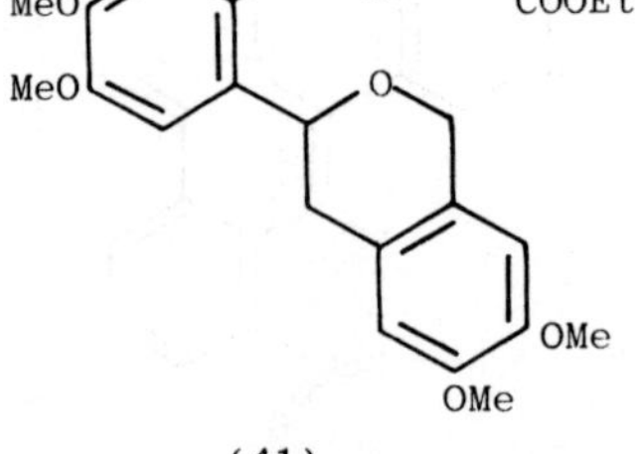

(41) 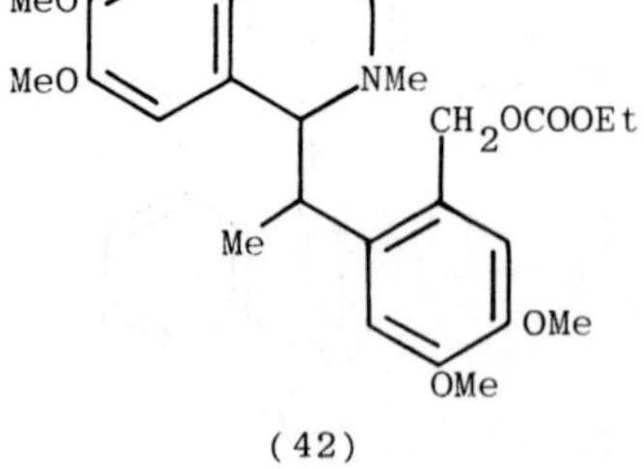   (42)

(43)   (44)

9 Protopines

Alkaloids of this sub-group have been isolated from the following:

Berberis darwinii[21]	protopine
Berberis empetrifolia[54]	protopine
Corydalis bulbosa[90]	protopine
Corydalis ledebouriana[91]	allocryptopine and cryptopine
Fumaria schrammii[94]	protopine
Fumaria vaillantii[22]	protopine
Papaver albiflorum[95]	allocryptopine and protopine
Papaver lecoquii[95]	allocryptopine, cryptopine, and protopine
Papaver litwinowii[4]	allocryptopine, cryptopine, and protopine
Papaver oreophilum[96]	allocryptopine and protopine

For the conversion of oxoallocryptopine into an indanobenzazepine see section 13.

10 Phthalideisoquinolines

Phthalideisoquinoline alkaloids have been isolated from the following:

Corydalis bulbosa[90]	(−)-adlumidine and (+)-bicuculline
Fumaria parviflora[93]	(+)-adlumidine and (−)-corlumine
Fumaria schrammii[94,126]	adlumine, adlumiceine, adlumidiceine, bicuculline, bicucullinidine, and fumschleicerine

| Fumaria vaillantii[22] | adlumine, N-methyladlumine chloride, adlumidine, adlumidiceine, and hydrastine |

N-Methyladlumine chloride and bicucullinidine (45) are new alkaloids and corlumine was previously found naturally only as the dextro-rotatory form.

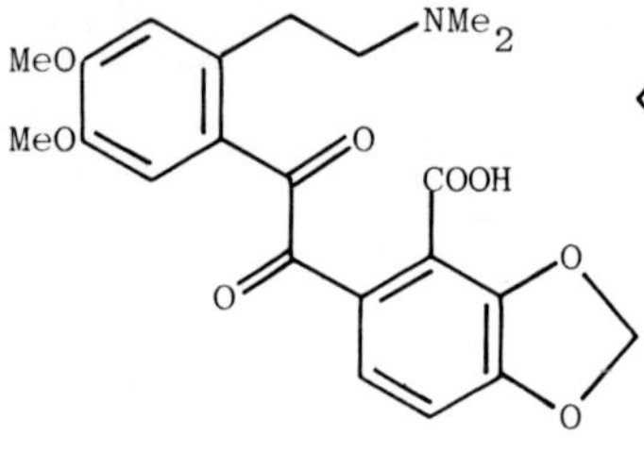

(45)

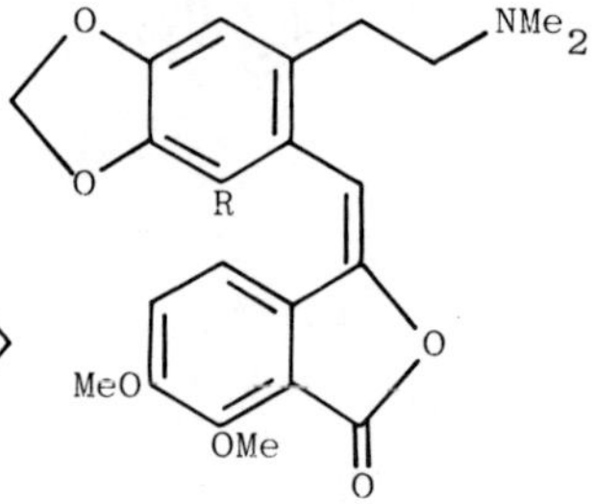

(46)

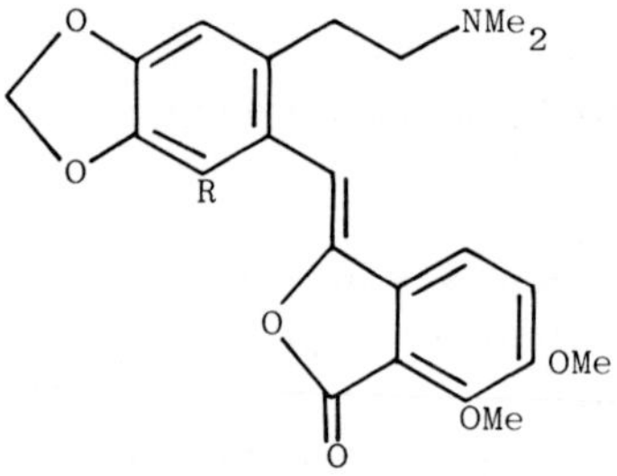

(47)

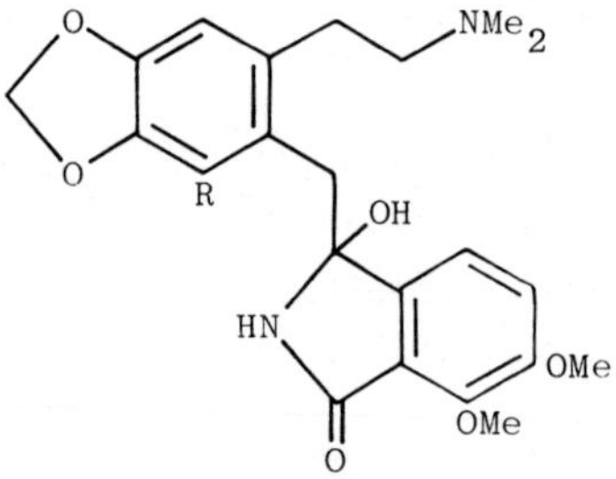

(48)

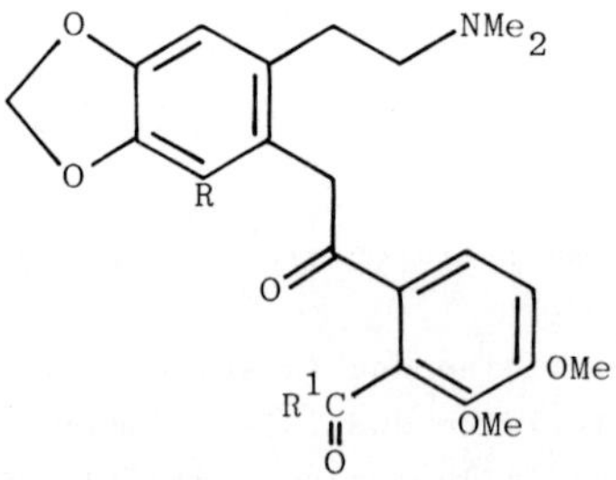

(49)

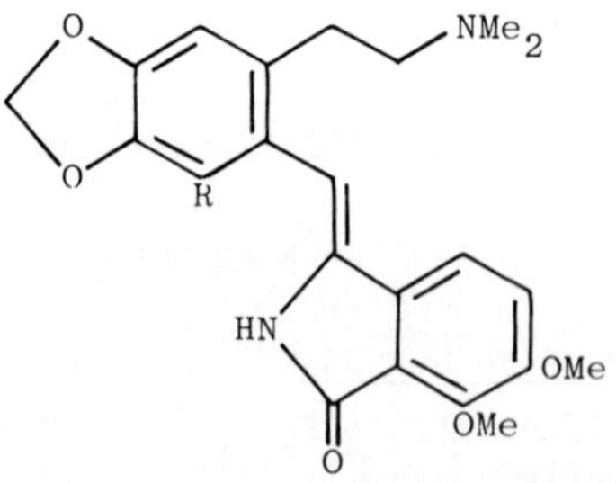

(50)

Degradation of N-methyl-α-hydrastinium salts has been shown to give the E-lactone (46; R = H) and of the isomeric β-hydrastinium salts to give the Z-lactone (47; R = H), which are converted by water or methanol into (49; R = H, R^1 = OH) or (49; R = H, R^1 = OMe) and by ammonia into the hydroxylactam (48; R = H), and the last of these is dehydrated by acids to the Z-enelactam (50; R = H), convertible into the isomeric E-enelactam by UV light.[127] Similar reactions with narcotine have afforded the compounds (47; R = OMe), (48; R = OMe), (49; R = OMe, R^1 = NH_2), and (50; R = OMe).[128] Reduction of α-narcotine with sodium di(2-methoxyethoxy)aluminium hydride gives

(51)   (52)

the hemi-acetal (51), which has been converted into the acetoxy-nitrile (52).[129] Oxidation of the erythro compound β-hydrastine to the iminium salt followed by reduction catalytically or with sodium borohydride or zinc and acetic acid gives only β-hydrastine and not the threo isomer.[130] The metabolism of narcotine in rats, rabbits, and humans has been shown to give the two isomeric hydroxy-methoxyphthalides.[131]

Cordrastine has been synthesised from the Reissert compound 2-benzoyl-1-cyano-6,7-dimethoxy-1,2-dihydroisoquinoline by reaction with the aldehydo-ester (53; R = Me) followed by N-methylation and hydrolysis; both erythro and threo forms were obtained.[132] Hydrastine has been prepared from the aldehydo-acid (53; R = H) by Passerini reaction with the appropriate phenethylisocyanide, to give the lactam lactone (54), followed by Bischler-Napieralsky ring-closure, N-methylation, and reduction.[133] Phthalide-isoquinolines have also been prepared from N-methylisoquinolinium salts and phthalides in the presence of sodium alkoxide followed by reduction of the 3,4-double bond[134] and from N-methyl-3,4-dihydroisoquinolinium salts and 3-halophthalides under reductive conditions in the presence of metallic zinc or zinc-copper couple.[135]

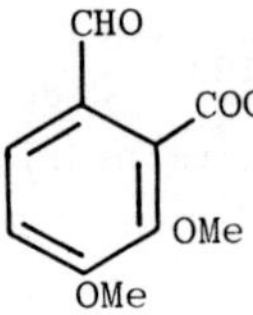

(53)

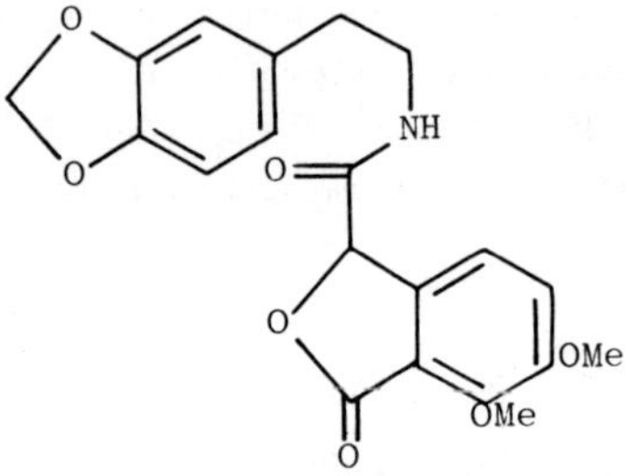

(54)

Some physiological effects of bicuculline have been studied.[136-139]

11 Spiro-benzylisoquinolines

Fumariline, ledecorine, and parfumine have been isolated from Fumaria vaillantii.[22] Several alkaloids previously assigned structures within this group have been reformulated as indenobenzazepines (see following section).

Fumariline (55; $RR^1 = R^2R^3 = CH_2$), parfumine (55; $R = Me$, $R^1 = H$, $R^2R^3 = CH_2$), and parfumidine (55; $R = R^1 = Me$, $R^2R^3 = CH_2$) have been converted into phenols in which $R^2 = Me$ and $R^3 = H$ by potassium hydroxide in methanol and homologues where $R^2 = Et$ and $R^3 = H$ with potassium hydroxide in ethanol, R and R^1 being unchanged in all cases. Oxidation of the bases (56; $RR^1 = CH_2$, $R^2 = H$), (56; $R = Me$, $R^1 = R^2 = H$), and (56; $R = R^1 = Me$, $R^2 = H$) with lead tetra-acetate gives the analogues in which $R^2 = OAc$, and these can be hydrolysed to the related alcohols, $R^2 = OH$. With potassium hydroxide the ester (57; $R = Me$, $R^1 = Ac$) gives (57; $R = R^1 = Me$) in methanol and (57; $R = R^1 = Et$) in ethanol.[140]

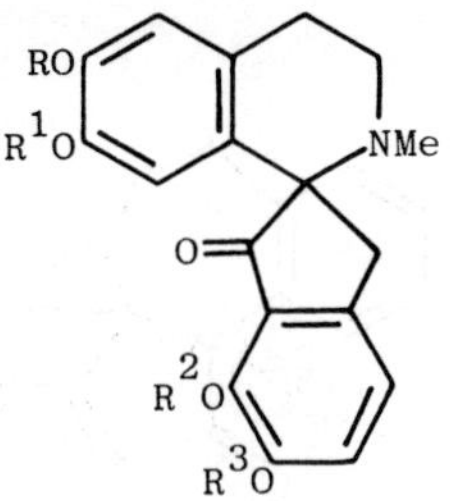

(55)

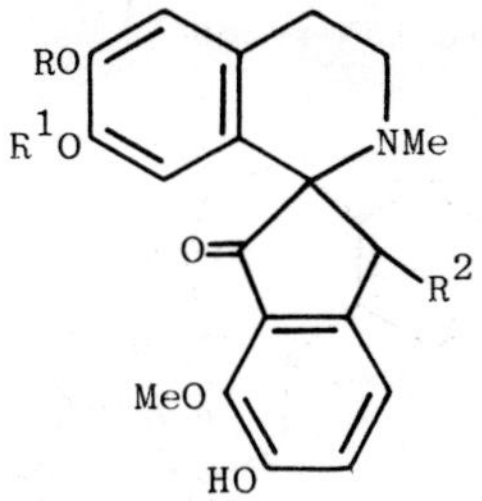

(56)

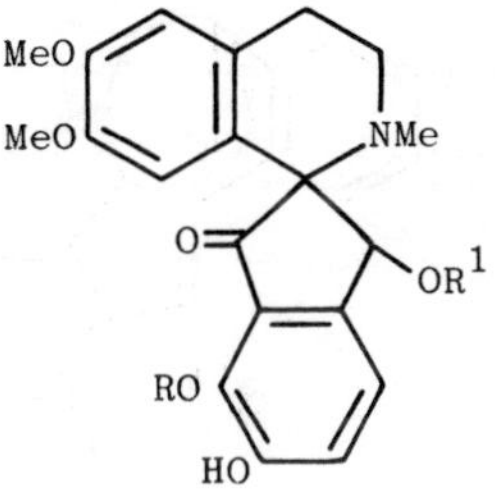

(57)

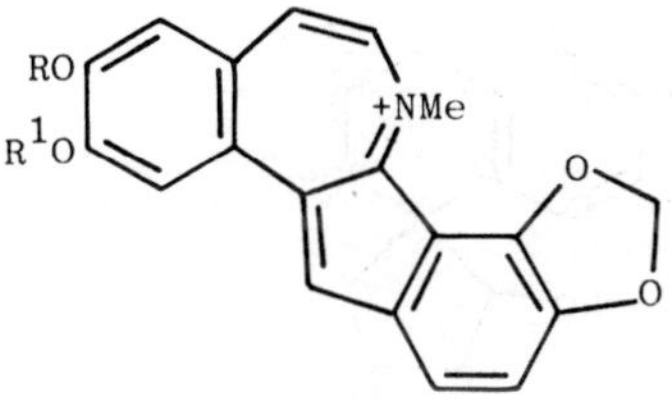

(58)

12 Indanobenzazepines

Two new alkaloids isolated from Fumaria parviflora, lahorine and lahoramine, have been shown to have the indenobenzazepine structures (58; RR^1 = CH_2) and (58; R = R^1 = Me) respectively by partial synthesis from fumariline (55; RR^1 = R^2R^3 = CH_2) and parfumidine (55; R = R^1 = Me, R^2R^3 = CH_2) by reduction to the secondary alcohols, dehydration with rearrangement of these, followed by oxidation with iodine.[141] Subsequently it was shown that rearrangement of dihydroparfumine and dihydroparfumidine with trifluoroacetic anhydride and quenching with methanol gave (59; R = H) and (59; R = Me), found to be identical with fumaritridine and fumaritrine respectively, alkaloids that had previously been assigned spiro-structures.[142] The structure of fumarofine has been reassigned as (60; R = H) following partial synthesis of its methyl ether (60; R = Me) by rearrangement of the isomer (61) of parfumidine to the olefin (62), followed by oxidation to the cis-glycol and further oxidation to the ketone.[143]

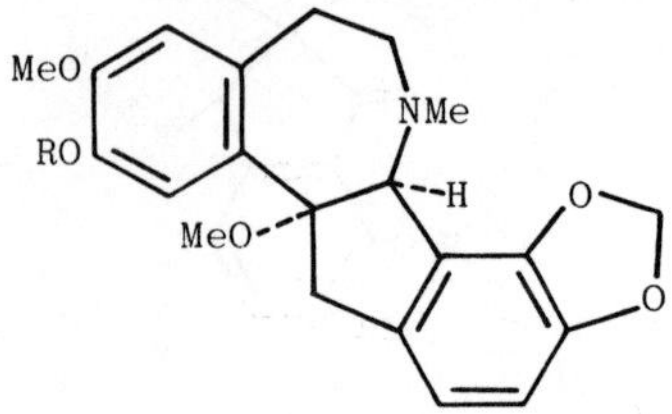

(59)

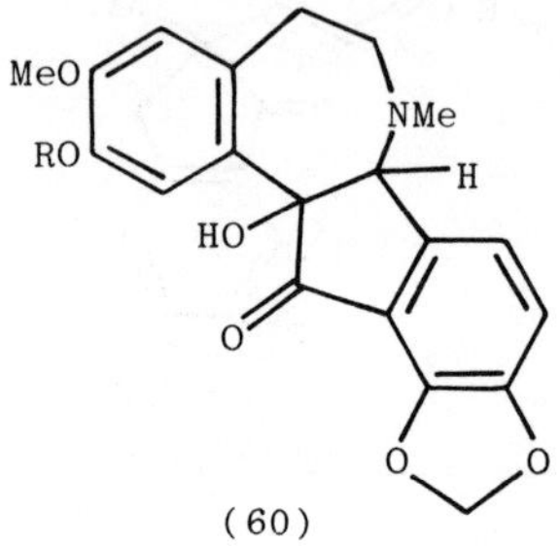

(60)

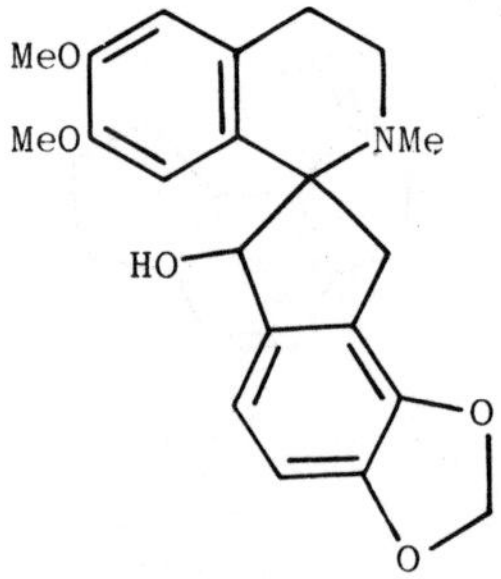

(61)

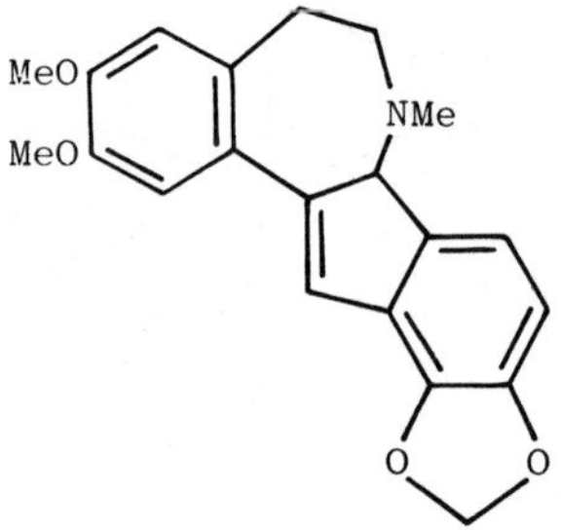

(62)

The reverse rearrangement of the glycols (63; R = H, R^1 = OH) and (63; R = OH, R^1 = H) to raddeanine (64; R = H, R^1 = OH) and yenhusomine (64; R = OH, R^1 = H) by trifluoroacetic anhydride and pyridine has also been accomplished. Raddeanine and the isomeric base (65) have also been prepared by rearrangement of the ketones (63; RR^1 = O) and (66) followed by reduction. These rearrangements appear to proceed equally easily with the trans fusion of the seven- and five-membered rings to give the same products.[144] The process presumably involves aziridinium salts as intermediates.

The aziridine (67), formed by irradiation of berberine betaine, on treatment with formaldehyde, gives the base (68), which on reduction with sodium cyanoborohydride affords a cis-glycol that is irreversibly isomerised by acid to the more stable trans-glycol (66). Periodate oxidation of the cis-glycol gives the keto-lactone (69).[145]

(63) (64) (65) (66) (67) (68) (69) (70)

The quaternary salts of the aziridine (67) react with hydrochloric acid in water to give the indanobenzazepine (66)[146,146a] and in methanol to give the corresponding methyl ether, and (66) may also be prepared from oxoallocryptopine (70) by irradiation in t-butanol in the presence of potassium t-butoxide.[146]

13 Rhoeadines

Alkaloids of this group have been isolated from the following:

Berberis empetrifolia[54]	1,14-dioxo-2-hydroxy-7,8-methylenedioxy-12,13-dimethoxyaporhoeadane
Papaver albiflorum[95]	rhoeadine, papaverrubines A, C, and E
Papaver lecoquii[95]	rhoeadine, papaverrubines A, C, and E
Papaver litwinowii[4]	papaverrubines A, C, D, and E
Papaver oreophilum[96]	rhoeadine, isorhoeadine, rhoeagenine, oreadine, oreogenine, and papaverrubines A, C, D, and E

Alpenigenine (71) has been oxidised to the N-formylnor-compound by Jones's reagent[147] and alpenigenine oxime has been dehydrated to the nitrile (72), the methiodide of which when subjected to Hofmann degradation gave the amide (73).[148] A positional isomer of xylopinine has been subjected to Hofmann degradation and the resulting olefin oxidised with osmium tetroxide to the diol (74), which was cleaved to the dialdehyde (75), photolysis of which gave 20-30% of cis- and 1% of trans-alpinigenine.[149] The lactone (69) has been reduced with sodium borohydride followed by acid treatment to give the lactone (75a), further reduction of which to the hemiacetal, followed by O-methylation, gave an analogue of rhoeadine.[145]

(71)

(72)

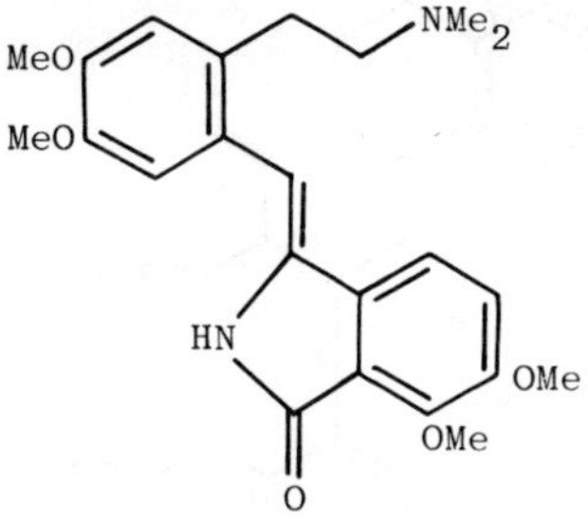

(73)

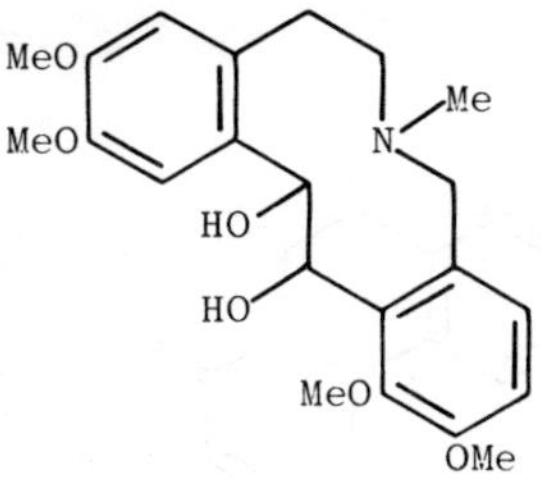

(74)

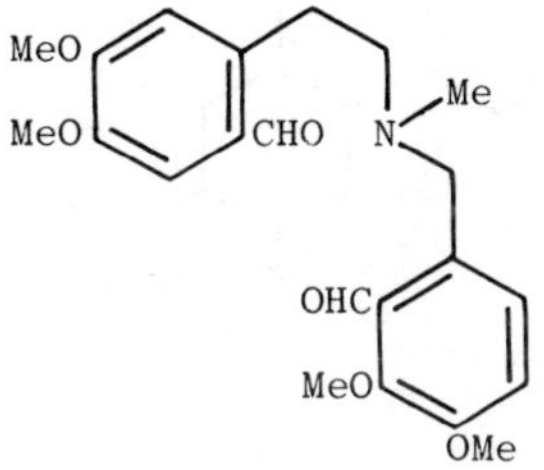

(75)

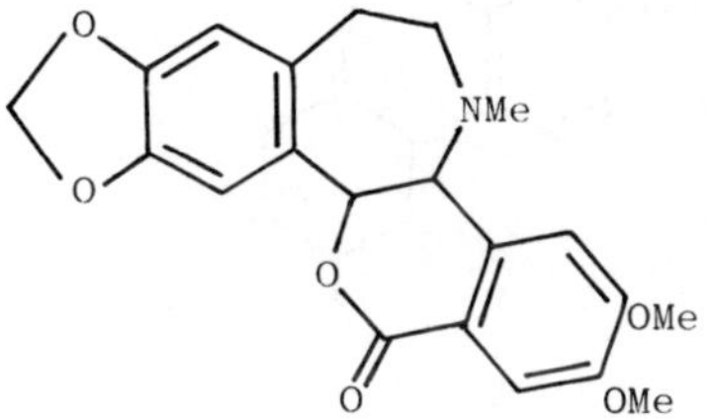

(75a)

14 Emetine and Related Alkaloids

Ankorine, cephaleine, and psychotrine have been isolated from Alangium salviifolium and ankorine from A. kurzii.[150] Full details of the synthesis of protoemetinol (and of protoemetine) from norcamphor (previously reported in Volume 10 pp. 107-108) have been published.[151] The lactam (77) has been prepared by the Michael addition of 6,7-dimethoxy-1-methyl-3,4-dihydroisoquinoline to the unsaturated ester (76) and the product converted into emetine by C-ethylation, hydrolysis, decarboxylation, and reduction.[152] The ester (79) has been prepared by the Bischler-Napieralsky route from the lactam (78) and converted by conventional methods into emetine.[153]

The antiviral effects of emetine[154] and the effects of the alkaloid on subcellular nucleic acid and protein metabolism[155] have been studied.

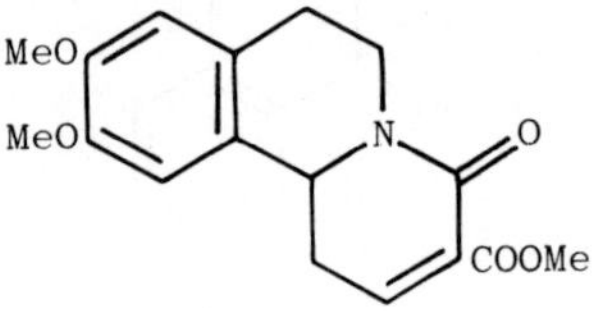

(76)

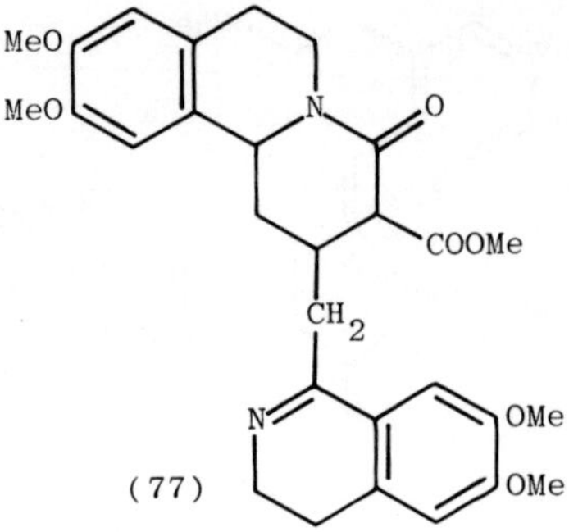

(77)

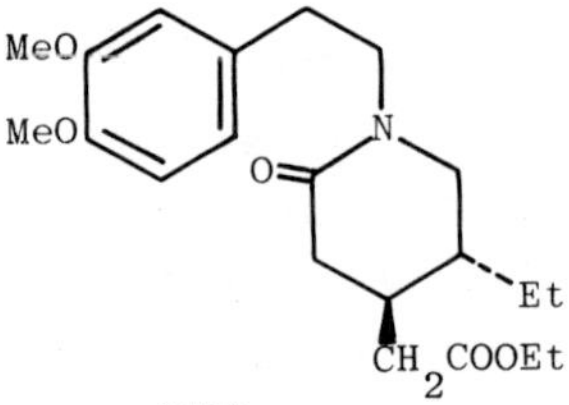

(78)

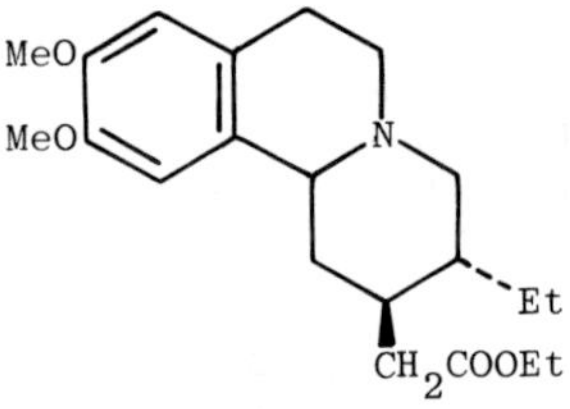

(79)

15 Morphine Alkaloids

Alkaloids of this group have been isolated from the following species:

Croton bonplandianum[156]	norsinoacutine
Croton salutaris[157]	(±)-salutaridine and norsalutaridine
Ocotea acutanglea[158]	pallidine, O-methylpallidine, pallidinine, and O-methylpallidinine
Papaver albiflorum[95]	thebaine
Papaver spicatum[159]	amurine, dihydronudaurine, and flavinantine
Papaver strictum[159]	amurine, dihydronudaurine, and flavinantine
Stephania longa[160]	longanone
Stephania sasakii[161]	1,1-bisaknadinine

In addition, the ability of tissue cultures of Papaver somniferum to produce morphine, codeine, and thebaine has been further demonstrated.[162] Of these alkaloids, norsalutaridine, O-methylpallidine, pallidine (80; R = H), O-methylpallidinine (80; R = Me), 8,14-

dihydronudaurine, longanone (81), and 1,1-bisaknadinine are new.

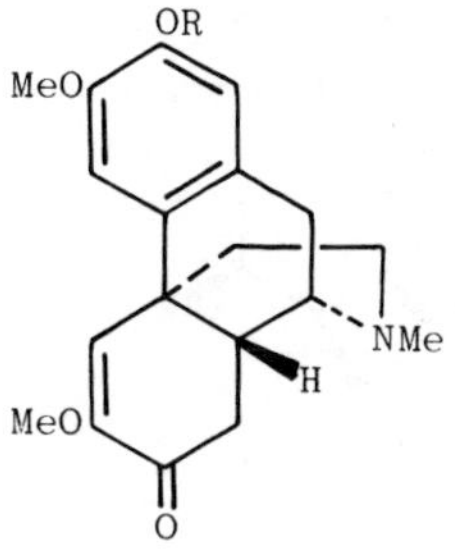

(80)

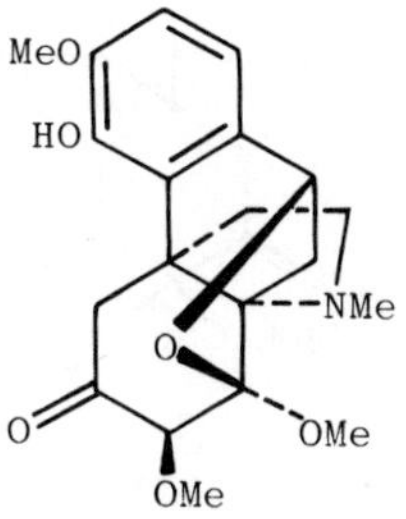

(81)

Codeinone has been shown to react with diazomethane to give the 7β,8β-methano-compound (82), which can be cleaved by hydrochloric acid to 8β-chloromethyldihydrocodeinone, reducible to 8β-methyldihydrocodeine.[163] Treatment of codeinone with alkaline hydrogen peroxide gives the 7β,8β-epoxide,[163-165] reducible by sodium borohydride to codeine epoxide.[166] Codeinone also undergoes Michael addition reactions leading to the bases (83; R = CH_2NO_2) and [83; R = $CH(CO_2Et)_2$], and 8β-vinyldihydrocodeinone (83; R = CH=CH_2) has been converted into the alcohols (83; R = CH_2CH_2OH) and [83; R = CH(OH)Me], both of which have been converted into the corresponding fluorides.[167]

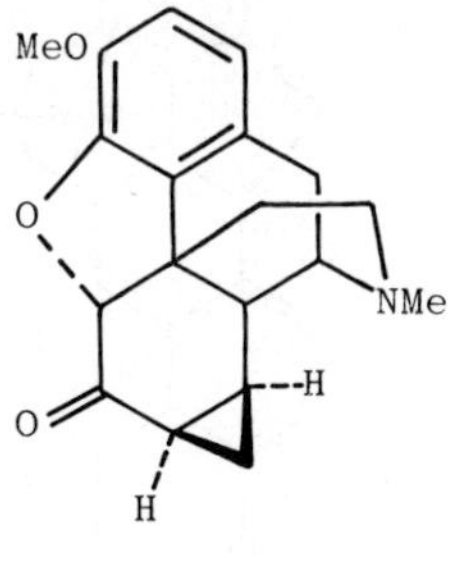

(82)

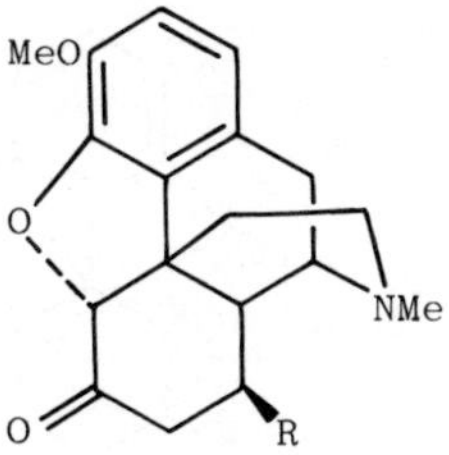

(83)

Dihydrocodeinone (83; R = H) and its 8β-methyl and 8β-ethyl analogues have been shown to react with formaldehyde in aqueous dioxan in the presence of calcium hydroxide to give the 7,7-di-

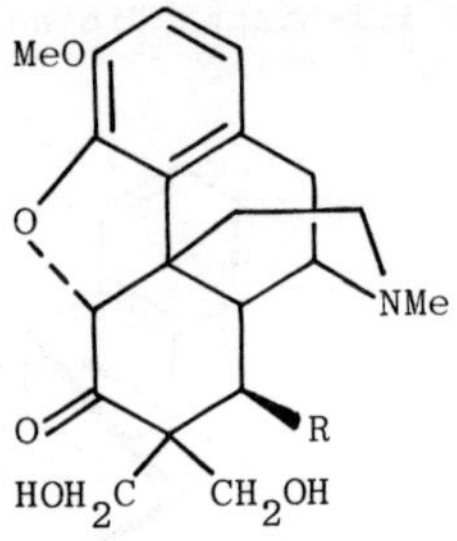

(84) 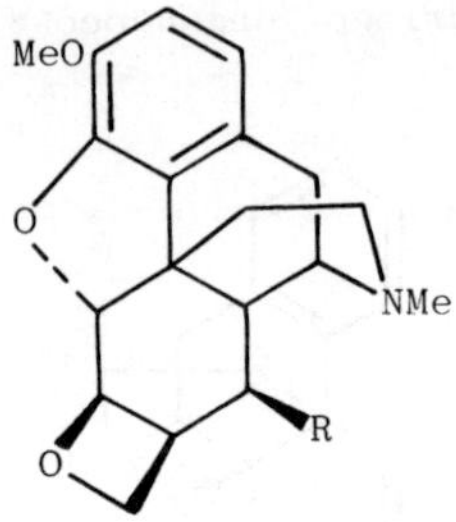   (85)

(hydroxymethyl)-compounds (84).[168,169] The bis-toluenesulphonyl ester of (84; R = H) is reduced by lithium triethylborohydride to 7,7-dimethyldihydroisocodeine, which can be oxidised to the ketone and then reduced to 7,7-dimethyldihydrocodeine. Reduction of the corresponding esters of (84; R = Me) and (84; R = Et) with the same reagent gives the bases (85; R = Me) and (85; R = Et), which may be further reduced by lithium aluminium hydride and aluminium chloride to the corresponding 7,7-dimethyl-8β-alkyl-dihydroisocodeines.[169] In the presence of dimethylamine, dihydrocodeinone reacts with formaldehyde to give the spiro-ether (86), presumably formed by cyclo-addition of the 7-methylene ketone to itself, rather than, as previously believed, 7,7-methylene-bis-dihydrocodeinone [part structure (87)].[170]

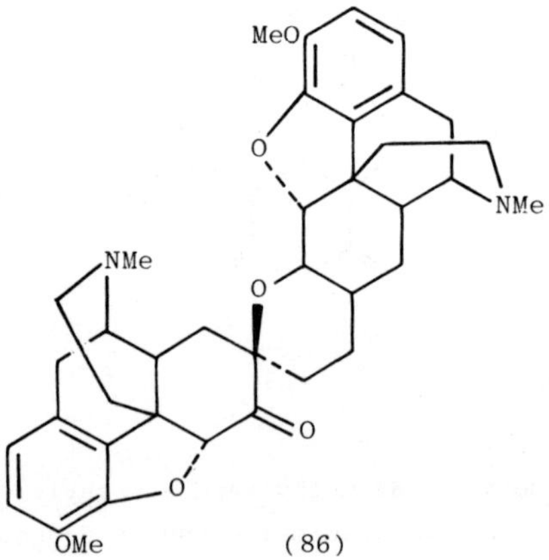

(86) 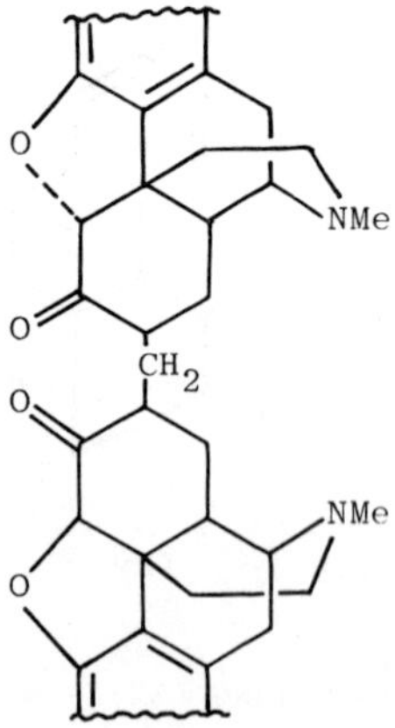   (87)

Photolysis of dihydrocodeinone in aqueous hydrochloric acid, methanol, ethanol, or acetic acid gives the ring-opened bases (88; R = OH), (88; R = OMe), (88; R = OEt), and (88; R = OAc) whereas under similar conditions 14β-methyldihydrocodeinone affords the 5β-compounds (89); in diethylamine, photolysis of dihydrocodeinone gives (90) and the 5β-diethylamino analogue of (88).[171]

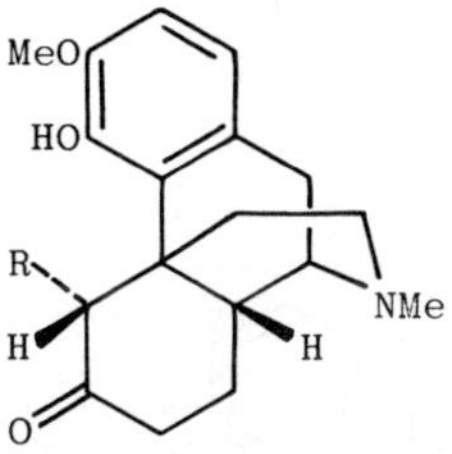

(88)

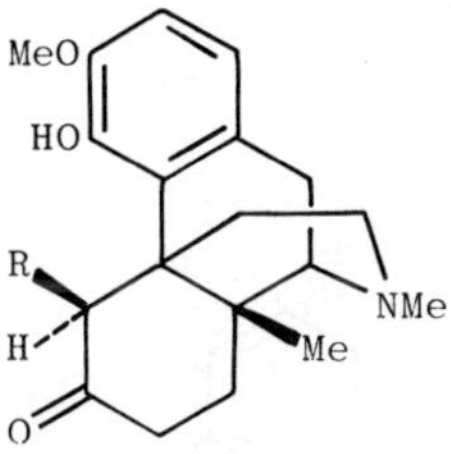

(89)

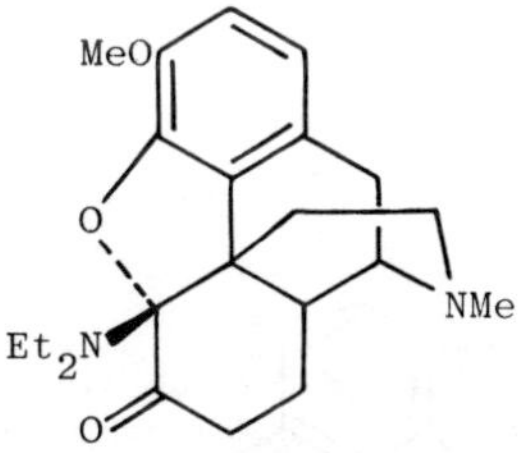

(90)

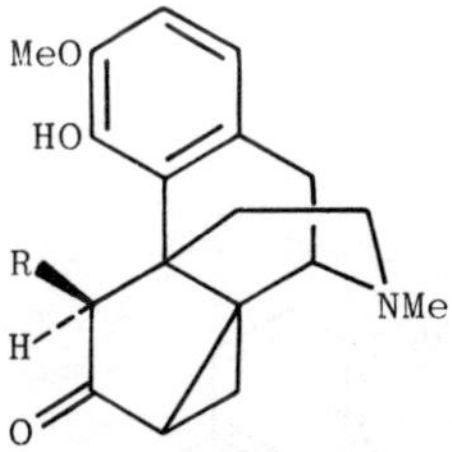

(91)

7,14-Cyclocodeinone behaves like 8β-methyldihydrocodeinone under the same conditions, giving the 5β-substituted compounds (91), where R is OH, OMe, OEt, OAc, or NEt_2.[172] 7,14-Cyclocodeinone reacts with diazomethane to give the ring-enlarged epoxides (92; n = 0), (92; n = 1), and (92; n = 2) and the methanomorphinanenone (93) resulting from rearrangement of (92; n = 0).[173]

Beckmann transformation of the oxime of 7,14-cyclocodeinone in methanol or ethanol gives the mixed acetals (94; R = Me) and (94; R = Et) respectively and in acetic acid the phenolic aldehyde (95), which on heating is converted into the cyclised lactam (96).[174]

(92)

(93)

(94)

(95)

(96)

(97)

Whereas the reaction of codeinone with lithium dialkylcuprates proceeds with production almost exclusively of 8β-alkyldihydrocodeinones, in the 14-hydroxycodeinone series appreciable amounts of 8α-compounds are formed, and increased steric hindrance at C-14 results in these being the principal product, e.g. (97; R = H) gives 56%, (97; R = SiMe$_3$) 80%, and (97; R = SiMe$_2$Bu^t) 100% of 8α-methyl, -ethyl, and -butyl compounds. The silicon protecting group can be removed by tetrabutylammonium fluoride.[175]

Claisen-Eschenmoser reaction of codeine with N,N-dimethylacetamide dimethylacetal affords the 8β-substituted deoxycodeine-C derivative (98; R = CONMe$_2$), which has been converted into (98; R = CH$_2$NMe$_2$), (98; R = CH$_2$OH), (98; R = CHO), (98; R = CH$_2$O-Tos), and (98; R = CH$_3$). The alcohol (98; R = CH$_2$OH) undergoes assisted ring-opening in 4M hydrochloric acid to give the phenolic ether (99).[176] N-Ethoxycarbonylnorcodeinone has been converted into the

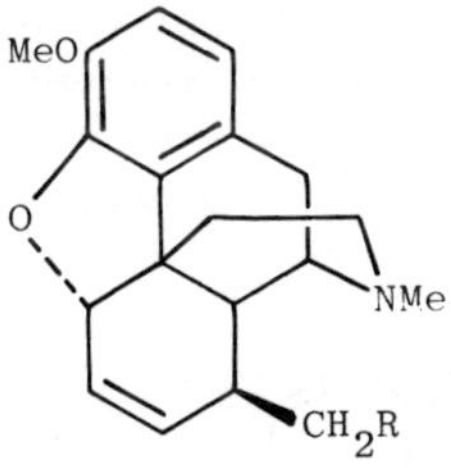

(98)

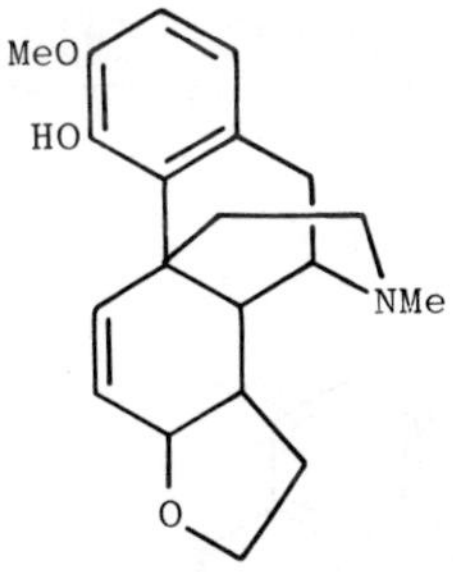

(99)

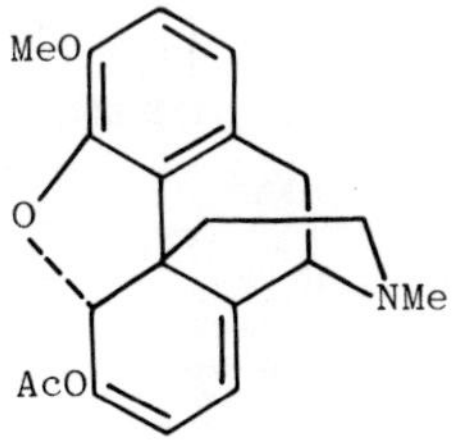

(100)

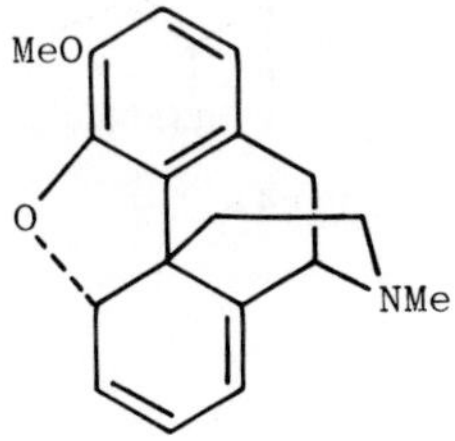

(101)

enol acetate (100), which can be oxidised by singlet oxygen to
N-ethoxycarbonyl-14-hydroxynorcodeinone, which can be demethylated
by boron tribromide to the corresponding morphinone; this
provides a useful route to 14-hydroxydihydronorcodeinone and
morphinone.[177]

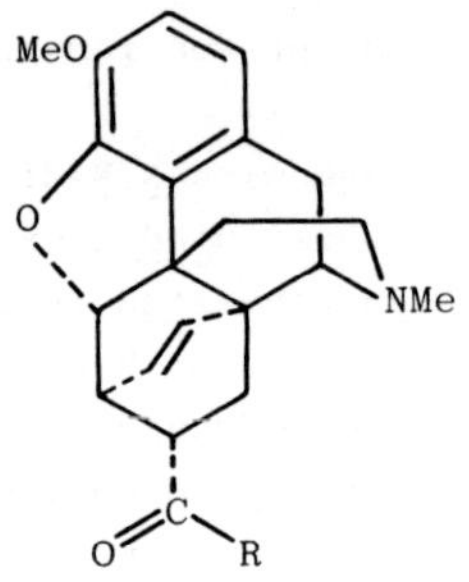

(102)

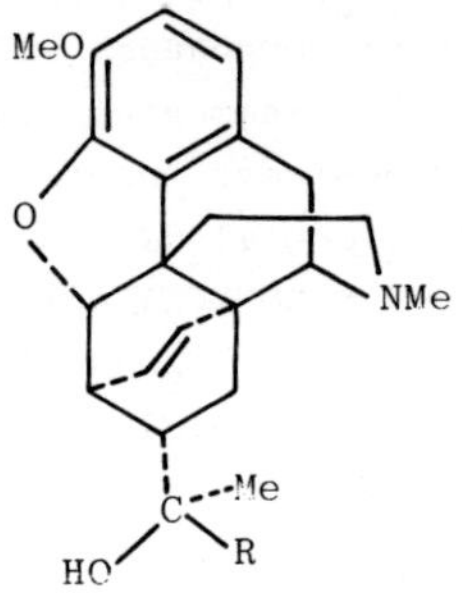

(103)

6-Demethoxythebaine (101) has been prepared from isocodeine by
elimination[178,179] and converted by Diels-Alder reaction into the
ester (102; R = OEt)[178] and the ketone (102; R = Me),[179] the last of
these being further converted into a number of tertiary alcohols
(103).[179]

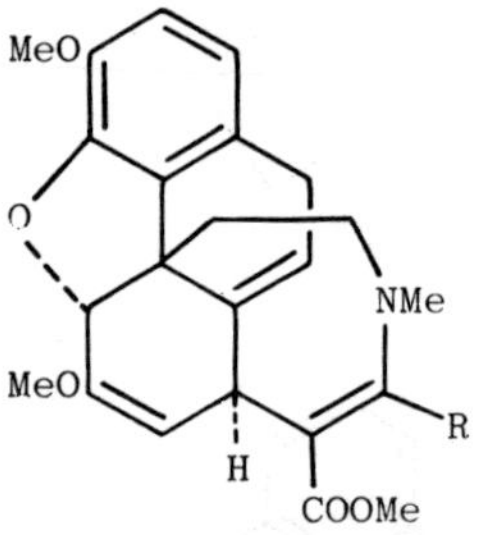

(104)

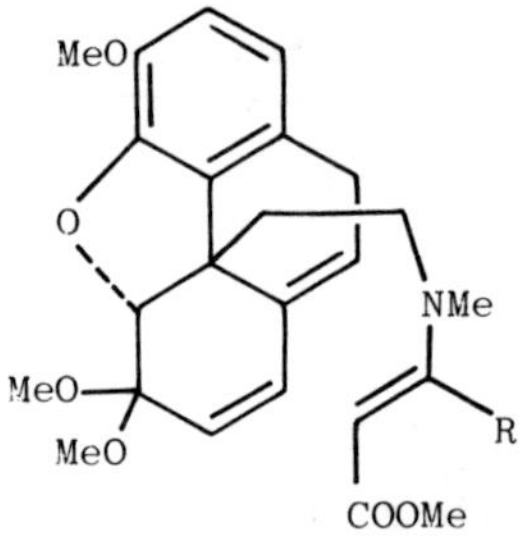

(105)

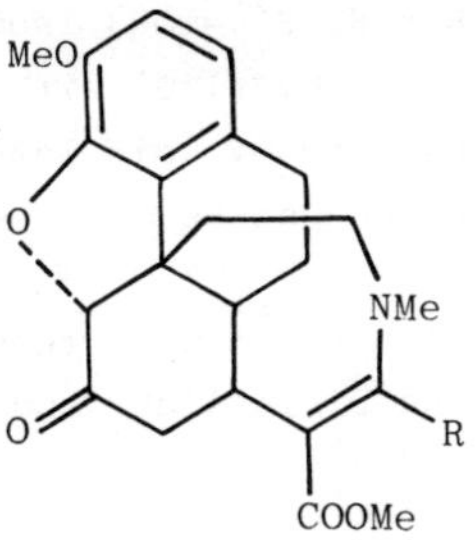

(106) 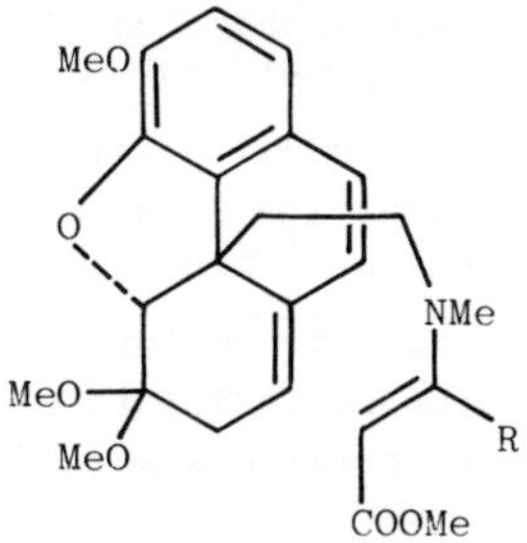 (107)

In polar solvents, thebaine has been found to react with esters of propiolic and acetylene-dicarboxylic acids to give the bases (104; R = H)[180,181] and (104; R = CO_2Me),[181] and in methanol the bases (105; R = H) and (105; R = CO_2Me) are also formed and may be converted into the corresponding (104) by acid.[181] Further treatment with acid converts (104) into (106; R = H)[180] and (106; R = CO_2Me),[181] and (105; R = H) is isomerised to (107; R = H) by base.[181]

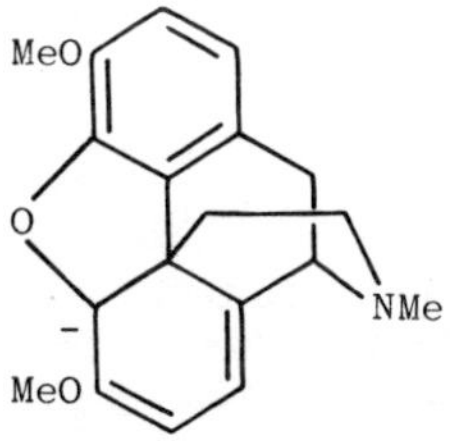

(108) 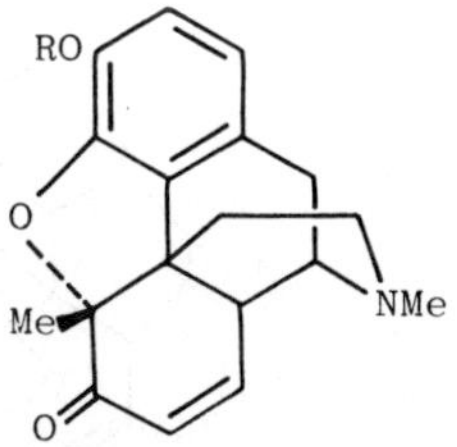 (109)

Treatment of thebaine with butyl-lithium at -78°C gives the anion (108), which when quenched with D_2O gives 5-deuteriothebaine and with methyl fluorosulphonate gives 5-methylthebaine, which has been converted into 5-methylcodeinone (109; R = Me), 5-methyldihydrothebaine, 5-methyldihydrocodeinone (109; R = Me; 7,8-dihydro), and metopon (109; R = H; 7,8-dihydro).[182]

Full details of the reaction of thebaine with tetranitromethane previously reported have been published. The peroxide (110), obtained in the presence of oxygen on treatment with sodium iodide and acetic acid, gives the diol (111; R = H, R^1 = OH), isomerised by base to (111; R = OH, R^1 = H). Reduction of (110) with triphenylphosphine gives the epoxide (112), converted by acid into the hydroxy-ketone (113; R = H, R^1 = OH), which can be oxidised to the diketone (113; RR^1 = O).[183]

N-Acetyl- and N-benzoylhydroxylamine react with thebaine and tetraethylammonium periodate to give the Diels-Alder adducts (114; R = Me) and (114; R = Ph).[184]

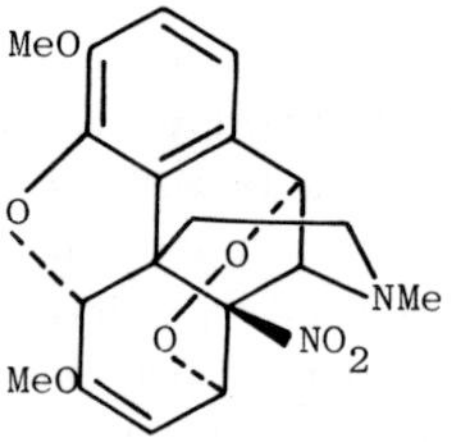

(110)

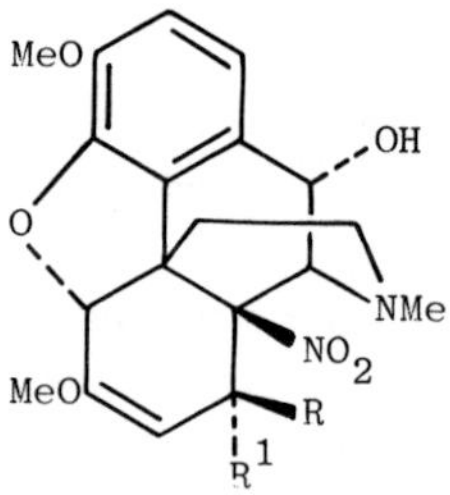

(111)

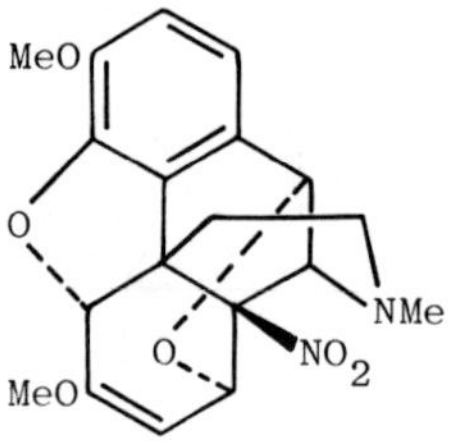

(112)

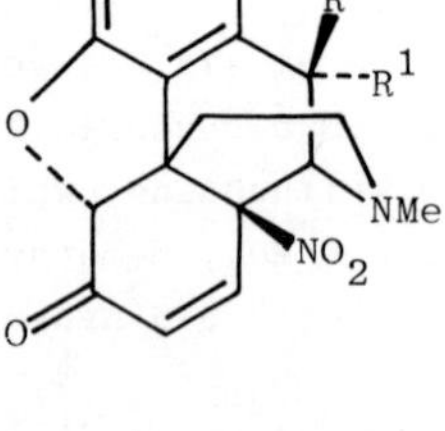

(113)

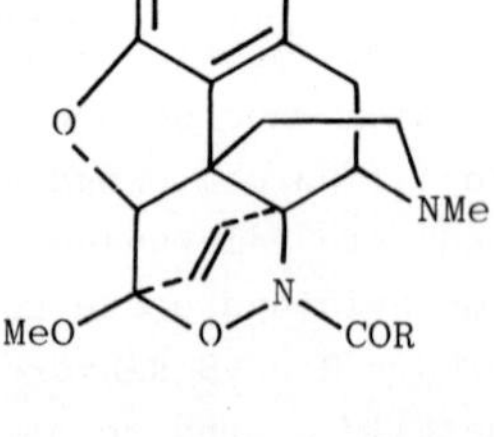

(114)

Morphine and dihydrocodeinone have been shown to react with diazonium salts at position 2, codeine at positions 2 and 8, and dihydrocodeinone at positions 2 and 7.[185] Naloxone, labelled with tritium in the N-allyl group,[186,187] [2-^{125}I]iodomorphine,[188] and ^{3}H- and ^{14}C-labelled ethylmorphine[189] have been prepared. Patents have been published for the conversion of thebaine into codeinone and neopinone,[190] for the preparation of 7-methyl-morphine and of 7-methyl-8-alkyldihydromorphine,[191] 6-fluoro- and 6,6-difluoro-compounds of the series (see Vol.12 p.122),[192] naloxone and its analogues[193] and 6-methyl-6-deoxyanalogues,[194] and N-cyclopropyl-methylnormorphine-6-sulphate.[195]

The conversion of thebaine into neopine and of codeinone into codeine by tissue cultures of Papaver somniferum,[196] the enzymic hydrolysis of morphine glucuronide,[197] and the metabolism of thebaine[198] and 14-hydroxycodeinone[199] have been studied.

(115) (116)

Hasubanonine (115) has been subjected to acetolysis, aromatisation, hydrolysis, and O-methylation to give 1,2,3,5,6-pentamethoxyphenanthrene.[200] Details of a synthesis of protostephanine reported previously[201] have been published[202] and this alkaloid has been shown to give the olefin (116) on Hofmann degradation.[200]

A review of approaches to the practical synthesis of morphine via reticuline and dihydrothebainone has been published,[203] as has a patent for the synthesis of dihydrothebainone and dihydrocodeinone by a previously described route.[204] Salutaridine has been obtained in 2.7% yield by the oxidation of reticuline with lead tetra-acetate[205] and from 4,6-diacetoxy-3-methoxymorphinan-5,8-diene via 4-acetoxy-6-hydroxy-3-methoxy- and 4-acetoxy-3,6-dimethoxymorphinandienone.[206]

Methods for the detection and estimation of morphine,[207-220] 3-acetylmorphine,[221,222] heroin,[223-226] codeine,[227] dihydrocodeinone,[228] and buprenorphine[211] have been published.

The analgesic,[229-234] emetic,[235] sedative,[236] lipolytic,[237] and toxic[238] effects and the pharmacodynamics[239] of morphine have been studied, as have the effects of the alkaloid on behaviour,[240-264] locomotor activity,[265,266] the intake of food[267-269] and water,[270] the output of urine,[271,272] the cardio-vascular system,[273-279] the secretion of prolactin,[280-284] growth hormone,[283,285] luteinising hormone,[286-289] thyroid stimulating hormone,[290] 5-hydroxytryptamine,[291,292] enkephalins,[293] cyclic-AMP,[294] and histamine,[295] body temperature,[268,296-298] the pituitary gland,[299,300] the brain,[301-305] neurones,[306-308] the spinal cord,[309] the cough reflex,[310] the gastro-intestinal tract,[311-316] respiration,[279,317,318] the eye,[319,320] the synthesis of peptides[321] and of serotonin,[322] the excretion of low-molecular-weight proteins,[323] phagocytosis,[324] and the activity of histamine transferase.[325]

The narcotic antagonist effects of naloxone have been studied[326-330] as have the effects of the compound on behaviour,[255,258,331-341] shock,[342,343] the intake of food[267,269,344,345] and water,[344,346-349] the cardio-vascular system,[279,350-353] the gastro-intestinal tract,[311,354-357] analgesia[358] and analgesic tolerance,[359] the brain,[360] neurones,[308] muscle tone,[361] locomotor activity,[362] respiration,[279] diazepam-induced narcosis,[363] lymphocytes,[364] platelets,[365] the adrenal cortex,[366] arthritis[367] and the secretion of cortisol,[368,369] prolactin,[369,370] growth hormone,[369] luteinising hormone,[286] and LH-releasing hormone.[371,372]

The pharmacological effects of the following have also been studied:
heroin,[259,373] codeine,[295,310,374,375] (+)-codeine,[375] dihydromorphinone,[376] 6-acetylmorphine,[373] 6-nicotinylmorphine,[377] morphine, codeine, and heroin 7,8-epoxides,[378] 6-azidodeoxymorphine,[379] etorphine,[380] buprenorphine,[381-394] nalorphine,[344,395] naltrexone,[360,362,396-400] nalbuphine,[401,402] 6-N-acylamino-6-deoxynaltrexone,[403] oripavine,[404,405] dihydrothebaine-φ,[404] and sinomenine.[406]

16 Benzophenanthridines

Alkaloids of this group have been isolated from the following species:

Corydalis ledebouriana[91]	dihydrochelerythrine, dihydrosanguinarine, and oxosanguinarine
Corydalis paniculigera[407]	pancorine (117)
Glaucium flavum[408]	6-iminosanguinarine (118)
Papaver oreophilum[96]	chelirubine and sanguinarine
Toddalia asiatica[409,410]	annotianamide, dihydrochelerythrine, 8-hydroxydihydrochelerythrine, 8-acetonyldihydrochelerythrine, and toddalidimerine (119).

In addition, sanguinarine has been isolated from cell cultures of Papaver bracteatum.[411]

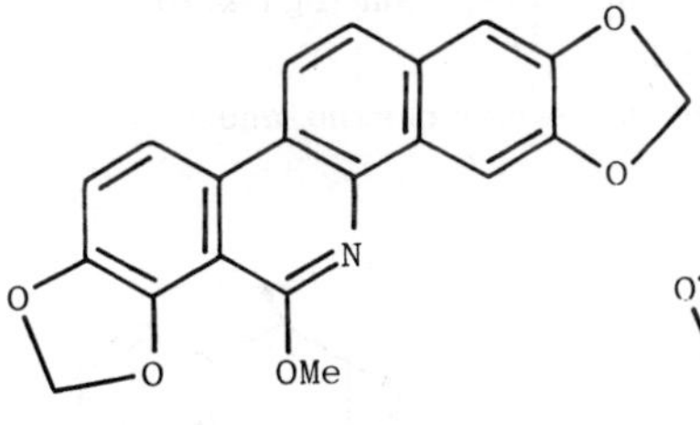

(117)

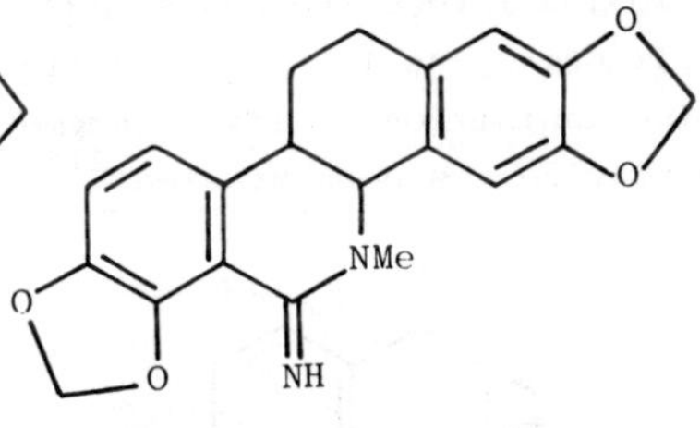

(118)

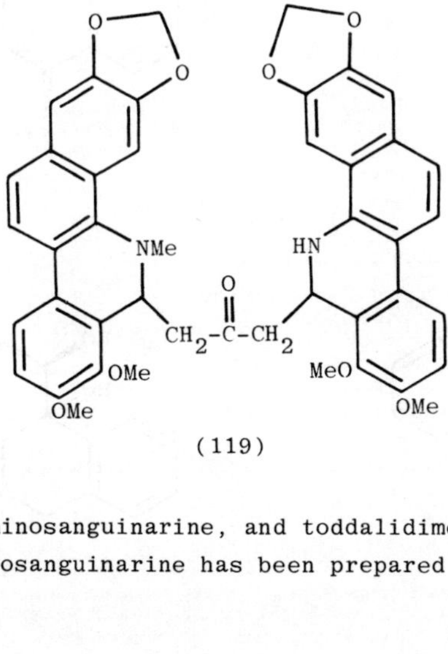

(119)

Pancorine, 6-iminosanguinarine, and toddalidimerine are new alkaloids; 6-iminosanguinarine has been prepared from oxosanguinarine.[408]

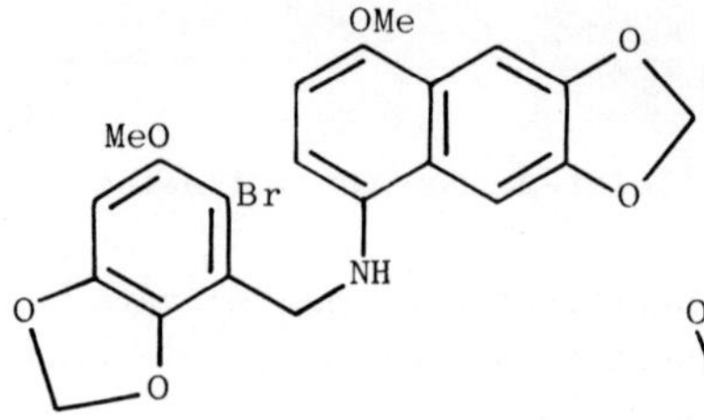

(120) 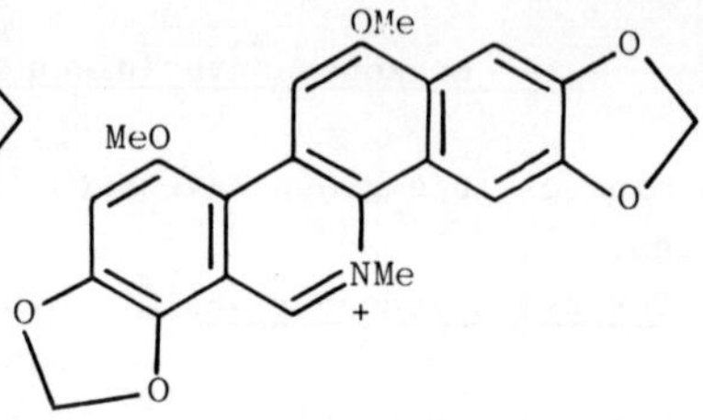 (121)

The structure of macarpine (121) has been confirmed by the synthesis of dihydromacarpine by photolytic cyclisation of the bromo-compound (120) followed by N-methylation.[412] The epoxide (122) has been reduced to the lactone (123), which on treatment successively with methylamine and lithium aluminium hydride, followed by catalytic reduction, dehydration, and further reduction, afforded the bases (124) and (125), analogues of epichelidonine and 10-hydroxychelidonine.[413]

The inhibition of liver enzymes by sanguinarine and chelerythrine has been studied.[414]

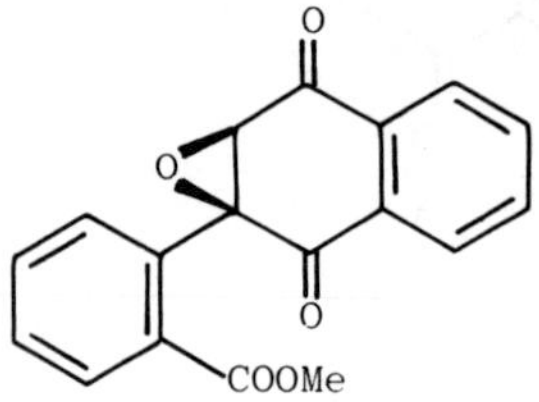

(122) 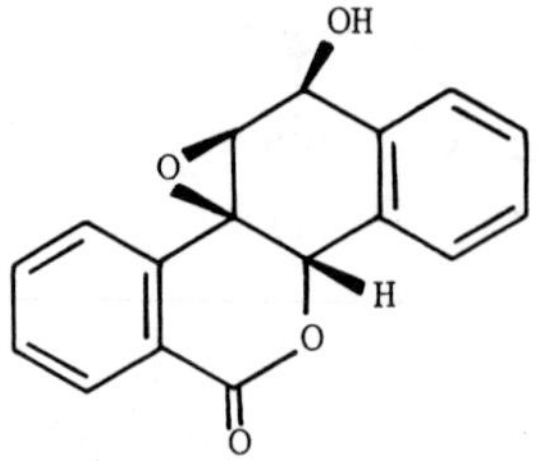 (123)

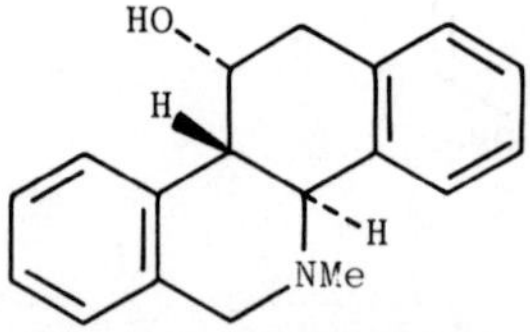

(124) 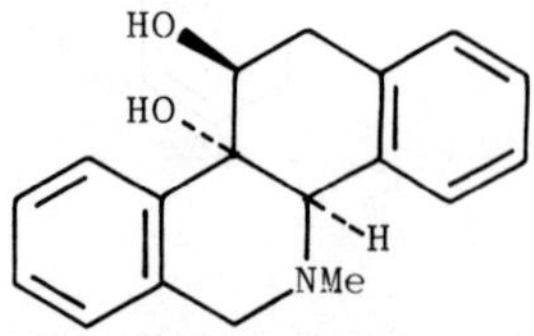 (125)

17 Colchicine and Related Bases

Demethylation of colchicine with concentrated sulphuric acid at 80-90°C gives 2-demethylcolchicine, and under the same conditions 1- and 3-O-demethylcolchicine give respectively the 1,2-O-demethyl- and 2,3-O-demethyl-compounds; the latter has been converted into (126), identical with natural cornigerine.[415] Deacetylcolchiceine, on methylation with diazomethane, gives a mixture of deacetyl-colchicine (128; R = H) and its isomer (127; R = H), and these, on treatment with trifluoroacetyloxyacetyl chloride in pyridine, give isocolchifoline (127; R = COCH$_2$OH) and colchifoline (128; R = COCH$_2$OH).[416]

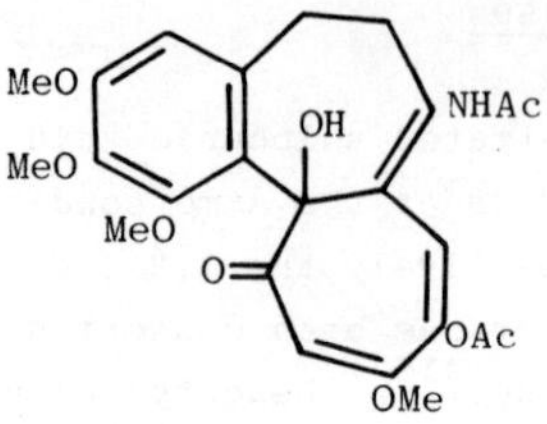

(132) 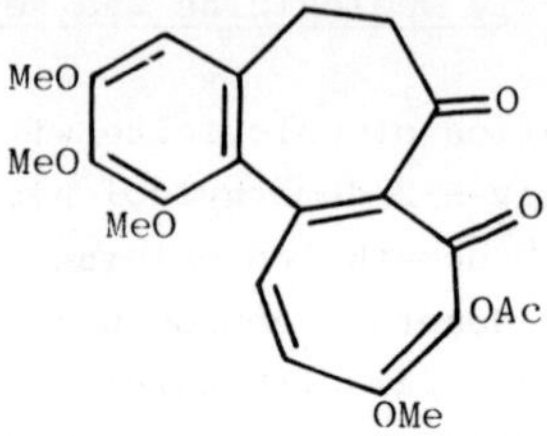   (133)

The enamide (129), on photolysis in benzene, affords the cyclised product (130),[417] but in the presence of oxygen the products are (131), (132), and (133).[418] Photolysis of (134; R = Me) in the presence of oxygen gives the carbinolamine (135; R = H) and the hydroperoxide (135; R = OH); treatment of the hydroperoxide with selenium dioxide gives the aldehyde (134; R = CHO).[418]

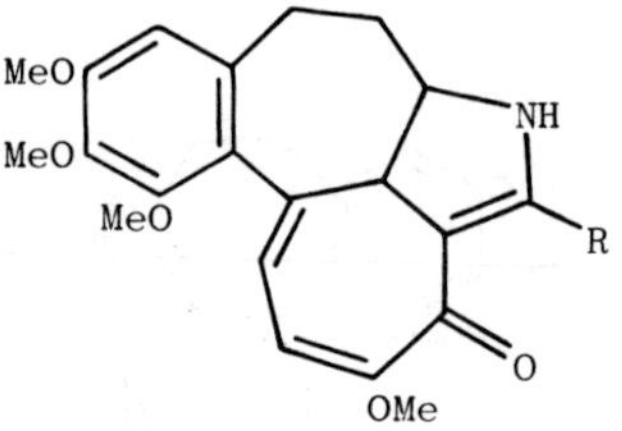

(134) 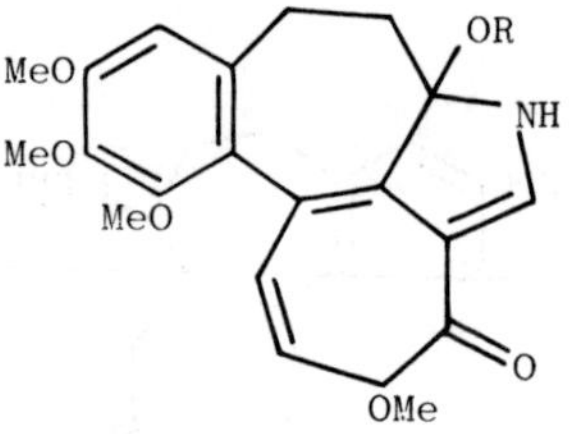   (135)

Following a route previously developed for the synthesis of deacetamidocolchicine, ($\pm$)-colchicine has been synthesised from (136; R = R^1 = OMe), which was hydrolysed by acid to (136; RR^1 = O) and this was cyclised by boron trifluoride etherate to the acid (137). Ring-expansion of the methyl ester of (137) with trifluoroacetic acid gave a mixture of (138) and the isomeric αβ-unsaturated ester, both of which, on oxidation, gave the tropolone (139), hydrolysis and decarboxylation of which gave deacetamido-isocolchicine. Hydrolysis of (139) and treatment with diphenyl-

(136) (137) (138) (139)

phosphoryl azide and triethylamine in t-butyl alcohol gave
(127; R = CO_2Bu^t), which was hydrolysed to deacetylcolchiceine, from
which colchicine (128; R = Ac) was prepared.[419]

Methods for the detection and determination of colchicine have
been reported[420,421] and reviewed.[422] The uptake of colchicine by
brain and liver has been studied[423] as have the effects of the
alkaloid on the generation of cyclic AMP in human leucocytes,[424]
the reactivity of lymphocytes,[425] the production of heteroploid
mulberry trees,[426] microtubule polymerisation,[427] the degradation[428]
and synthesis[429] of proteins, cell agglutination in experimental hepatoma,[430] the intra-axonal transport of tubulin,[431] spermato-
genesis,[432] the chick retina,[433] long-term memory,[434] the structure
of seminiferous tubules,[435] neuroblastoma cells,[436] the control of
frog skeletal muscle,[437] endothelial cell migration and replication
of epithelium,[438] the oriental hornet,[439] clone mutation,[440] cell
culture,[441] adenylcyclase activity in cirrhotic rats,[442] the active

Arthus reaction in rabbits,[443] humoral antibody responses,[444] the regeneration of nerves,[445] the distribution of water in rat liver and hepatomas,[446] the synthesis of collagen in liver,[447] polymorphonuclear adhesion,[448] intra-ocular pressure,[449] and degeneration of sciatic nerve,[450] of colchemid on sarcoma 180 tumour,[451] and of lumicolchicine on cell culture[441] and nerve regeneration.[445]

References

1. S. Pummangura and J.L. McLaughlin, J. Nat. Prod., 1981, 44, 498.
2. F. Chen and H. Wang, Zhongcaoyao, 1981, 12, 389.
3. S. Pummangura, J.L. McLaughlin and R. Schifferdecke, J. Nat. Prod., 1981, 44, 614.
4. J. Slavik and L. Slavikova, Coll. Czech. Chem. Commun., 1981, 46, 1534.
5. H. Petershafer-Halbmayer, O. Kubelka, J. Jurenitsch and W. Kubelka, Sci. Pharm., 1982, 50, 29.
6. R. Mata and J.L. McLaughlin, Rev. Latinoam. Quim., 1981, 12, 95.
7. C.A. Russo and E.G. Gros, J. Labelled Compds.Radiopharm., 1981, 18, 1185.
8. B. Ruscic and L. Klasnic, Kem. Ind., 1981, 30, 443.
9. P. Machairas and G. Tsatsas, Prakt. Akad. Athenon, 1981, 55 (A-B), 119
10. R.N. Shull, R.J. Sbordone and D.A. Gorelick, Physiol. Psychol., 1981, 9, 208.
11. J. Siwon, R. Verpoorte, T. Van Beek, H. Meerburg and A.B. Svendsen, Phytochemistry, 1981, 20, 323.
12. H. Hara, O. Hoshino and B. Umezawa, Heterocycles, 1981, 15, 911.
13. H. Hara, A. Tsunashima, H. Shinoki, O. Hoshino and B. Umezawa, Heterocycles, 1982, 17 (Special Issue), 293.
14. R.B. Tirodkar and R.N. Usgaonkar, Indian J. Chem.,Sect. B, 1981, 20, 813.
15. G.A. Smythe, M.W. Duncan and J.E. Bradshaw, IRCS Med. Sci., Libr.Compend., 1981, 9, 472.
16. J. Faust, R. Ripperger, D. Sandoval and K. Schreiber, Pharmazie, 1981, 36, 718.
17. M. Leboeuf, C. Lequeut, A. Cave, J.F. Desconclois, P. Forgacs and H. Jacquemin, Planta Med., 1981, 42, 37.
18. G. Blasko, M. Shamma, A.A. Ansari and Atta-ur-Rahman, Heterocycles, 1982, 19, 257.
19. M.H.A. Zarga, G.A. Miana and M. Shamma, Heterocycles, 1982, 18(Special Issue), 63.
20. M.H.A. Zarga, G.A. Miana and M. Shamma, Tetrahedron Lett., 1981, 22, 541.
21. L.H. Villaroel and M.R. Torres, Bol. Soc. Chil. Quim., 1982, 27, 273.
22. M. Alimova and I.A. Israilov, Khim. Prir. Soedin., 1981, 602.
23. M. Lavault, M.M. Debray and J. Bruneton, Planta Med., 1981, 42, 50.
24. R. Hocquemiller, A. Cave and A. Raharisololalao, J. Nat. Prod., 1981, 44, 551.
25. P. Buchs and A. Brozzi, Helv. Chim. Acta, 1981, 64, 681.
26. F.R. Hewgill and M.C. Pass, Tetrahedron Lett., 1981, 22, 2125.
27. A. Buzas and G. Lavielle, Ger. Offen. 3,027,325 (12 Feb. 1981) Chem. Abs., 1981, 95, 25378d
28. A. Buzas and G. Lavielle, Ger. Offen. 3,027,338 (19 Feb. 1981) Chem. Abs., 1981, 95, 150978a.
29. K. Yamada, M. Takeda and T. Iwakuma, Tetrahedron Lett., 1981, 22, 3869.
30. P. Janicki, A. Czlonkowski, B. Osipiak, U. Myszkowska, W. Gumulka, J. Libich, A. Chodkowski and B. Gutowska, Pol. J. Pharmacol. Pharm., 1980, 32, 141.
31. C. Wang, Yao Hsueh T'ung Pao, 1981, 16, 51.
32. L. Huang, D. Zhong and C. Wang, Yaoxue Xuebao, 1981, 16, 931.
33. C. Szantay, G. Dornyei, G. Blasko, M. Barczai-Beke and P. Pechy, Arch. Pharm. (Weinheim), 1981, 314, 983.

34 W. Wiegrebe, S. Mahboobi, G. Dannhardt, K.K. Mayer and E. Eibler, Chimia, 1981, 35, 288.
35 R.B. Sharma and R.S. Kapil, Indian J. Chem.,Sect. B, 1982, 21B, 141.
36 G.R. Lenz and C.-M. Woo, J. Heterocyclic Chem., 1981, 18, 691.
37 H. Otomatsu, K. Higashiyama, T. Hondo and T. Kametani, Heterocycles, 1982, 19, 353.
38 J.B. Stenlake, R.D. Waigh, G.H. Dewar, R. Hughes, D.J. Chapple and G.C. Coker, Eur. J. Med. Chem. - Chim. Ther., 1981, 16, 515.
39 J.B. Stenlake, R.D. Waigh, J. Urwin, G.H. Dewar, R. Hughes and D.J. Chapple, Eur. J. Med. Chem. - Chim. Ther., 1981, 16, 503.
40 R. Hughes and J.P. Payne, Int. Congr. Ser. - Excerpta Med., 1981, 538, 197.
41 H. Nawrath, J. Pharmacol. Exp. Ther., 1981, 218, 544.
42 J.C. Bowen, J.C. LeDoux, J.L. Ochsner, M.G. Ochsner and J.G. Payne, Surgery (St. Louis), 1981, 90, 41.
43 T. Shiba, T. Uruno, K. Kubota and K. Takagi, Jpn. J. Pharmacol., 1981, 31, 553.
44 L.I. Kazak, Fiziol. Zh. (Kiev), 1981, 27, 706.
45 N. Oudart and R.G. Boulu, Pathophysiol. Pharmacother. Cerebrovasci Disord. Satell. Symp. 2nd., 1980, 52.
46 Z. Cheng, L. Liu, T. Zhou, Q.Li and H. Wang, Yaoxue Xuebao, 1981, 16, 721.
47 I.S. Hoffman and L.X. Cubeddu, J. Pharmacol. Exp. Ther., 1982, 220, 16.
48 H. Watanabe, M. Ikeda, K. Watanabe and T. Kikuchi, Planta Med., 1981, 42, 213.
49 Y.-P. Feng, Y. Zhang, H.-S. Zhan and G.-Y. Zeng, Chung-kuo Yao Li Hsueh Pao, 1981, 2, 114.
50 H.-W. Han, C.-C. Wan, F.-L. Sun and K.-Y. Tseng, Yao Hsueh T'ung Pao, 1980, 15, 41.
51 X.-Z. Zhong and F.-H. Wu, Chung Ts'ao Yao, 1981, 12, 30.
52 G. Hoogewijs, Y. Michotte, J. Lambrecht and D.L. Massart, J. Chromatogr., 1981, 226, 423.
53 V. Fajardo, M. Garrido and B.K. Cassels, Heterocycles, 1981, 15, 1137.
54 V. Fajardo, A. Leon, M.C. Loncharic, V. Elango, M. Shamma and B.K. Cassels, Bol. Soc. Chil. Quim., 1982, 27, 159.
55 S.F. Hussain, L. Khan, K.K. Sadozal and M. Shamma, J. Nat. Prod., 1981, 44, 274.
56 T. Suess and F.R. Stermitz, J. Nat. Prod., 1981, 44, 680.
57 K.H. Baser, Doga,Seri A, 1981, 5, 163.
58 S. Mukhamedova, S. Kh. Maekh and S. Yu.Yunusov, Khim. Prir. Soedin., 1981, 250.
59 P. Wiriyachitra and B. Phuriyakorn, Aust. J. Chem., 1981, 34, 2001.
60 L. Koike, A.J. Marsaioli and F. de A.M. Reis, J. Org. Chem., 1981, 46, 2385.
61 D. Neuhaus, H.S. Rzepa, R.N. Sheppard and I.R.C. Bick, Tetrahedron Lett., 1981, 22, 2933.
62 I. Hitoshi, A. Harada, K. Ichino and K. Ito, Heterocycles, 1981, 16, 1275.
63 E.P. Nakova, O.N. Tol'kachev, R.P. Evstigneeva, Khim. Prir. Soedin., 1981, 457.
64 O.N. Tol'kachev, O.N. Tol'kacheva and R.P. Evstigneeva, USSR Pat. 554,674 Chem. Abs., 1981, 95, 133216h.
65 N.V. Korobov, Farmakol. Toksikol. (Moscow), 1982, 45, 115.
66 H. Gajdzinska and J. Wedrychowski, Arch. Med. Sadowej Kriminol., 1980, 30, 229.
67 A. Meulmans, J. Chromatogr., 1981, 226, 255.
68 V.K. Bhargava, D. Bhargava, D.N. Sharma and K.D. Tripathi, Neurosci. Lett., 1981, 23, 175.
69 M.T. Lin and A. Chandra, Experientia, 1981, 37, 986.
70 N. Shaker, A.T. Eldefrawi, L.G. Aguayo, J.E. Warnick, E.X. Albuquerque and M.E. Eldefrawi, J. Pharmacol. Exp. Ther., 1982, 220, 172.
71 A.A. Shoro, Methods Find. Exp. Clin. Pharmacol., 1981, 3, 149.
72 A.A. Selyanko, V.A. Derkach and V.I. Skok, Adv. Physiol. Sci. Proc. Int. Congr. 28th., 1980, 23, 285.
73 W.P. Brotherton and R.S. Matteo, Anesthesiology, 1981, 55, 273.

74. K. Aono, S. Morimoto, K. Hashimoto, K. Sato, I. Joja, S. Kimoto, H. Ezoe, Y. Takeda and M. Miyake, Okayama Igakkai Zasshi, 1980, 92, 1015.
75. K. Aono, N. Shiraishi, T. Arita, B. Inouye, T. Nakazawa and K. Utsumi, Physiol. Chem. Phys., 1981, 13, 137.
76. T. Kuramochi, T. Shimada, M. Inouchi, S. Kojima, M. Murayama and M. Ishida, Sei Marianna Ika Daigaku Zasshi, 1981, 9, 253.
77. Y. Tagashira, R. Fujiwara, M. Ono, K. Ohashi, Y. Kamikawa, N. Tomaka, H. Miwa, T. Mannami and E. Konaga, Gan to Kagaku Ryoho, 1981, 8, 234.
78. Y. Lu, D. Xu, Y. Mao, X. Wei and Z. Yang, Zhongguo Yaoli Xuebao, 1981, 2, 223.
79. W. Yao, D. Fang, G. Xia, L. Qu and M. Jang, Wuhan Yixueyan Xuebao, 1981, 10, 81.
80. J. Ke, S. Wang, G. Zhang, J. Wang and R. Fu, Zhongguo Yaoli Xuebao, 1981 2, 235.
81. R.Z. Wang and Q.Q. Pan, Chung-Lua Chung Liu Tsa Chih, 1981, 3, 86.
82. I. Chen and F. Lu, Yaoxue Tongbao, 1981, 16, 60.
83. C.-C. Ma, Yao Hsueh T'ung Pao, 1980, 15, 46.
84. M. Zhu and X. Tang, Yao Hsueh T'ung Pao, 1981, 16, 3.
85. K. Kido and Y. Watanabe, Chem. Pharm. Bull., 1981, 29, 861.
86. R. Elliott, F. Hewgill, E. McDonald and P. McKenna, Tetrahedron Lett., 1980, 21, 4633.
87. H. Hara, O. Hoshino and B. Umezawa, Nippon Kagaku Kaishi, 1981, 813.
88. M.J. Campello, L. Castedo, J.M. Saa, R. Suau and M.C. Vidal, Tetrahedron Lett., 1982, 23, 239.
89. H. Otsuka, H. Fujimura, T. Sawada and M. Goto, Yakugaku Zasshi, 1981, 101, 883.
90. Kh. Kiryakov, E. Iskrenova, B. Kuzmanov and L. Evstatieva, Planta Med., 1981, 43, 51.
91. Kh. Sh. Khusainova and Yu. D. Sadykov, Khim. Prir. Soedin., 1981, 671.
92. C. Tani, N. Nagakura, S. Saeki and M.T. Kao, Planta Med., 1981, 41, 403.
93. G. Blasko, S.F. Hussain and M. Shamma, J. Nat. Prod., 1981, 44, 349.
94. Kh. Kiryakov, Z. Mardisoyan and P. Panov, Dokl. Bolg. Akad. Nauk, 1981, 34, 43.
95. J. Slavik, L. Slavikova and L. Dolejs, Coll. Czech.Chem. Commun., 1981, 46, 2587.
96. F. Veznik, E. Taborska and J. Slavik, Coll. Czech. Chem. Commun., 1981, 46, 926.
97. X.A. Dominguez, O.R. Franco, C.G. Cano, S. Garcia and R.S. Tamez, Rev. Latinoam. Quim., 1981, 12, 61.
98. M.H.A. Zarga and M. Shamma, Tetrahedron Lett., 1980, 21, 3739.
99. K. Iwasa and M. Cushman, Heterocycles, 1981, 16, 901.
100. K. Iwasa, Y.P. Gupta and M. Cushman, J. Org. Chem., 1981, 46, 4744.
101. K. Iwasa, Y.P. Gupta and M. Cushman, Tetrahedron Lett., 1981, 22, 2333.
102. K. Iwasa and M. Cushman, J. Org. Chem., 1982, 47, 545.
103. T.R. Govindachari, B.R. Pai, S. Rajeswari, S. Natarajan, S. Chandrasekaram, M.S. Premila, R. Charubala, K. Venkatesan and M.M. Bhadbhade, Heterocycles, 1981, 15, 1463.
104. L.I. Petlichnaya, N.M. Turkevich and A.F. Mynka, Ukr. Khim. Zh. (Russian Edn.), 1981, 47, 864.
105. S. Prior, W. Wiegrebe and G. Sariyar, Arch. Pharm. (Weinheim), 1982, 315, 273.
106. M.L. Contreras and R. Rozas, Heterocycles, 1981, 16, 1735.
107. M. Hanaoka, M. Inoue, M. Takahashi and S. Yasuda, Heterocycles, 1982, 18, 31.
108. L.I. Petlichnaya, S.V. Ivasivka and A.I. Potopal'skii, Khim.-Farm. Zh., 1981, 15, 46.
109. T. Kametani, N. Takagi, M. Toyota, T. Honda and K. Fukumoto, Heterocycles, 1981, 16, 591.
110. T. Kametani, N. Takagi, M. Toyota, T. Honda and K. Fukumoto, J. Chem. Soc., Perkin Trans. 1, 1981, 2830.

111 T. Naito, Y. Tada, C. Hashimoto, K. Katsumi, T. Kiguchi and I. Ninomiya, Tennen Yuki Kagobutsu Toronkai Koen Yoshishu 24th., 1981, 460.
112 T. Naito, Y. Tada and I. Ninomiya, Heterocycles, 1981, 16, 1141.
113 G.D. Pandey and K.P. Tiwari, Tetrahedron, 1981, 37, 1213.
114 A. Chatterjee and S. Ghosh, Synthesis, 1981, 818.
115 N.S. Narasimhan, A.C. Ranade and B.H. Bhide, Indian J. Chem., Sect. B, 1981, 20B, 439.
116 P. Kerekes, Acta. Chim. Acad. Sci. Hung., 1981, 106, 303.
117 T. Naito, O. Miyata, I. Ninomiya and S.C. Pakrashi, Heterocycles, 1981, 16, 725.
118 S.T. Ko and D.Y. Lim, Arch. Pharmacol. Res., 1980, 3, 23.
119 E.A. Swabb, Y.-H. Tai and L. Jordan, Am. J. Physiol., 1981, 241, G248.
120 Y.H. Tai, J.F. Feser, W.G. Marname and J.F. Desjeux, Am. J. Physiol., 1981, 241, G253.
121 R.B. Slack and J.L. Froehlich, Infect. Immun., 1982, 35, 471.
122 L.I. Petlichnaya, B.S. Zimenkovskii and A.F. Mynka, Farm. Zh. (Kiev), 1981, 34.
123 X. Chen, Yaoxue Tongbao, 1981, 16, 19.
124 X. Jiang, Q. Wu, H. Shi, W. Chen, S. Chang, S. Zhao, X. Tiau, L. Zhou, S. Guo and Y. Li, Yaoxue Xuebao, 1982, 17, 61.
125 W. Wiegrebe, S. Prior and K.K. Mayer, Arch. Pharm. (Weinheim), 1982, 315, 262.
126 H.G. Kiryakov, Z. Mardirossian, D.B. MacLean and J.P. Ruder, Phytochemistry, 1981, 20, 1721.
127 G. Blasko, V. Elango, B. Sener, A.J. Freyer and M. Shamma, J. Org. Chem., 1982, 47, 880.
128 Z. Koblicova, J. Kreckova and J. Trojanek, Cesk. Farm., 1981, 30, 177.
129 H. Schmidhammer, Sci. Pharm., 1981, 49, 304.
130 A. Chatterjee, S. Bhattacharyya and S. Bhattacharya, Indian J. Chem., Sect. B, 1981, 20, 74.
131 N. Tsunoda and H. Yoshimura, Xenobiotica, 1981, 11, 189.
132 P. Kerekes, G. Gaal, R. Bognar, T. Toro and B. Costisella, Acta Chim. Acad. Sci. Hung., 1980, 105, 283.
133 J.R. Falck and S. Manna, Tetrahedron Letters, 1981, 22, 619.
134 R.H. Prager, J.M. Tippett and A.D. Ward, Aust. J. Chem., 1981, 34, 1085.
135 C.E. Slemon, L.C. Hellwig, J.P. Ruder, E.W. Hoskins and D.B. MacLean, Can. J. Chem., 1981, 59, 3055.
136 W.R. Buckett, J. Pharmacol. Methods, 1981, 5, 35.
137 E.J. Heyer, L.M. Nowak and R.L. MacDonald, Brain Res., 1982, 232, 41.
138 D.J. Nutt, P.J. Cowen, C.C. Batts, D.G. Grahame-Smith and A.R. Green, Psychopharmacology (Berlin), 1982, 76, 84.
139 P.J. Cowen, D.J. Nutt, C.C. Batts, A.R. Green and D.J. Heal, Psychopharmacology (Berlin), 1982, 76, 88.
140 G. Blasko, S.F. Hussain and M. Shamma, J. Amer. Chem. Soc., 1982, 104, 1599.
141 G. Blasko, S.F. Hussain, A.J. Freyer and M. Shamma, Tetrahedron Lett., 1981, 22, 3127.
142 G. Blasko, N. Murugesan, A.J. Freyer, R.D. Minard and M. Shamma, Tetrahedron Lett., 1981, 22, 3143.
143 G. Blasko, N. Murugesan, R.D. Minard, M. Shamma, B. Sener and M. Tanker, Tetrahedron Lett., 1981, 22, 3135.
144 G. Blasko, N. Murugesan, A.J. Freyer, D.J. Gula, B. Sener and M. Shamma, Tetrahedron Lett., 1981, 22, 3139.
145 N. Murugesan, G. Blasko, R. Minard and M. Shamma, Tetrahedron Lett., 1981, 22, 3131.
146 G. Blasko, V. Elango, N. Murugesan, M. Shamma, J. Chem. Soc. Chem. Commun., 1981, 1246.
146a M. Hanaoka, M. Inoue, K. Nagami, Y. Shimada and S. Yasuda, Heterocycles, 1982, 19, 313.
147 D. Lavie, H. Berger-Josephs, T. Yehezkel, H.E. Gottlieb and E.C. Levy, J. Chem. Soc., Perkin Trans. 1, 1981, 1019.

148 H. Roensch, Tetrahedron, 1981, 37, 371.
149 S. Prabhakar, A.M. Lobo, M.R. Tavares and I.M.C. Oliveira, J. Chem. Soc., Perkin Trans. 1, 1981, 1273.
150 L. Hou, M. Chen and H. Zhu, Zhongcaoyao, 1981, 12, 352.
151 S. Takano, S. Hatakeyama, Y. Takahashi and K. Ogasawara, Heterocycles, 1982, 12 (Special Issue), 263.
152 T. Kametani, S.A. Surgenor and K. Fukumoto, J. Chem. Soc., Perkin Trans. 1, 1981, 920.
153 S. Takano, M. Sato and K. Ogasawara, Heterocycles, 1981, 16, 799.
154 F. Dubini, R. Mattina and M. Falchi, Chemioter. Antimicrob., 1980, 3, 5.
155 A.K. Chatterjee and A.D. Ray, Jpn. J. Exp. Med., 1982, 52, 27.
156 K.P. Tiwari, R.N. Choudhary and G.D. Pandey, Phytochemistry, 1981, 20, 863.
157 R.A. Barnes and O.H. Soeiro, Phytochemistry, 1981, 20, 543.
158 V. Vecchietti, C. Casagrande, G. Ferrari, B. Danieli and G. Palmisano, J. Chem. Soc., Perkin Trans. 1, 1981, 578.
159 G. Sariyar and A. Oztekin, Planta Med. Phytother., 1981, 15, 160.
160 A. Lao, Z. Tang and R. Xu, Yaoxue Xuebao, 1981, 16, 940.
161 J. Kunimoto, Y. Murakami, M. Oshikata, T. Shingu, M. Akasu, S.-T. Lu and I.-S. Chen, Phytochemistry, 1980, 19, 2735.
162 K.K. Kano, W. Kimoto, A.-F. Hsu, P.G. Mahlberg and D.D. Bills, Phytochemistry, 1982, 21, 219.
163 M.P. Kotick, J. Med. Chem., 1981, 24, 722.
164 Kowa Co. Ltd., Jpn. Kokai Tokkyo Koho 81 30980 (28.3.81).
          Chem. Abs., 1981, 95, 89117c.
165 Kowa Co. Ltd., Jpn. Kokai Tokkyo Koho 82 02288 (7.1.82).
          Chem. Abs., 1982, 96, 218087n.
166 Kowa Co. Ltd., Jpn. Kokai Tokkyo Koho 81 30981 (28.3.81).
          Chem. Abs., 1981, 95, 98116b.
167 M.P. Kolick and J.O. Polazzi, J. Heterocyclic Chem., 1981, 18, 1029.
168 M.P. Kolick and D.L. Leland, U.S. Pat. 4275205 (23.6.81.).
          Chem. Abs., 1982, 96, 35613m.
169 D.L. Leland and M.P. Kolick, J. Med. Chem., 1981, 24, 717.
170 J.O. Polazzi, J. Org. Chem., 1981, 46, 4262.
171 M. Boess and W. Fleischhacker, Liebig's Ann. Chem., 1981, 2002.
172 M. Boess and W. Fleischhacker, Liebig's Ann. Chem., 1981, 1994.
173 W. Fleischhacker and E. Leitner, Monatsh., 1980, 111, 1307.
174 M. Boess and W. Fleischhacker, Liebig's Ann. Chem., 1982, 112.
175 D.L. Leland, J.O. Polazzi and M.P. Kotick, J. Org. Chem., 1981, 46, 4012.
176 W. Fleischhacker and B. Richter, Chem. Ber., 1980, 113, 3866.
177 M.A. Schwarz and R.A. Wallace, J. Med. Chem., 1981, 24, 1525.
178 P.R. Crabbendam, L. Maat and H.C. Beyerman, Recl. J.R. Neth. Chem. Soc., 1981, 100, 293.
179 C.W. Hutchins, G.K. Cooper, S. Puerro and H. Rappoport, J. Med. Chem., 1981, 24, 773.
180 K. Hayakawa, S. Motohiro, I. Fujii and K. Kanematsu, J. Amer. Chem. Soc., 1981, 103, 4605.
181 A. Singh, S. Archer, K. Hoogsteen and J. Hirshfield, J. Org. Chem., 1982, 47, 752.
182 R.M. Boden, M. Gates, S.P. Ho and P. Sundarasaman, J. Org. Chem., 1982, 47, 1347.
183 R.M. Allen, G.W. Kirby and D.J. McDougall, J. Chem. Soc., Perkin Trans. 1, 1981, 1143.
184 G.W. Kirby and J.G. Sweeny, J. Chem. Soc., Perkin Trans. 1, 1981, 3250.
185 H. Anterhoff and W. Tittjung, Arch Pharm. (Weinheim), 1980, 313, 985.
186 J.-L. Gong and T. Zhu, Shang-hai Ti 1 I Hsueh Yuan Hsueh Pao, 1981, 8, 133.
187 C.N. Filer, D. Ahern, R. Fazio and R.J. Seguin, J. Org. Chem., 1981, 46, 4968.
188 P.A. Mason, B. Law and R.E. Ardrey, J. Labelled Compd. Radiopharm., 1981, 18, 1497.

189 Y. Yost and J.L. Holtzman, J. Labelled Compd. Radiopharm., 1981, 18, 1793.
190 W.G. Dauben, C.P. Baskin and H.C.H.A. Van Riel, U.S. Pat. Appl. 115 411. (27.3.81); Chem. Abs., 1981, 95, 115834d.
191 D.L. Leland, M.P. Kolick, R.N. Schut and J.O. Polazzi, Eur. Pat. Appl. 23576 (11.2.81.); Chem. Abs., 1981, 95, 98113y.
192 G.A. Boswell and R.M. Henderson, U.S. Pat. 4241065 (23.12.80) Chem. Abs., 1981, 95, 150977z.
193 D.J. Maffer and D.F. Loncrini, PCT Int. Appl. 81 00409 (19.2.81) Chem. Abs., 1981, 95, 150979b.
194 E.C. Hermann, K.T. Lee and M.J. Myers, Eur. Pat. Appl. EP 36066 (4.11.81) Chem. Abs., 1982, 96, 199973e
195 Sankyo Co. Ltd., Jpn. Kokai Tokkyo Koho 82 04991 Chem. Abs., 1982, 96, 218088p.
196 W.H.J. Tam, W.G.W. Kurz, F. Constabel and K.B. Chatson, Phytochemistry, 1982, 21, 253.
197 J. Combie, J.W. Blake, T.E. Nugent and T. Tobin, Clin. Chem. (Winston-Salem N.C.) 1982, 28, 83.
198 Y. Yamazoe, H. Numata and T. Yanagita, Jpn. J. Pharmacol., 1981, 31, 433.
199 T. Ishida, K. Oguri and H. Yoshimura, J. Pharmacobio.-Dyn., 1982, 5, 134.
200 A.R. Battersby, R.C.F. Jones, R. Kazlauskas, A.P. Ottridge, C. Poupat and J. Staunton, J. Chem. Soc.,Perkin Trans. 1, 1981, 2010.
201 A.R. Battersby, A.K. Bhatnagar, P. Hackett, C.W. Thornber and J. Staunton, Chem. Commun., 1968, 1214.
202 A.R. Battersby, A.K. Bhatnagar, P. Hackett, C.W. Thornber and J. Staunton, J. Chem. Soc., Perkin Trans. 1, 1981, 2002.
203 A. Brossi, Proc. Asian Symp. Med. Plants Spices, 4th., 1980, 1, 261.
204 K.C. Rice, U.S. Pat. Appl. 165690 (19.12.80) Chem. Abs., 1981, 95, 98121z.
205 C. Szantay, M. Barczai-Beke, P. Pechy, G. Blasko and G. Dornyer, J. Org. Chem., 1982, 47, 594.
206 G. Horvath and S. Makleit, Acta Chim. Acad. Sci. Hung., 1981, 106, 37.
207 Yu. A. Chichnev, Sud.-Med. Ekspert., 1981, 24, 53.
208 G. Vanzetti, V. Benedetti, M. Cassani, G. Scaring and D. Valente, Gazz. Ital. Chim. Clin., 1980, 5, 403.
209 S.H. Loh, Proc. Malays. Biochem. Soc. Conf. 6th., 1980, 119.
210 K. Khupulsup and B. Petchclai, J. Med. Assoc. Thailand, 1981, 64, 72.
211 J.W. Villiger, R.A. Boas and K.M. Taylor, Life Sci., 1981, 29, 229.
212 T. I. Ul'yankina, V.V. Lakin and D.D. Enikeeva, Lab. Delo, 1981, 436.
213 A. Huhtikangas, K. Wickstrom and T. Vartiainen, Prog. Clin. Pharm., 1981, 3, 89.
214 J. Combie, J.W. Blake, B.E. Rainey and T. Tobin, Am. J. Vet. Res., 1981, 42, 1523.
215 L. Von Meyer, G. Kanert and G. Drasch, Beitr. Gerichtl. Med., 1981, 39, 113.
216 D. Valente, M. Cassini, M. Pigliapochi and G. Vansetti, Clin.Chem. (Winston-Salem N.C.),1981, 27, 1952.
217 P.E. Nelson, S.L. Nolam and K.R. Bedford, J. Chromatogr., 1982, 234, 407.
218 J.B. Lopez, N.M. Desa, M. Zaini-Rahman and G.F. De Will, Lab. Pract., 1981, 30, 1223.
219 D.R. Stanski, L. Paalzow and P.O. Edlund, J. Pharm. Sci., 1982, 71, 314.
220 I.E. Kovalev, T.N. Robakidze, O. Yu. Polevaya and I.V. Agapitova, Khim.-Farm. Zh., 1982, 16, 39.
221 D. Bernhauer, E.F. Fuchs, M. Gloger and H. Neumann, Arch. Kriminol., 1981, 168, 139.
222 H. Neumann and G. Vordermaier, Arch. Kriminol., 1981, 167, 33.
223 T.A. Gough and P.B. Baker, J. Chromatogr. Sci., 1981, 19, 227.
224 G. Machata and W. Vycudilik, J. Anal. Toxicol., 1980, 4, 318.
225 S.T. Chow, J. Forensic. Sci., 1982, 27, 32.
226 P. Demedts, M. Van den Heede and A. Heyndrickx, J. Anal. Toxicol., 1982, 6, 30.
227 G. Lhoest and M. Mercier, Pharm. Acta Helv., 1980, 55, 316.

228 V.V. Mikhno, *Farm. Zh. (Kiev)*, 1981, 49.
229 J.E. Sherman, J.W. Lewis, R.E. De Wetter and J.C. Liebeskind, *Proc. West. Pharmacol. Soc.*, 24th, 1981, 327.
230 J.W. Lewis, E.H. Chudler and J.T. Cannon, *Proc. West. Pharmacol. Soc.*, 24th 1981, 323.
231 T.L. Yash, S.V.R. Reddy, *Anesthesiology*, 1981, 54, 451.
232 L.F. Jarvik, J.H. Simpson, D. Guthrie and E.H. Liston, *Psychopharmacology (Berlin)*, 1981, 75, 124.
233 T. Doi and I. Jurna, *Brain Res.*, 1982, 234, 399.
234 M.D. Hughes and R.W. Fuller, *Drug Dev. Res.*, 1982, 2, 33.
235 D.B. Belselin, S.K. Krstic, K. Stefanovic-Denic, M. Strbac and D. Micic, *Brain Res. Bull.*, 1981, 6, 451.
236 H. Marcais, J.J. Bonnet and J. Costentin, *Life Sci.*, 1981, 28, 2737.
237 P. Nencini and E. Paroli, *Pharmacol. Res. Commun.*, 1981, 13, 535.
238 D.G. Lange, J.M. Fujimoto and R.B. Franklin, *Res. Commun. Subst. Abuse*, 1981, 2, 97.
239 J.C. Umans and C.E. Inturessi, *J. Pharmacol. Exp. Ther.*, 1981, 218, 409.
240 D.M. Grilly, *Life Sci.*, 1981, 28, 1883.
241 P. Slater, *Pharmacol. Biochem. Behav.*, 1981, 14, 625.
242 R.J. Bodnar, J.H. Kordower, M.M. Wallace and H. Tamir, *Pharmacol. Biochem. Behav.*, 1981, 14, 645.
243 T. Nakaki, M. Saito, T. Nadakate, Y. Tokemaga and R. Kato, *Psychopharmacology (Berlin)*, 1981, 73, 215.
244 G.B. Stefano, L. Hiripi, K. Saghy-Rozsa and J. Salanki, *Adv. Physiol. Sci. Proc. Int. Congr.*, 28th., 1980, 23, 285.
245 R.C. Drugan, J.W. Grau, S.F. Maier, J. Madden and J.D. Barchas, *Pharmacol. Biochem. Behav.*, 1981, 14, 677.
246 W.T. Nelson, S.S. Steiner, M. Brutus, R. Farrell and S.J. Ellman, *Psychopharmacology (Berlin)*, 1981, 74, 58.
247 L. Switzman, B. Fishman and Z. Amit, *Psychopharmacology (Berlin)*, 1981, 74, 149.
248 W.J. Jacobs, D.A. Zellner, V.H. LoLordo and A.L. Riley, *Pharmacol. Biochem. Behav.*, 1981, 14, 779.
249 B.E. Jones and J.A. Prada, *Drug Alcohol Depend.*, 1981, 7, 203.
250 M.L. Kirby, *Exp. Neurol.*, 1981, 73, 430.
251 M.E. Olds and J.L. Fobes, *Pharmacol. Biochem. Behav.*, 1981, 15, 167.
252 H.S. Wheeling, S. Casson and C. Kornetsky, *Subst. Alcohol Actions/Misuse*, 1981, 2, 107.
253 S.G. Smith, T.E. Warner and W.M. Davis, *Drug Alcohol Depend.*, 1981, 7, 305.
254 H.D. Kimmel and M.M. Budronis, *Pavlovian J. Biol. Sci.*, 1981, 16, 163.
255 A.F.T. Arnstein, D.S. Segal, S.E. Loughlin and D.C.S. Roberts, *Brain Res.*, 1981, 222, 351.
256 L.S. Brady and S.G. Holtzman, *J. Pharmacol. Exp. Ther.*, 1981, 219, 344.
257 R. Blair and Z. Amit, *Pharmacol. Biochem. Behav.*, 1981, 15, 651.
258 D.M. Thompson and J.M. Moerschbaecher, *J. Exp. Anal. Behav.*, 1981, 36, 371.
259 L. Switzman, T. Hunt and Z. Amit, *Pharmacol. Biochem. Behav.*, 1981, 15, 755.
260 R.L. Preshaw, H. Zenick and R.M. Stutz, *Pharmacol. Biochem. Behav.*, 1982, 16, 81.
261 L.D. Byrd, *J. Pharmacol. Exp. Ther.*, 1982, 220, 139.
262 L.P. Spear, G.P. Horowitz and J. Lipovsky, *Behav. Brain Res.*, 1982, 4, 279.
263 M.L. Kirby and S.G. Holtzman, *Pharmacol. Biochem. Behav.*, 1982, 16, 263.
264 M.D. Mank, J.F. Warren and R.F. Thompson, *Science (Washington)*, 1982, 216, 434.
265 R.F. Mucha, C. Volkovskis and H. Kalant, *J. Comp. Physiol. Psychol.*, 1981, 95, 351.
266 L.S. Brady and S.G. Holtzman, *J. Pharmacol. Exp. Ther.*, 1981, 218, 613.
267 R.R. Griffiths, R.M. Wurster and J.V. Brady, *J. Exp. Anal. Behav.*, 1981, 35, 335.
268 F.S. Tepperman, M. Hirst and C.W. Gowdey, *Life Sci.*, 1981, 28, 2459.
269 Y. Shimomura, J. Oku, Z. Glick and G.A. Bray, *Physiol. Behav.*, 1982, 28, 441.

270   J.M. Stapleton, A. Castiglioni and J.C. Liebeskind, Proc. West. Pharmacol. Soc., 24th., 1981, 161.
271   J. Huidoboro-Toro, J. Pablo and F. Huidoboro, J. Pharmacol. Exp. Ther., 1981, 217, 579.
272   J.L. Montastruc, F. Morales-Olivas and P. Montastruc, Arch. Farmacol. Toxicol., 1980, 6, 287.
273   G. Lee, R.I. Low, E.A. Amsterdam, A.N. De Maria, P.W. Huber and D.T. Mason, Clin. Pharmacol. Ther., 1981, 29, 576.
274   J.Y. Hwa and S.H.H. Chan, Brain Res., 1981, 214, 205.
275   I. Kalmar, M. Szazados, J. Soos, M. Olajos, F. Remji-Vamos, L. Papp and Z. Szabo, Anasthesiol. Intenziv. Ther., 1980, 10, 241.
276   M. Weinstock, E. Erez and D. Roll, J. Pharmacol. Exp. Ther., 1981, 218, 504.
277   L.L. Priano and S.F. Vatner, Anesthesiology, 1981, 55, 236.
278   W. Feldberg and E. Wei, Perspect. Cardiovasc. Res., 1981, 6, 229.
279   M.A. Hurle, A. Mediavilla and J. Florey, J. Pharmacol. Exp. Ther., 1982, 220, 642.
280   L.F. Fanjul, C.M. Ruiz de Galarreta and J. Meites, Proc. Soc. Exp. Biol. Med., 1981, 166, 542.
281   A. Ratner, W. Woo, D. Yelvington and K. Torres, Proc. West. Pharmacol. Soc., 24th, 1981, 335.
282   I.S. Login, I. Nagy and R.M. MacLeod, Neuroendocrinology, 1981, 33, 101.
283   A.E. Panerai, F. Casanueva, A. Mortini, P. Mantegazza and A.M. DiGiulio, Endocrinology (Baltimore), 1981, 108, 2400.
284   W.B. Wehrenberg, D. McNicol, S.L. Wardlaw, A.G. Frantz and M. Ferrin, Endocrinology (Baltimore), 1981, 109, 544.
285   E. Eriksson, S. Eden and K. Modigh, Neuroendocrinology, 1981, 33, 91.
286   S.P. Kalra and J.W. Simpkins, Endocrinology (Baltimore), 1981, 109, 776.
287   J. Lakoski and G.F. Gebhart, Neuroendocrinology, 1981, 33, 105.
288   J. Lakoski and G.F. Gebhart, Brain Res., 1982, 232, 231.
289   J.H. Johnson, G.T. Maughan and L. Anderhub, Life Sci., 1982, 30, 1473.
290   B. Sharp, J.E. Morley, H.E. Carlson, J. Gordon, J. Briggs, S. Melmed and J.M. Hershman, Brain Res., 1981, 219, 335.
291   J. Weil-Fugazza, F. Godefroy, D. Coudert and J.M. Besson, Brain Res., 1981, 214, 440.
292   A.K. Sanyal, D.N. Srivastava and S.K. Bhattacharya, Indian J. Med. Res., 1981, 73, 787.
293   S.P. Sholl, M.A. Connors, A.K. Killian and B.H. Wainer, Res. Commun. Subst. Abuse, 1981, 2, 367.
294   P. Slater and C. Blundell, J. Neurosci. Res., 1981, 6, 701.
295   N. Grosman, Agents Actions, 1981, 11, 196.
296   H. Bhargava, Life Sci., 1981, 29, 1015.
297   R. Numan and H. Lal, Prog. Neuro-Psychopharmacol., 1981, 5, 363.
298   C.E. Rosow, J.M. Miller, J. Poulsen-Burke and J.E. Cochin, J. Pharmacol. Exp. Ther., 1982, 220, 464.
299   R.W. James, R. Heywood and D. Crook, Toxicol. Lett., 1980, 7, 61.
300   R. Hemmings, G. Fox and G. Tolis, Fertil. Steril., 1982, 37, 389.
301   D. Le Bars, D. Clutour, E. Kraus, A.M. Clot, A.H. Dickenson and J.M. Besson, Brain Res., 1981, 215, 257.
302   J.P. Rosenfeld and S. Stocco, Brain Res., 1981, 215, 342.
303   Y.F. Jacquet and G. Wolf, Brain Res., 1981, 219, 214.
304   C.-M. Lee and L.L. Iversen, Brain Res., 1981, 219, 355.
305   A. Roby, B. Bussel and J.C. Willer, Brain Res., 1981, 222, 209.
306   I. Jurna, Naunyn-Schmiedberg's Arch. Pharmacol., 1981, 316, 149.
307   L.A. Gromov, S.V. Krivorotov and R.N. Skryma, Neirofiziologiya, 1981, 13, 332.
308   A. Schurr, B.M. Rigor, B.T. Ho and N. Dafny, Brain Res. Bull., 1981, 6, 473.
309   C.J. Woolf and M. Fitzgerald, Neurosci. Lett., 1981, 25, 37.
310   S. Yanaura, T. Hosokawa, H. Kitagawa and M. Misawa, Jpn. J. Pharmacol., 1981, 31, 529.

311 S.N. Sullivan, M. Champion and R. Darwish, Am. J. Gastroenterol., 1981, 76, 44.
312 D.E. Burleigh, J.J. Galligan and T.F. Burks, Eur. J. Pharmacol., 1981, 75, 283.
313 L. Bueno, J. Fioramonti and M. Ruckebusch, Eur. J. Pharmacol., 1981, 75, 239.
314 E. Scheufler and G. Zetler, Naunyn-Schmiedberg's Arch.Pharmacol., 1981, 318, 66.
315 Q. Lin, J. Yu and C. Wang, Shengli Xuebao, 1981, 33, 379.
316 Q. Lin, C. Meng, J. Yu and C. Wang, Shengli Xuebao, 1981, 33, 404.
317 S. Piepenbrock, M. Zenz, G.W. Sybrecht and B. Otten, Reg.-Anaesth., 1981, 4, 32.
318 R.L. Knill, J.L. Clement and W.R. Thompson, Can. Anaesth. Soc. J., 1981, 28, 537.
319 J.T. Huang, Res. Commun. Subst. Abuse, 1981, 2, 221.
320 M.C. Wallenstein, Am. J. Physiol., 1981, 241, R130.
321 M. Hachisu and C. Takashige, Showa Igakkai Zasshi, 1980, 40, 701.
322 J. Weil-Fugazza, F. Godefroy and D. Clutour, C.R. Seances Acad. Sci.,Ser.3, 1981, 293, 89.
323 I. Akiba, H. Endou, T. Susuki, S. Yamaura and F. Sakai, Life Sci., 1981, 29, 1057.
324 G.C. Mougdil, Can. Anaesth. Soc. J., 1981, 28, 597.
325 G.D. Prell and I.M. Mazurkiewicz-Kwilecki, Prog. Neuro-psycho. Pharmacol., 1981, 5, 581.
326 R.D. Kaufman, M.L. Gabathuler and J.W. Bellville, J. Pharmacol. Exp. Ther., 1981, 219, 156.
327 E. Masini, P. Blandina and P.F. Maunioni, Clin. Toxicol., 1981, 18, 1021.
328 R.F. Mucha and H. Kalant, Psychopharmacology (Berlin), 1981, 75, 132.
329 F. Poreca, A. Cowan and R.J. Tallarida, Eur. J. Pharmacol., 1981, 76, 55.
330 C.L. Wong, M.K. Wai and M.B. Roberts, Clin. Exp. Pharmacol. Physiol., 1982, 9, 69.
331 R.M. Quock and T.B. Welsh, J. Pharm. Pharmacol., 1981, 33, 111.
332 C. Castellano, Psychopharmacology (Berlin), 1981, 73, 152.
333 P. De Witte, Psychopharmacology (Berlin), 1981, 73, 391.
334 N.L. Ostrowski, R.G. Noble and L.D. Reid, Pharmacol. Biochem. Behav., 1981, 14, 881.
335 L.H. Gorris and J.H.F. Abeelen, Psychopharmacology (Berlin), 1981, 74, 355.
336 C.V. Vorhees, Neurobehav. Toxicol. Teratol., 1981, 3, 295.
337 S.E. Filer and T. Silverstone, Psychopharmacology (Berlin), 1981, 74, 353.
338 D.A. Gorelick, M.L. Elliott and R.J. Sbordone, Res. Commun. Subst. Abuse, 1981, 2, 419.
339 R.U. Esposito, W. Perry and C. Kornetsky, Pharmacol. Biochem. Behav., 1981, 15, 905.
340 J.L. Nelson, N.L. Ostrowski, R.G. Noble and L.D. Reid, Physiol. Psychol., 1981, 9, 367.
341 C. Fabre-Nys, R.E. Meller and E.B. Keverne, Pharmacol. Biochem. Behav., 1982, 16, 653.
342 T.A. Salerno, B. Milne and K.H. Jhamandas, Surg. Gynecol. Obstet., 1981, 152, 773.
343 L. Rios and J. Jacob, Arch. Inst. Pasteur Tunis, 1981, 58, 313.
344 N.L. Ostrowski, N. Rowland, T.L. Foley, J.L. Nelson and L.D. Reid, Pharmacol. Biochem. Behav., 1981, 14, 549.
345 J. Birk and R.G. Noble, Life Sci., 1981, 29, 1125.
346 S.M. Siviy, F. Bermudez-Rattoni, G.A. Lockwood, C.M. Dargie and L.D. Reid, Pharmacol. Biochem. Behav., 1981, 15, 257.
347 G.A. Rockwood, S.M. Siviy and L.D. Reid, Pharmacol. Biochem. Behav., 1981, 15, 319.
348 N. Rowland, Pharmacol. Biochem. Behav., 1982, 16, 87.
349 R.D. Olson, R.C. Fernandez, A.J. Kastin, G.A. Olson, S.W. Delatte, T.K. Von Almen, D.G. Erickson, D.C. Hastings and D.H. Coy, Pharmacol. Biochem. Behav., 1981, 15, 921.

β-Phenylethylamines and the Isoquinoline Alkaloids 169

350   C. Farsang, M.D. Ramirez-Gonzales and K. Gyorgy, Kiserl. Orvostud, 1981, 33, 181.
351   A.I. Faden, T.P. Jacobs, G. Feurstein and J.W. Holaday, Brain Res., 1981, 213, 415.
352   J.P. Huidoboro-Toro and J.M. Musacchio, Arch. Int. Pharmacodyn., Ther., 1981, 251, 310.
353   D.A. Bennett, J.J. De Feo, E.E. Elko and H. Lal, Drug. Dev. Res., 1982, 2, 175.
354   M.H. Wood, R.V. Reddy, I. Broome and E.J. Dawe, Surg. Forum, 1981, 32, 141.
355   C.L. Maas, Eur. J. Pharmacol., 1982, 77, 71.
356   R. Caldora, M. Cambelli, E. Masci, M. Guslandi, C. Barbieri and C. Ferrari, Gut, 1981, 22, 720.
357   M. Feldman and Y.M. Cowley, Dig. Dis. Sci., 1982, 27, 308.
358   T. Shimizu, T. Koja, T. Fujisaki and T. Fukuda, Brain Res., 1981, 208, 463.
359   S. Ramaswamy, N.P. Pillai, V. Gopalakrishnan and M.N. Ghosh, Indian J. Exp. Biol., 1981, 19, 555.
360   G.J. Schaefer and R.P. Michael, Psychopharmacology (Berlin), 1981, 74, 17.
361   T. Duka, E.P. Bonetti, G.P. Bondiolotti and M. Wuester, Experientia, 1981, 37, 881.
362   C. Castellano and S. Puglisi-Allegra, Pharmacol. Biochem. Behav., 1982, 16, 561.
363   B. Kraynaek, J. Guitantas, T. Hughston and D. De Shan, Proc. West. Pharmacol. Soc., 24th., 1981, 189.
364   T.M. Loseva and G.I. Kolyaskina, Zh. Nevropatol. Psikhiatr. im. S.S. Korsakova, 1981, 81, 1006.
365   P. Almqvist, M. Kuenzig and S.I. Schwartz, Surg. Forum, 1981, 32, 304.
366   J.R. Lymangrover, L.A. Dokas, A. Kong, R. Martin and M. Saffran, Endocrinology (Baltimore), 1981, 109, 1132.
367   V. Kayser and G. Guilbaud, Brain Res., 1981, 226, 344.
368   L.J. Judd, D.S. Janowsky, A. Zettner, L.Y. Huey, Y. Leighton and K.I. Takahashi, Psychiatry Res., 1981, 4, 277.
369   D. Naber, D. Pickar, G.C. Davis, R.M. Cohen, D.C. Junerson, M.A. Elchisak, E.G. Defraites, N.H. Kalin, S.C. Risch and M.S. Buchsbaum, Psychopharmacology (Berlin), 1981, 74, 125.
370   L. Steardo, M. Maj, P. Monteleone, S. Zizolfi, G. Buscaino and D. Kemali, Acta Neurol., 1981, 36, 379.
371   M.M. Wikes and S.S.C. Yen, Life Sci., 1981, 28, 2355.
372   D.A. Van Vugt, C.F. Aylsworth, P.W. Sylvester, F.C. Leung and J. Meites, Neuroendocrinology, 1981, 33, 261.
373   J.G. Humans and C.E. Inturessi, J. Pharmacol. Exp. Ther., 1981, 218, 409.
374   R. Liljequist, Eur. J. Clin. Pharmacol., 1981, 20, 99.
375   T.T. Chan and L.S. Harris, NIDA Res Monogr., 1980, 34, 58.
376   J.J. Vallner, J.T. Stewart, J.A. Kotzan, E.B. Kirsten and I.L. Honigberg, J. Clin. Pharmacol., 1981, 21, 152.
377   W. Linder and M. Raab, Arch. Pharm. (Weinheim), 1981, 314, 595.
378   I. Takayanagi, N. Miyata, K. Uba, K. Watanabe and M. Hirobe, Jpn. J. Pharmacol., 1981, 31, 573.
379   J. Knoll, L. Kerescen, G.T. Somogyi and A. Kovacs, Arch. Int. Pharmacodyn. Ther., 1981, 251, 52.
380   D.C. Perry, J.S. Rosenbaum, M. Kurowski and W. Sadee, Mol. Pharmacol., 1982, 21, 272.
381   J. Dum, J. Blaesig and A. Herz, Eur. J. Pharmacol., 1981, 70, 293.
382   U. Filibeck, C. Castellano and A. Oliverio, Psychopharmacology (Berlin), 1981, 73, 134.
383   N.K. Mello, M.P. Bree and J.H. Mendelson, NIDA Res. Monogr., 1980, 34, 36.
384   N.K. Mello, M.P. Bree and J.H. Mendelson, Pharmacol. Biochem. Behav., 1981, 15, 215.
385   G. De Klerk, H. Mathie and J. Spierdijk, Acta Anaesthesiol. Belg., 1981, 32, 131.

386. H. Iizuka, A. Shimada and T. Yanagita, Jitchuken Zenrinsho Kenkyuho, 1981, 7, 279.
387. K. Takada, T. Kawaguchi and T. Yanagita, Jitchuken Zenrinsho Kenkyuho, 1981, 7, 323.
388. H. Numata, T. Tsuda, H. Atai, M. Tanaka and T. Yanagita, Jitchuken Zenrinsho Kenkyuho, 1981, 7, 342.
389. T. Yanagita, S. Kato, Y. Wakasa and N. Oinuma, Jitchuken Zenrinsho Kenkyuho, 1981, 7, 337.
390. J.W. Villiger and K.M. Taylor, Life Sci., 1981, 29, 2699.
391. J.H. Mendelson, J. Ellingboe, N.K. Mello and J. Kuehnie, J. Pharmacol. Exp. Ther., 1982, 220, 252.
392. S. Shintani, M. Umezato, Y. Toba, Y. Yamaji, K. Kitaura, T. Tani, H. Ishizama, T. Kikuchi and T. Mori, Nippon Yakurigaku Zasshi, 1982, 79, 173.
393. T. Hiyama, S. Shintani, M. Tsutsui and Y. Yasuda, Nippon Yakurigaku Zasshi, 1982, 79, 147.
394. Y. Yasuda, Y. Shioya, S. Nakai, S. Shintani and T. Hiyama, Nippon Yakurigaku Zasshi, 1982, 79, 163.
395. E. Rolandi, G. Magnani, A. Sannia and T. Barreca, Eur. J. Clin. Pharmacol., 1981, 21, 23.
396. R.J. Valentino, S. Herling, J.H. Woods, F. Medzihradsky and H. Merz, J. Pharmacol. Exp. Ther., 1981, 217, 652.
397. R.J. Katz, Pharmacol. Biochem. Behav., 1981, 15, 131.
398. N.K. Mello, J.H. Mendelson and M.P. Bree, J. Pharmacol. Exp. Ther., 1981, 218, 558.
399. M. Apfelbaum and A. Mandenoff, Pharmacol. Biochem. Behav., 1981, 15, 89.
400. C. Advokat, Pharmacol. Biochem. Behav., 1981, 15, 677.
401. G.F. Stemfels, G.A. Young and N. Khazan, NIDA Res. Monogr., 1980, 34, 138.
402. C.A. Di Fazio, J.C. Moscicki and M.R. Magruder, Anesth. Analg. (Cleveland), 1981, 60, 629.
403. S.J. Ward, P.S. Portoghese and A.E. Kakemori, J. Pharmacol. Exp. Ther., 1982, 220, 494.
404. S.Y. Yeh, Arch. Int. Pharmacodyn. Ther., 1981, 254, 223.
405. WHO Advisory Group, Bull. Narc., 1981, 23, 29.
406. M. Zhang, Shanxi Xinyiyao, 1981, 10, 57.
407. M. Alumova, I.A. Israilov, M.S. Yunusov and S. Yu. Yunusov, Khim. Prir. Soedin., 1981, 671.
408. L. Castedo, D. Dominguez, M. Pereira, J. Saa and R. Suau, Heterocycles, 1981, 16, 533.
409. P.N. Sharma, A. Shoeb, R.S. Kapil and S.P. Popli, Phytochemistry, 1981, 20, 2781.
410. P.N. Sharma, A. Shoeb, R.S. Kapil and S.P. Popli, Phytochemistry, 1982, 21, 252.
411. G.B. Lockwood, Phytochemistry, 1981, 20, 1463.
412. N. Takao, M. Kamigauchi, M. Sugiura, I. Ninomiya, O. Miyata and T. Naito, Heterocycles, 1981, 16, 221.
413. Y. Harigaya, K. Yotsumoto, S. Takamatsu, H. Yamaguchi and M. Onda, Chem. Pharm. Bull., 1981, 29, 2557.
414. D. Walterova, J. Ulrichova, V. Preininger, V. Simanek, J. Lenfeld and J. Laskovsky, J. Med. Chem., 1981, 24, 1100.
415. M. Roesner, F.-L. Hsu and A. Brossi, J. Org. Chem., 1981, 46, 3686.
416. M.A. Iorio, M. Molinari and A. Brossi, Can. J. Chem., 1981, 59, 283.
417. R. Hunter, J.J. Bonet and A. Blade-Font, Afinidad, 1981, 38, 120.
418. R. Hunter, J.J. Bonet and A. Blade-Font, Afinidad, 1981, 38, 122.
419. D.A. Evans, S.P. Tanis and D.J. Hart, J. Amer. Chem. Soc., 1981, 103, 5813.
420. K.V. Kovatsis, M.N. Christianopoulou and V. Papageorgiou, Int. Congr. Assoc. Poison Control Cent. 9th., 1980, 220.

421	R. Pontikis, J.M. Schermann, N.H. Nan, L. Bondet and L. Pichat, J. Immunoassay, 1980, 1, 449.
422	D.K. Wyatt, L.T. Grady and S. Sun, Anal. Profiles Drug Subst., 1981, 10, 139.
423	E.L. Bennett, M.H. Alberti and J.F. Flood, Pharmacol. Biochem. Behav., 1981, 14, 863.
424	S.A. Rudolph and S.E. Malawista, Janssen Res Found. Ser., 1980, 3, 481.
425	D. Milosevic, I. Sani, N. Stanojevic-Bakic and I. Spuzic, Experientia, 1981, 37, 519.
426	M.O. Aliev, Izv. Akad. Nauk Az. SSR, Ser. Biol. Nauk, 1981, No. 6, p.31.
427	R.A.B. Keates and G.B. Mason, Can. J. Biochem., 1981, 59, 361.
428	B. Grinde and P.O. Seglen, Hoppe-Seyler's Z. Physiol. Chem., 1981, 362, 549.
429	V.V. Il'chukov, S.M. Shchiparev and I.V. Kuznetsova, Vestn. Leningr. Univ., Biol., 1981, 103.
430	Z.-W. Wang and P.-Q. Zhou, Chung-kuo Yao Li Hsueh Pao, 1981, 2, 135.
431	M. Kurokawa and Y. Komiya, Integr. Control Funct. Brain, 1980, 3, 44.
432	K. Fukutani, H. Ishida, M. Shinohara, S. Minowada, T. Niijima, K. Hijikata and Y. Izawa, Fertil. Steril., 1981, 36, 76.
433	I.G. Morgan, Neurosci. Lett., 1981, 24, 255.
434	J.F. Flood, D.W. Landry, E.L. Bennett and M.E. Jarvik, Pharmacol. Biochem. Behav., 1981, 15, 289.
435	L.D. Russell, J.P. Malone and D.S. MacCurdy, Tissue Cell, 1981, 13, 349.
436	T. Suzuki, T. Teiichi and N. Ishiwara, Igaku no Ayumi, 1981, 118, 23.
437	I.S. Valitov, E.M. Volkov, G.I. Politaev and Kh. S. Khamitov, Byull. Eksp. Biol. Med., 1981, 91, 670.
438	S.C. Selden, P.S. Rabinovitch and S.M. Schwartz, J. Cell. Physiol., 1981, 108, 195.
439	J.S. Ishay, T.B. Shimoy, O.S. Schecter and M.B. Brown, Toxicology, 1981, 21, 129.
440	H. Yamamoto, M. Ishida and T. Tomimori, Shoyakugaku Zasshi, 1981, 35, 15.
441	K. Masden, Arch. Biochem. Biophys., 1981, 211, 368.
442	M. Mourelle, M. Rojikind and B. Rubalcava, Toxicology, 1981, 21, 213.
443	Y. Miyachi, K. Danno and S. Imamura, Br. J. Dermatol., 1981, 105, 279.
444	R.W. Trottier and T.J. Fitzgerald, Drug Dev. Res., 1981, 1, 241.
445	R.E. Davis and S. Benloucif, Neurotoxicology (Park Forest South, Ill.), 1981, 2, 419.
446	M.A. Lea, Cancer Lett. (Shannon, Eire), 1981, 14, 317.
447	M.S. Tanner, D. Jackson and A.P. Mowat, J. Pathol., 1981, 135, 179.
448	J.N. Fordham, J. Kirwan, J. Cason and H.L.F. Currey, Ann. Rheum. Dis., 1981, 40, 605.
449	R.N. Williams and P. Bhattacherjee, Eur. J. Pharmacol., 1982, 77, 17.
450	Y. Komiya, Shinkei Kagaku, 1981, 20, 248.
451	T. Nomura and J.E. Plager, Cancer Treat. Rep., 1981, 65, 283.

9
Aporphinoid Alkaloids

BY M. SHAMMA AND H. GUINAUDEAU

1 Proaporphines

The new proaporphine (-)-hexahydromecambrine-B (1), stereoisomeric at C-10 with the known (-)-N-methyllitsericine, has been found in Papaver lecoquii.[1] Isooridine (2) is present in Papaver oreophilum.[2,3]

Known proaporphines that have been reisolated from plants are:

(-)-Mecambrine:	Ref.	Oridine:	Ref.
Papaver lecoquii	1	Papaver oreophilum	7
P. albiflorum ssp. austromoravicum	1	Pronuciferine:	
P. litwinowii	4,5		
P. spicatum ssp. spicatum	6	Xylopia buxifolia	8
P. spicatum ssp. luschanii	6	Berberis hakeoides	9
P. strictum	6	Stepharine:	
P. oreophilum	7		
		Annona muricata	10
		Cocculus laurifolius	11

The tetrahydrobenzylisoquinolines (±)-norcoclaurine, (±)-coclaurine, and (±)-N-methylcoclaurine can act as precursors to the proaporphine N-methylcrotsparine. The latter may in turn undergo dienone-phenol rearrangement to aporphines of the nuciferine type.[12] There is always a clear preference for aryl over alkyl migration in the transformation of a proaporphine into an aporphine.[13]

2 Aporphines

A study of two Xylopia species has led to the characterization of five new aporphines. Norisodomesticine (3), xyloguyelline (4), and danguyelline (5) were obtained from X. danguyella, and norstephalagine (6) and buxifoline (7) were found in X. buxifolia.[8] The structural assignments for xyloguyelline and danguyelline are presently only tentative.

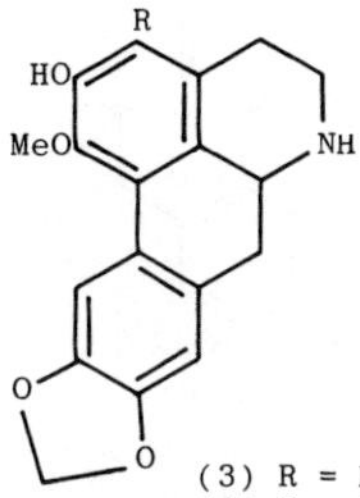

(3) R = H
(4) R = OMe

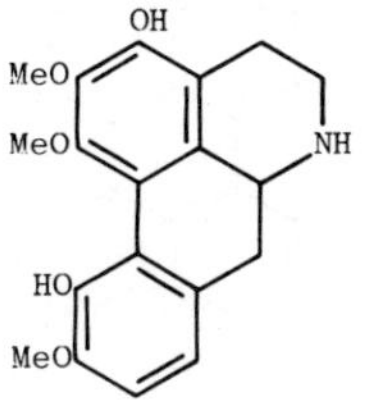

(5)

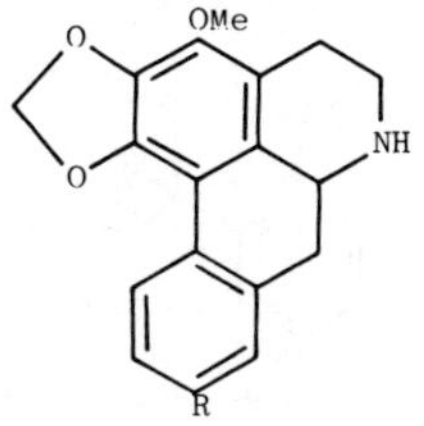

(6) R = H
(7) R = OMe

N,O,O-Trimethylsparsiflorine (8) has been isolated from Thalictrum foliolosum[14] and Laurelia philippiana has furnished 4-hydroxyanonaine (9).[15a,b]

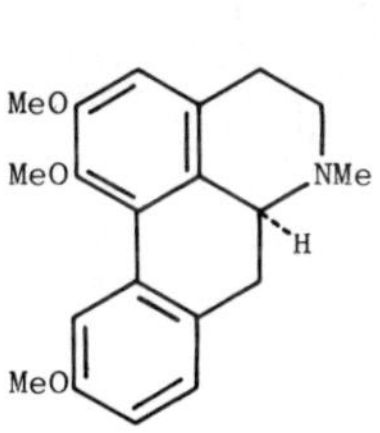

(8)

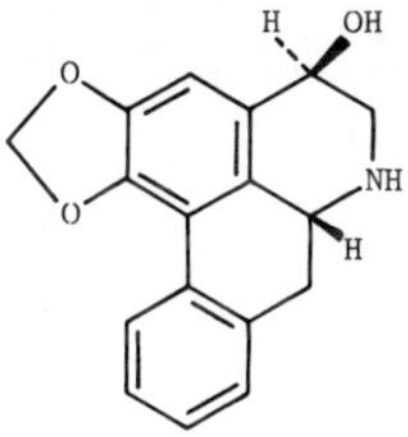

(9)

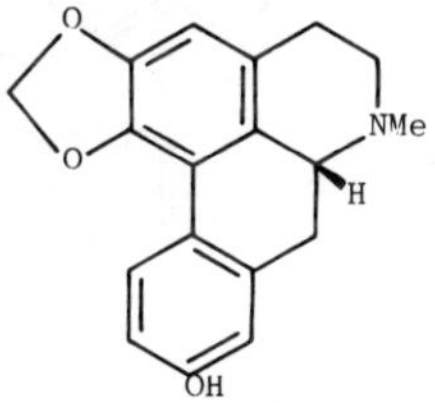

(10)

In connection with a continuing investigation of the alkaloidal content of Formosan Stephania sasakii, (-)-roemeroline (10) and (-)-4-hydroxycrebanine (11) have been isolated and characterized.[16]

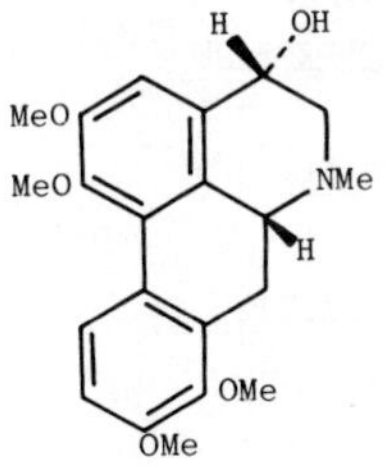

(11)

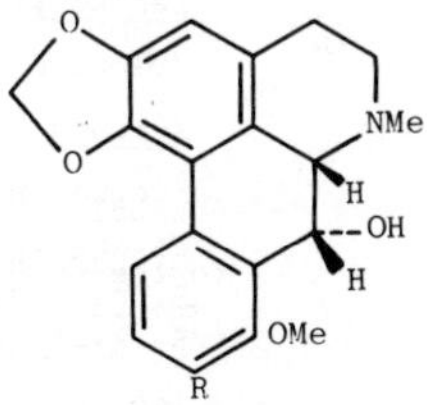

(12) R = H
(13) R = OMe

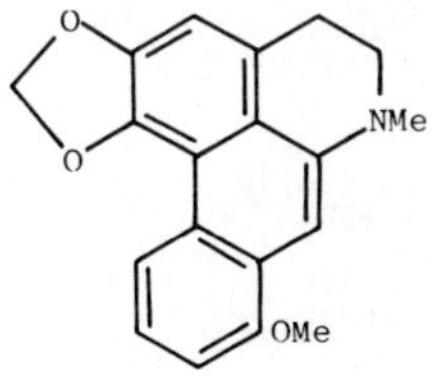

(14)

Two new 7-hydroxylated aporphines, (-)-ayuthianine (12) and (-)-sukhodianine (13), have been located in Stephania venosa.[17] Dehydrostephanine (14) has been obtained from Stephania cepharantha.[18]

Studies of a variety of Annonaceae species have furnished some most unusual aporphines. The two 7-dimethylated aporphines melosmine (15) and melosmidine

(16) have been isolated from Guatteria melosma,[19] while the 6a-methylated aporphines (+)-guattescine (17) and (-)-guattescidine (18) occur in G. scandens.[20]

(15) R = H
(16) R = Me

(17)

(18)

Two other interesting aporphines, duguecalyne (19) and duguenaine (20), which incorporate an extra oxazine ring, have been discovered in Duguetia calycina.[21]

(19)

(20)

Aporphines that have recently been reisolated, together with their natural sources, are listed below.

Asimilobine:	Ref.
Laurelia philippiana	15a,b
N-Methylasimilobine:	
Xylopia buxifolia	8
Stephania cepharantha	18
Nornuciferine:	
Xylopia buxifolia	8
Nuciferine:	
Papaver oreophilum	3
Anonaine:	
Xylopia buxifolia	8
Laurelia philippiana	15b
Roemerine:	
Papaver strictum	6
P. spicatum ssp. luschanii	6
P. spicatum ssp. spicatum	6
P. albiflorum ssp. albiflorum	1
P. albiflorum ssp. austromoravicum	1
P. litwinowii	5

Roemrefidine:	Ref.
Papaver litwinowii	4,5
N-Acetylanonaine:	
Zanthoxylum bungeanum	22
Xylopine:	
Xylopia buxifolia	8
Stephanine:	
Stephania micrantha	23
S. cepharantha	18
S. kwansiensis	24
Isothebaine:	
Papaver orientale	25
P. pseudo-orientale	25
Obovanine:	
Laurelia novae-zelandiae	15a,b
Crebanine:	
Stephania cepharantha	18

Stesakine:	Ref.	Nornantenine:	Ref.
Stephania cepharantha	18	Xylopia danguyella	8
Isoboldine:		Laurelia philippiana	15b
		Hernandia cordigera	32
Guatteria melosma	19	Nantenine:	
Xylopia danguyella	8		
Corydalis marschalliana	26	Corydalis bulbosa	27
C. bulbosa	27	C. slivenensis	28
C. slivenensis	28	C. marschalliana	26
Fumaria vaillantii	29		
Neolitsea fuscata	30	Actinodaphnine:	
Thalictrum foetidum	31	Hernandia cordigera	32
Hernandia cordigera	32		
Thaliporphine:		N-Methylactinodaphnine:	
Thalictrum foetidum	31	H. cordigera	32
Corydalis bulbosa	27	Neolitsine:	
Mahonia repens	33	H. cordigera	32
(-)-Domesticine:		Corytuberine:	
Corydalis bulbosa	27	Papaver lecoquii	1
C. slivenensis	28	P. albiflorum ssp. austromoravicum	1
C. marschalliana	26	P. albiflorum ssp. albiflorum	1
C. suaveolens	34	P. litwinowii	5
		Corydalis suaveolens	34
Laurolitsine:		Magnoflorine:	
Monimia rotundifolia	35		
Boldine:		Stephania elegans	39
		Cyclea barbata	40
M. rotundifolia	35	Thalictrum minus var. microphyllum	41
Predicentrine:		T. foetidum	31
		Heptacyclum zenkeri	42
Corydalis bulbosa	27	Mahonia repens	33
C. slivenensis	28	M. aquifolium	43
Laurotetanine:		Berberis koreana	44
		Papaver oreophilum	3
Xylopia danguyella	8	P. albiflorum ssp. austromoravicum	1
Monimia rotundifolia	35	P. lecoquii	1
Hernandia cordigera	32	Norcorydine:	
N-Methyllaurotetanine:		Xylopia danguyella	8
Monimia rotundifolia	35	Laurelia philippiana	15a,b
Hernandia cordigera	32	Corydine:	
Papaver strictum	6		
Norglaucine:		Aconitum leucostomum	45
		Xylopia danguyella	8
Monimia rotundifolia	35	Corydalis bulbosa	27
Glaucine:		C. slivenensis	28
		C. marschalliana	26
Aconitum yezoense	36	Glaucium flavum	46
Thalictrum foetidum	31	G. leiocarpum	38
Corydalis bulbosa	27	Papaver lecoquii	1
C. ambigua	37	P. albiflorum ssp. austromoravicum	1
Papaver strictum	6	P. litwinowii	5
P. spicatum ssp. luschanii	6	P. oreophilum	3
P. spicatum ssp. spicatum	6	Mahonia repens	33
Glaucium leiocarpum	38	Dicranostigma leptopodium	47
Mahonia repens	33	Guatteria moralessi	48
		G. cubensis	48

Norisocorydine:	Ref.	Bulbocapnine:	Ref.
Hernandia cordigera	32	Corydalis bulbosa	27
Xylopia danguyella	8	C. ledebouriana	49
		C. slivenensis	28
(−)-Isocorydine:		C. marschalliana	26
Corydalis slivenensis	28	Dehydroroemerine:	
Isocorydine:		Stephania micrantha	23
		Papaver spicatum ssp. luschanii	6
C. ledebouriana	49	P. spicatum ssp. spicatum	6
Glaucium flavum	46		
G. leiocarpum	38	Dehydrostephanine:	
Papaver lecoquii	1	Stephania micrantha	23
P. albiflorum ssp. austromoravicum	1		
P. litwinowii	5	Dehydronantenine:	
P. oreophilum	3	Corydalis bulbosa	27
Stephania micrantha	23	C. slivenensis	28
Mahonia repens	33	C. marschalliana	26
Dicranostigma leptopodium	47		
		Dehydrocrebanine:	
Menisperine:		Stephania cepharantha	18
Magnolia grandiflora	50	Ushinsunine:	
Papaver oreophilum	3	S. venosa	17

Isocorytuberine (21) has been prepared via the benzylisoquinoline (22).[51]

(21)   (22)   (23)

Treatment of benzylisoquinoline (23) with lead tetraacetate furnished quinol acetate (24), which was cyclized with trifluoroacetic acid to isothebaine (25). The three related aporphines (26), (27), and (28) were synthesized by a similar pathway.[52]

(24)

(25) R^1= OMe, R^2= R^3= H, R^4= OH
(26) R^1= R^3= H, R^2= OMe, R^4= OH
(27) R^1= R^2= H, R^3= OMe, R^4= OH
(28) R^1= R^3= R^4= H, R^2= OMe

Oxidation of reticuline (29) with m-chloroperoxybenzoic acid in methanol containing cuprous chloride afforded the 1,2,10,11-oxygenated aporphine corytuberine (30).[53] Alternatively, treatment of reticuline with lead tetraacetate in the presence of trichloroacetic acid led to isoboldine (31) in 14% yield.[54]

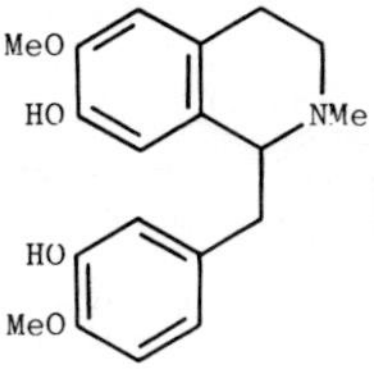

(29) 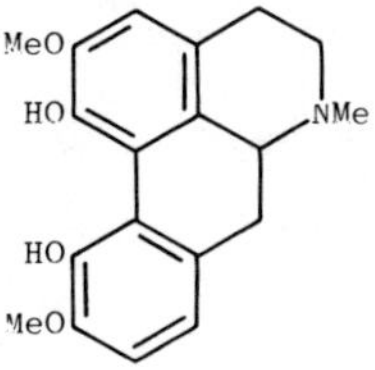   (30) 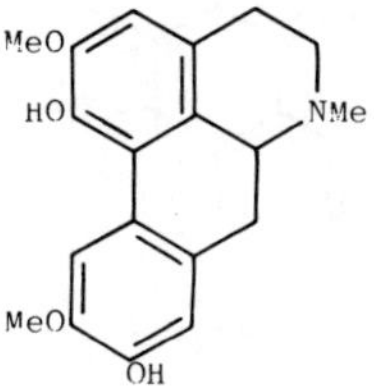   (31)

The 4-hydroxylated aporphine steporphine (32) was prepared as a racemate from acetal (33) and benzylisoquinoline (34). This work represents the first total synthesis of a 4-hydroxylated aporphine.[55]

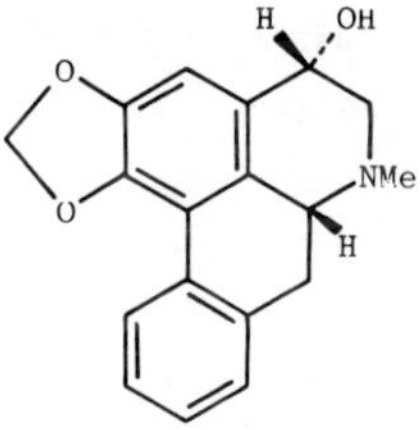

(32) 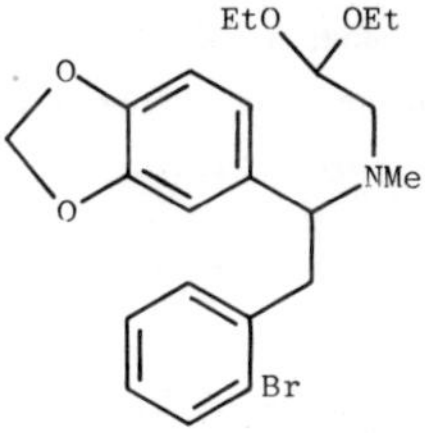   (33) 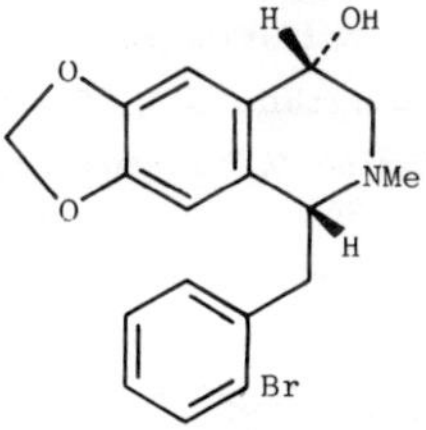   (34)

The synthesis of a variety of aporphinoids, based upon the intermolecular Diels-Alder-type cyclization between benzyne and appropriate 1-methyleneisoquinolines, has been described. Thus, treatment of methyleneisoquinoline (35) with benzyne produced N-acetyldehydronornuciferine (36).[56]

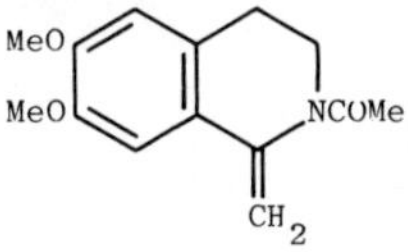

(35) 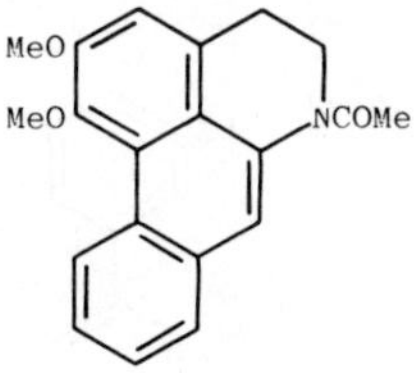   (36)

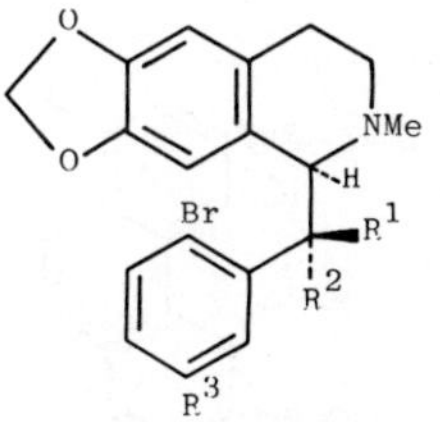

(37) $R^1 = R^3 = H$, $R^2 = OH$
(38) $R^1 = OH$, $R^2 = H$, $R^3 = OMe$

A method for the photooxidation of dehydroaporphines in benzene solution into didehydroaporphines has appeared.[57] Photoirradiation of the α-hydroxylated tetrahydrobenzylisoquinolines (37) and (38) in cyclohexane solution supplied oliveroline (39) and oliveridine (40), respectively.[58]

A report has appeared stating that reaction of glaucine (41) and nuciferine (42) with dichlorocarbene leads to phenanthrenes (43) and (44), respectively.[59]

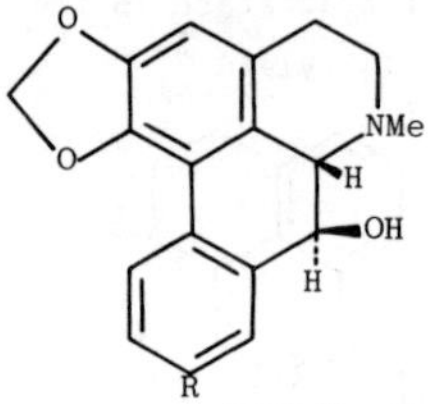

(39) R = H
(40) R = OMe

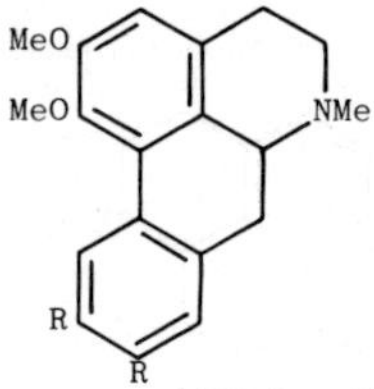

(41) R = OMe
(42) R = H

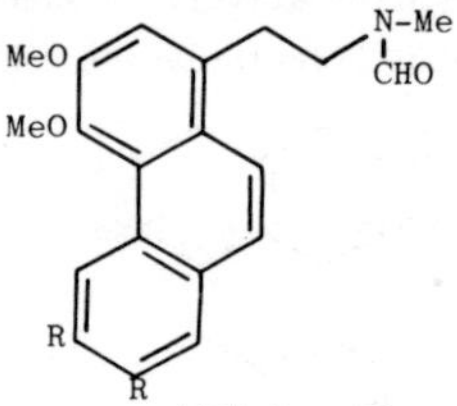

(43) R = OMe
(44) R = H

Dehydroglaucine (45) may be readily alkylated or acylated at C-7, and 7-benzoyldehydroglaucine has been converted into 7-benzylideneglaucine. Reduction of 7-formyldehydroglaucine with lithium aluminum hydride and aluminum trichloride leads to 7-methyldehydroglaucine. Alternatively, when dehydroglaucine was heated with formaldehyde, the dimeric species 7,7'-bisdehydroglaucinemethane was isolated.[60]

Aerial oxidation of glaucine leads to dehydroglaucine (45) and 1,2,9,10-tetramethoxyoxoaporphine (46). On the other hand, photo irradiation gave rise to dihydropontevedrine (47), pontevedrine (48), glaucine N-oxide, and corunnine (49).[61]

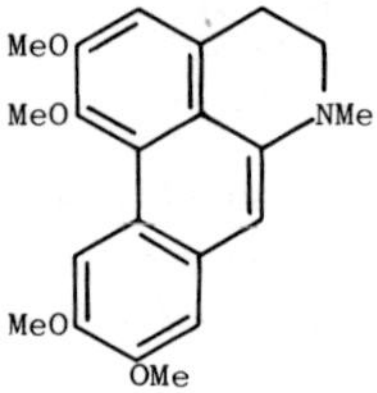

(45) 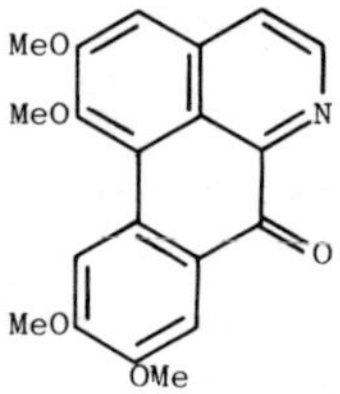 (46) 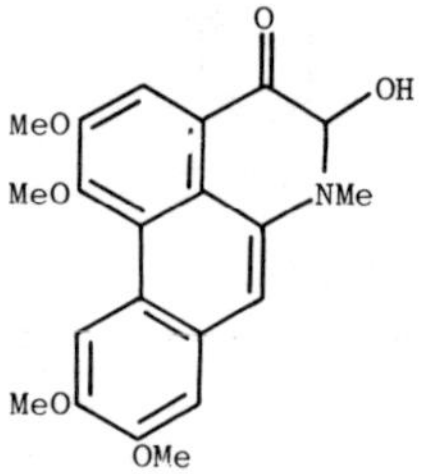 (47)

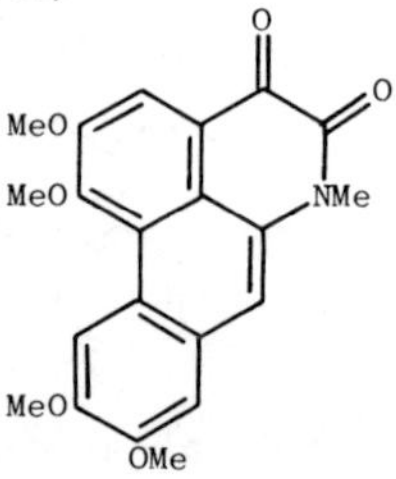

(48) 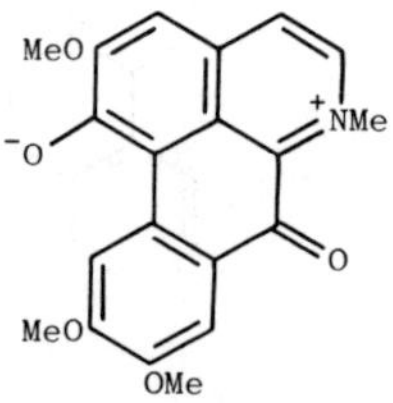 (49)

A potentially useful new synthetic route to the aporphines involves reduction of a properly substituted nitrotetrahydrobenzylisoquinoline with zinc in a mixture of trifluoroacetic acid and trifluoromethanesulfonic acid. In this fashion, (50) was cyclized to the aporphine (51), which in turn was converted into N-methyllaurotetanine. 1,2,10-Trimethoxynoraporphine and the corresponding aporphine were similarly prepared.[62]

(50)

(51)

A comparative ^{13}C n.m.r. study of various phenolic aporphines in neutral and in basic solutions has been carried out.[63]

A practical method has appeared for the preparation of (-)-apomorphine and (-)-N-n-propylnorapomorphine from thebaine which involves the use of a phenyl tetrazolyl ether. The method is also applicable to the conversion of the aporphine (+)-bulbocapnine into (+)-apomorphine.[64] The dopamine receptor interactions of a series of trihydroxyaporphines have been evaluated,[65] and the role of apomorphine and related aporphines as probes of the dopamine receptor has been reviewed.[66] The apomorphine-induced stereotypical cage climbing in mice has been investigated.[67] The general topic of dopaminergic agonists has been discussed in a review.[68]

(+)-N-n-Butyllaurotetanine has shown antiarrhythmic properties.[69] In a study of the interaction between horse liver alcohol dehydrogenase and aporphine alkaloids, it was concluded that the aporphines bind to the active center of the enzyme in a 1:1 ratio.[70]

3 Dimeric Aporphines

(+)-Istanbulamine (52), the first aporphine-benzylisoquinoline dimer formed from a reticuline-type unit linked to an N-methylcoclaurine moiety, has been found in Turkish Thalictrum minus var. microphyllum. The same plant also contains the new dimers (+)-bursanine (53) and (+)-iznikine (54), which belong to the thalicarpine and fetidine series, respectively;[71] as well as the known (+)-adiantifoline.[41,71]

The known dimers thalicarpine and pakistanine have been obtained from Thalictrum foliolosum[14] and Berberis empetrifolia,[72,73] respectively. Thalictrum dimers incorporate at least one reticuline-type moiety, whereas Berberis dimers consist of two N-methylcoclaurine-type units.

(52)

(53) R^1 = H, R^2 = OH
(54) R^1 = OH, R^2 = H

4 Oxoaporphines

The new oxoaporphines oxoanolobine (55a) and peruvianine (55b) have been isolated from <u>Guatteria melosma</u> [74] and <u>Telitoxicum peruvianum</u>,[75] respectively. <u>Stephania venosa</u> has yielded the first two non-phenolic N-methyloxoaporphinium salts known, namely uthongine (56a) and thailandine (56b). These salts underwent facile N-demethylation on a tlc plate to the corresponding oxoaporphine free bases. This result indicates that some of the natural oxoaporphines characterized in the past may actually exist in the plant as the corresponding N-metho salts.[76]

(55a)   (55b)   (56) a; R = OMe
                       b; R = H

Known oxoaporphines that have been recently reisolated, and their natural sources, are listed below.

Liriodenine:	Ref.	Atheroline:	Ref.
Annona acuminata	77	Monimia rotundifolia	35
A. cristalensis	78	Oxonantenine:	
Guatteria cubensis	48		
Xylopia buxifolia	8	Corydalis bulbosa	27
Siparuna gilgiana	79	C. marschalliana	28
Lysicamine:		Siparuna gilgiana	79
Annona acuminata		Nandazurine:	
Oxoputerine:		Corydalis bulbosa	27
Laurelia novae-zelandiae	15a,b	Corunnine:	
Homomoschatoline:		Thalictrum foetidum	31
Annona acuminata	77	Thalicminine:	
Lanuginosine:		Thalictrum dioicum	80
Xylopia buxifolia	8		

Photocyclization of a series of substituted 2,3-diaryl-Δ^2-pyrroline-4,5-diones of type (57) afforded the corresponding phenanthrenes (58).[81] Oxoglaucine, lysicamine, and dicentrinone were then prepared by Fremy's salt oxidation of the photocyclized products.[81] Oxidation of 1,2,9,10-tetramethoxydidehydroaporphine, again using Fremy's salt, has led to oxoglaucine and corunnine.[55] Similarly, Fremy's salt oxidation of norglaucine leads to oxoglaucine.[82]

$R^1 = R^2 = $ OMe
$R^1 = $ OMe, $R^2R^2 = $ OCH$_2$O
$R^1 = $ H, $R^2 = $ OMe

(57)   (58)

5 4,5-Dioxoaporphines

Aristolochia tuberosa has yielded two new 4,5-dioxoaporphines, tuberosinone (59) and tuberosinone $\underline{N}$-β-D-glucoside (60).[83] Condensation of benzyne with the masked diene (61) generated norcepharadione-B (62) in 40% yield;[56] aporphines can also be prepared by a related approach (see Section 2 above).

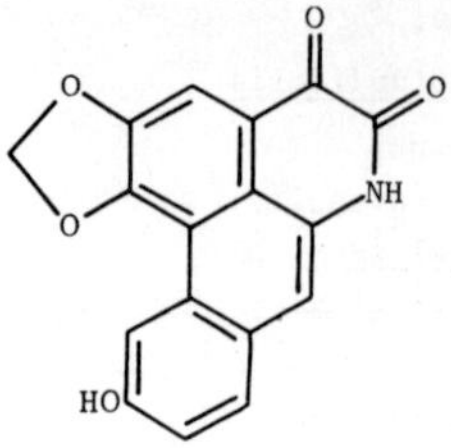

(59)

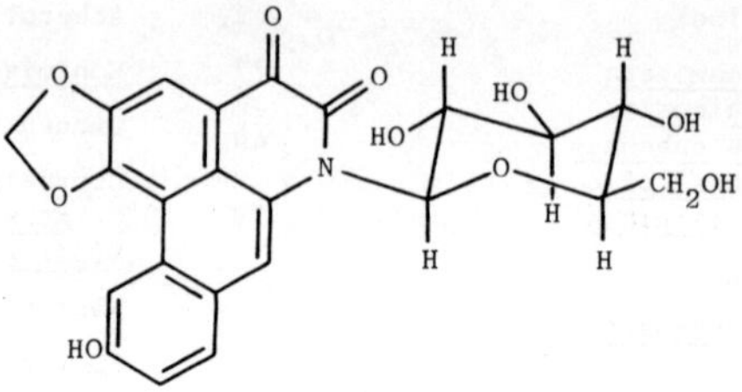

(60)

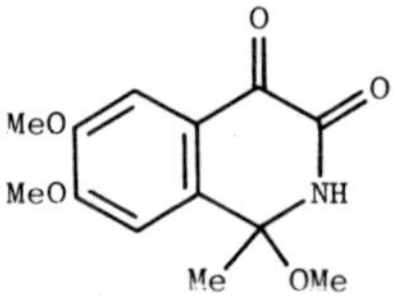

(61)

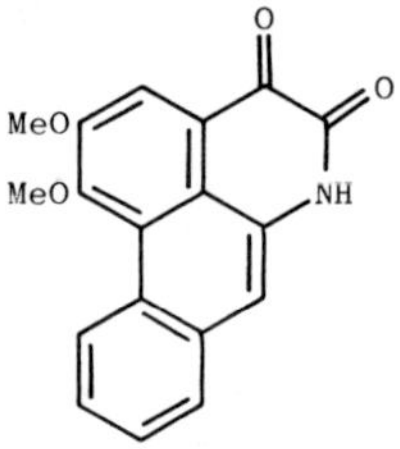

(62)

6 Phenanthrenes

The known atherosperminine and thaliglucinone have been reisolated from Annona muricata and Thalictrum minus var. microphyllum, respectively.[10,41]

7 Aristolochic Acids and Aristolactams

Three new aristolochic acids of special interest have been isolated. Two are substituted at C-7 and -8, instead of at the usual C-6 and C-8 sites; these are 7-hydroxyaristolochic acid-A (63) and 7-methoxyaristolochic acid-A (64). Both occur in Aristolochia debilis.[84] The third compound is the O-glycoside aristoloside (65), produced by A. manshuriensis.[85] Additionally, aristolochic acid-IV methyl ester (66) has been found in A. kwangsiensis,[86,87] and aristolactam-C N-β-D-glucoside (67) has been isolated from A. indica.[88]

Hofmann elimination of the N-metho salts of N-methyllaurotetanine and of boldine, followed by nitration, afforded the non-natural aristolochic acids (68) and (69). Catalytic reduction and cyclization then led to the corresponding aristolactams.[89]

Perkin condensation of o-iodophenylacetic acid (70) with m-hemipinic anhydride (71), followed by treatment with ammonium acetate and acetic acid, gave rise to the enelactam (72), whose photolysis provided cepharanone-B (73).[90]

An alternate synthesis of an aristolactam involves condensation of benzyne with the enelactam (74). The resulting aristolactam (75) was isolated in 22% yield.[56]

Aerial oxidation of the dehydroaporphine (76) in the presence of an alkali catalyst produced a mixture, one component of which was the non-natural aristolactam (77).[91]

(76)   (77)

References

1 J. Slavík, L. Slavíková, and L. Dolejš, Collect. Czech. Chem. Commun., 1981, 46, 2587.
2 F. Věžník, P. Sedmera, V. Preininger, V. Šimánek, and J. Slavík, Phytochemistry, 1981, 20, 347.
3 F. Věžník, E. Taborska, and J. Slavík, Collect. Czech. Chem. Commun., 1981, 46, 926.
4 J. Slavík, K. Picka, L. Slavíková, E. Táborská, and F. Věžník, Collect. Czech. Chem. Commun., 1980, 45, 914.
5 J. Slavík and L. Slavíková, Collect. Czech. Chem. Commun., 1981, 46, 1534.
6 G. Sariyar and A. Oztekin, Plant. Med. Phytother., 1981, 15, 160.
7 D.A. Murav'eva and V.V. Melik-Guseinov, Farmatsiya (Moscow), 1980, 29, 23.
8 R. Hocquemiller, A. Cavé, and A. Raharisololalao, J. Nat. Prod., 1981, 44, 551.
9 L.H. Villarroel and M.R. Torres, Bol. Soc. Chil. Quim., 1982, 27, 273.
10 M. Lebœuf, C. Legueut, A. Cavé, J.F. Desconclois, P. Forgacs, and H. Jacquemin, Planta Med., 1981, 42, 37.
11 D.S. Bhakuni and S. Jain, Tetrahedron, 1980, 36, 3107.
12 D.S. Bhakuni, S. Jain, and R. Chaturvedi, Tetrahedron, 1979, 35, 2323; D.S. Bhakuni and S. Jain, Tetrahedron, 1981, 37, 3175.
13 B.F. Bowden, R.W. Read, and W.C. Taylor, Aust. J. Chem., 1981, 34, 799.
14 D.S. Bhakuni and R.S. Singh, J. Nat. Prod., 1982, 45, 252.
15 (a) A. Urzúa and B.K. Cassels, Bol. Soc. Chil. Quim., 1982, 27, 165;
   (b) A. Urzúa and B.K. Cassels, Phytochemistry, 1982, 21, 773.
16 J.-I. Kunitomo, M. Oshikata, and Y. Murakami, Chem. Pharm. Bull., 1981, 29, 2251.
17 H. Guinaudeau, M. Shamma, B. Tantisewie, and K. Pharadai, J. Nat. Prod., 1982, 45, 355.
18 J. Kunitomo, M. Oshikata, and M. Akasu, Yakugaku Zasshi, 1981, 101, 951. See also Z.D. Min and S.M. Zhong, Yao Hsueh Hsueh Pao, 1980, 15, 532.

19 V. Zabel, W.H. Watson, C.H. Phoebe, J.E. Knapp, P.L. Schiff, Jr., and D.J. Slatkin, J. Nat. Prod., 1982, 45, 94.
20 R. Hocquemiller, S. Rasamizafy, and A. Cavé, Tetrahedron, 1982, 38, 911.
21 F. Roblot, R. Hocquemiller, and A. Cavé, C.R. Seances Acad. Sci., Ser. 2, 1981, 293, 373.
22 L. Ren and F. Xie, Yaoxue Xuebao, 1981, 16, 672; Chem. Abstr., 1982, 96, 48974e.
23 Z. Min, X. Liu, and W. Sun, Yaoxue Xuebao, 1981, 16, 557; Chem. Abstr., 1982, 97, 3595m.
24 K.-J. Cheng, K.-C. Wang, and Y.-H. Wang, Yao Hsueh T'ung Pao, 1981, 16, 49; Chem. Abstr., 1981, 95, 138460t.
25 J.D. Phillipson, A. Scutt, A. Baytop, N. Ozhatay, and G. Sariyar, Planta Med., 1981, 43, 261.
26 Kh. Kiryakov, E. Iskrenova, B. Kuzmanov, and L. Evstatieva, Planta Med., 1981, 41, 298.
27 Kh. Kiryakov, E. Iskrenova, B. Kuzmanov, and L. Evstatieva, Planta Med., 1981, 43, 51.
28 Kh. Kiryakov, E. Iskrenova, E. Daskalova, B. Kuzmanov, and L. Evstatieva, Planta Med., 1982, 44, 168.
29 M. Alimova and I.A. Israilov, Khim. Prir. Soedin., 1981, 602; Chem. Natural Compd. (Engl. Transl.), 1981, 17, 437.
30 A.A.L. Gunatilaka, S. Sotheeswaran, S. Sriyani, and S. Balasubramaniam, Planta Med., 1981, 43, 309.
31 S. Mukhamedova, S. Kh. Maekh, and S. Yu. Yunusov, Khim. Prir. Soedin., 1981, 251.
32 M. Lavault, M.M. Debray, and J. Bruneton, Bull. Mus. Natl. Hist. Nat., Sect. B, 1980, 2, 387.
33 R.T. Suess and F.R. Stermitz, J. Nat. Prod., 1981, 44, 680.
34 W.F. Xin and M. Lin, Chung Ts'ao Yao, 1981, 12, 1; Chem. Abstr., 1981, 95, 209450c.
35 P. Forgacs, G. Buffard, J.F. Desconclois, A. Jehanno, J. Provost, R. Tiberghien, and A. Touche, Plant. Med. Phytother., 1981, 15, 80.
36 H. Takayama, A. Tokita, M. Ito, S. Sakai, F. Kurosaki, and T. Okamoto, Yakugaku Zasshi, 1982, 102, 245.
37 D.Y. Zhu, C.-Q. Song, Y.-L. Gao, and R.-S. Xu, Hua Hsueh Hsueh Pao, 1981, 39, 280; Chem. Abstr., 1981, 95, 93849h.
38 T. Gözler and S. Ünlüyol, Doga, Seri C, 1982, 6, 21.
39 R.S. Singh, P. Kumar, and D.S. Bhakuni, J. Nat. Prod., 1981, 44, 664.
40 G. Klughardt and F. Zymalkowski, Arch. Pharm. (Weinheim,Ger.), 1982, 315, 7.
41 K.H.C. Başer, Doga, Seri A, 1981, 5, 163.
42 F.K. Duah, P.D. Owusu, J.E. Knapp, D.J. Slatkin, and P.L. Schiff, Jr., Planta Med., 1981, 42, 275.
43 D. Kostalova, B. Brazdovicova, and J. Tomko, Chem. Zvesti, 1981, 35, 279.
44 D. Kostalova, B. Brazdovicova, and H.-Y. Jin, Farm. Obz., 1982, 51, 213; Chem. Abstr., 1982, 97, 36105e.
45 V.N. Plugar, Ya.V. Rashkes, M.G. Zhamierashvili, V.A. Tel'nov, M.S. Yunusov, and S. Yu. Yunusov, Khim. Prir. Soedin., 1982, 80; Chem. Natural Compd. (Engl. Transl.), in press.
46 T. Gözler and S. Ünlüyol, Doga, Seri C, 1981, 5, 25.
47 H.-J. Chang, H.-H. Wang, and K.-E. Ma, Yao Hsueh T'ung Pao, 1981, 16, 52.
48 M. Diaz, C. Schreiber, and H. Ripperger, Rev. Cub. Farm., 1981, 15, 93.
49 Kh. Sh. Khusainova and Yu. D. Sadykov, Dokl. Akad. Nauk Tadzh. SSR, 1981, 24, 489; Khim. Prir. Soedin., 1981, 670.
50 K.V. Rao and T.L. Davis, J. Nat. Prod., 1982, 45, 283.
51 T.R. Suess and F.R. Stermitz, J. Nat. Prod., 1981, 44, 688.
52 H. Hara, O. Hoshino, T. Ishige, and B. Umezawa, Chem. Pharm. Bull., 1981, 29, 1083.
53 Sendai Heterocyclic Chemical Research Foundation, Jpn. Kokai Tokkyo Koho 80 139 362 (Cl C07D221/18), 31 Oct., Appl. 79/44 293, 13 Apr. 1979.
54 C. Szantay, M. Barczai-Beke, P. Pechy, G. Blaskó, and G. Dörnyei, J. Org. Chem., 1982, 47, 594.

55 J. Kunitomo, M. Oshikata, K. Suwa, K. Nakayama, and Y. Murakami, Heterocycles, 1982, 19, 45.
56 L. Castedo, E. Guitian, J.M. Saá, and R. Suau, Tetrahedron Lett., 1982, 23, 457.
57 L. Castedo, T. Iglesias, A. Puga, J.M. Saá, and R. Suau, Heterocycles, 1982, 19, 245.
58 S.V. Kessar, Y.P. Gupta, and T. Mohammad, Indian J. Chem., Sect. B, 1981, 20, 984.
59 L. Castedo, J.L. Castro, and R. Riguera, Heterocycles, 1982, 19, 209.
60 S. Philipov, O. Petrov, and N. Mollov, Arch. Pharm. (Weinheim, Ger.), 1981, 314, 1034.
61 V. Chervenkova, N. Mollov, and S. Paszyc, Phytochemistry, 1981, 20, 2285.
62 T. Ohta, R. Machida, K. Takeda, Y. Endo, K. Shudo, and T. Okamoto, J. Am. Chem. Soc., 1980, 102, 6385.
63 L. Castedo, R. Riguera, and F.J. Sardina, An. Quim., Ser. C, 1981, 77, 138.
64 V.J. Ram and J.L. Neumeyer, J. Org. Chem., 1981, 46, 2830.
65 J.L. Neumeyer, G.W. Arana, S.J. Law, J.S. Lamont, N.S. Kula, and R.J. Baldessarini, J. Med. Chem., 1981, 24, 1440.
66 J.L. Neumeyer, S.J. Law, and J.S. Lamont in 'Apomorphine and Other Dopaminomimetics', Vol. I, G.L. Gessa and G.U. Corsini, ed., Raven Press, New York, 1981, pp. 209-218.
67 R.V. Smith, A.E. Klein, R.E. Wilcox, and W.H. Riffee, J. Pharm. Sci., 1981, 70, 1144.
68 J.G. Cannon, J. Med. Chem., 1981, 24, 1113.
69 M. Leboeuf, A. Cavé, J. Provost, R. Tiberghien, and P. Forgacs, Ann. Pharm. Fr., 1980, 38, 537.
70 D. Walterová, J. Kovář, V. Preininger, and V. Šimánek, Collect. Czech. Chem. Commun., 1982, 47, 296.
71 H. Guinaudeau, A.J. Freyer, R.D. Minard, and M. Shamma, Tetrahedron Lett., 1982, 23, 2523.
72 V. Fajardo, C. Prats, and M. Garrido, Contrib. Cient. Tecnol., 1981, 11, 61.
73 V. Fajardo, A. Leon, M.C. Loncharic, V. Elango, M. Shamma, and B.K. Cassels, Bol. Soc. Chil. Quim., 1982, 27, 159.
74 C.H. Phoebe, P.L. Schiff, Jr., J.E. Knapp, and D.J. Slatkin, Heterocycles, 1980, 14, 1977.
75 M.D. Menachery and M.P. Cava, J. Nat. Prod., 1981, 44, 320.
76 H. Guinaudeau, M. Shamma, B. Tantisewie, and K. Pharadai, J. Chem. Soc., Chem. Commun., 1981, 1118.
77 I. Borup-Grochtmann and D.G.I. Kingston, J. Nat. Prod., 1982, 45, 102.
78 J. Faust, A. Ripperger, D. Sandoval, and K. Schreiber, Pharmazie, 1981, 36, 713.
79 S.Y.C. Chiu, R.H. Dobberstein, H.H.S. Fong, and N.R. Farnsworth, J. Nat. Prod., 1981, 45, 229.
80 X.A. Dominguez, O.R. Franco, C.G. Cano, S. Garcia, and R.S. Tamez, Rev. Latinoam. Quim., 1981, 12, 61.
81 L. Castedo, C. Saá, J.M. Saá, and R. Suau, J. Org. Chem., 1982, 47, 513.
82 L. Castedo, A. Puga, J.M. Saá, and R. Suau, Tetrahedron Lett., 1981, 22, 2233.
83 D. Zhu, B. Wang, B. Huang, and R. Xu, Y. Qiu, and X. Chen, Heterocycles, 1982, 17, 345.
84 Z.-L. Chen, B.-S. Huang, D.-Y. Zhu, and M.-L. Yin, Hua Hsueh Hsueh Pao, 1981, 39, 237; Chem. Abstr., 1981, 95, 156428t.
85 T. Nakanishi, K. Iwasaki, M. Nasu, I. Miura, and K. Yoneda, Phytochemistry, 1982, 21, 1759.
86 F.-H. Chou, P.-Y. Liang, S.-C. Chu, and C. Wen, Yao Hsueh T'ung Pao, 1981, 16, 56; Chem. Abstr., 1981, 95, 175618w.
87 F. Zhou, P. Liang, C. Qu, and J. Wen, Yaoxue Xuebao, 1981, 16, 638; Chem. Abstr., 1982, 96, 31686q.
88 B. Achari, S. Chakrabarty, and S.C. Pakrashi, Phytochemistry, 1981, 20, 1444.
89 P. Gorecki and H. Otta, Monatsh. Chem., 1981, 112, 1077.
90 L. Castedo, E. Guitián, J.M. Saá, and R. Suau, Heterocycles, 1982, 19, 279.
91 J.I. Kunitomo, Y. Murakami, M. Oshikata, T. Shingu, M. Akasu, S.-T. Lu, and I.-S. Chen, Phytochemistry, 1980, 19, 2735.

10
Amaryllidaceae Alkaloids

BY M. F. GRUNDON

1 Isolation and Structural Studies

The bulbs and roots of <u>Hippeastrum equestre</u> have been shown to contain galanthine, hippeastrine, lycorine, and tazettine[1] and extraction of <u>H. bicolor</u> yielded haemanthamine and lycorine.[2]

A number of new crinine-type alkaloids have been identified this year. Thus, further investigation of <u>H. ananuca</u> (cf. Vol. 10, p.135) resulted in the isolation of the new alkaloid (1) as well as haemanthamine; the structure of compound (1) was deduced mainly

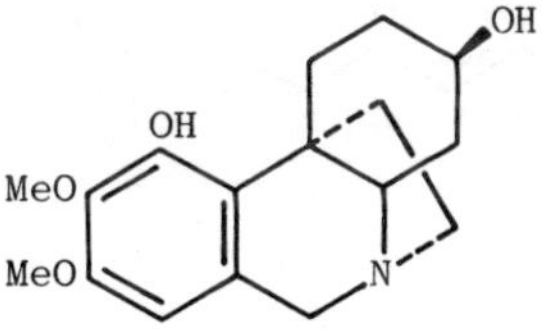

(1)

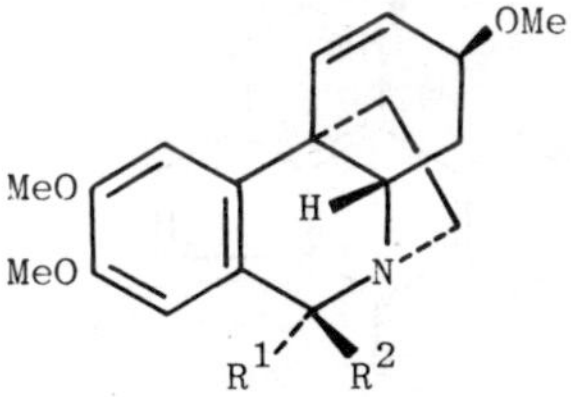

O-Methylmaritidine (2) $R^1 = R^2 = H$

Panyramine $\begin{cases} (3a)\ R^1 = H,\ R^2 = OH \\ (3b)\ R^1 = OH,\ R^2 = H \end{cases}$

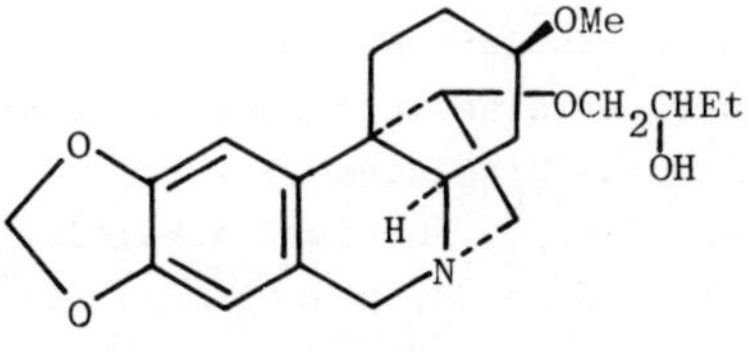

Narcimarkine (4)

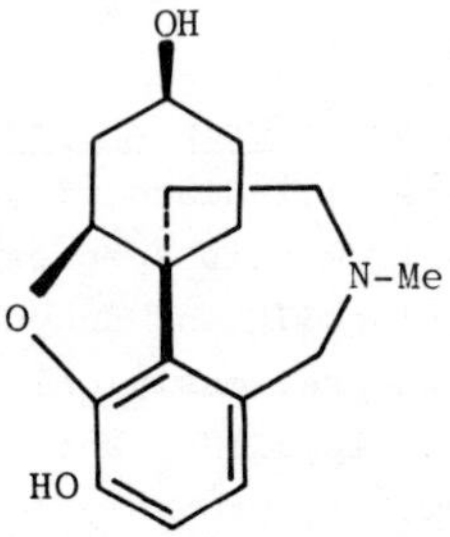

O-Demethyl-lycoramine (5)

from the u.v., i.r., and ^1H n.m.r. spectra.[3] The structure and stereochemistry of <u>O</u>-methylmaritidine (2), which was isolated for the first time from <u>Narcissus tazetta</u>, was also determined by

spectroscopy.[4] The related new alkaloid papyramine (3) occurs in
Narcissus papyraceus with galanthamine, lycoramine, lycorine,
maritidine, pseudolycorine, and tazettine;[5] a variable-temperature
n.m.r. study showed that papyramine exists as a mixture of epimers
(3a) and (3b) at ambient temperature and that the epimer ratio
depends on solvent polarity.[6] Structure (4) for narcimarkine from
Narcissus poeticus was proposed mainly on the basis of its mass
spectrum.[7]

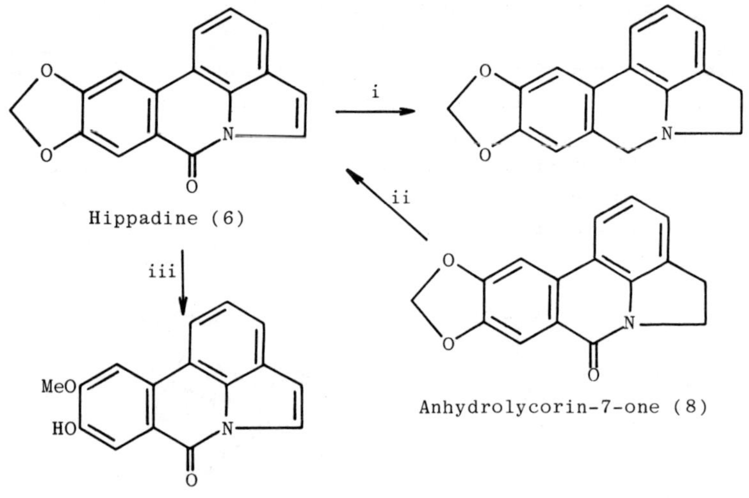

Hippadine (6)

Anhydrolycorin-7-one (8)

Pratorinine (7)

Reagents: i, LiAlH$_4$, Et$_2$O-THF, reflux; ii, 2,3-dichloro-5,6-dicyanobenzoquinone,
PhH, reflux; iii, NaOMe, Me$_2$SO, at 150°C

Scheme 1

A study of **Crinum pratense** led to the isolation of the alkaloids
ambelline, 1,2-diacetyllycorine, and narcissidine, not obtained previously from this species, and the alkaloids hippadine (6),
pratorinine (7), and anhydrolycorin-7-one (8);[8] the last alkaloid
was known previously only as a synthetic compound, pratorinine is
new, and hippadine, which was also obtained from **Hippeastrum
vittatum**,[9] is probably identical with 'N-3' isolated earlier from
Lycoris sanguinea Maxim. (cf. Vol. 5, p.170). The structures of
hippadine and pratorinine were established by spectroscopy and by
interconversions (Scheme 1). Since hippadine is insoluble in the
aqueous acidic medium used for extraction of Amaryllidaceae

alkaloids, it may be more widely distributed and have escaped isolation.[8]

Leucotamine (9) R^1 = H, R^2 = $COCH_2CH(OH)Me$

O-Methyl-
    leucotamine (10) R^1 = Me, R^2 = $COCH_2CH(OH)Me$

(11) R^1 = Me, R^2 = H

Clivisyalin (12)

Three new alkaloids of the galanthamine group have been isolated. The structure of O-demethyllycoramine (5) from <u>Lycoris radiata</u> was apparent from its formation from the reaction of lycoramine with pyridine hydrochloride at 190°C.[10] <u>Leucojum aestivum</u> contains leucotamine (9), which was converted with diazomethane into a second new alkaloid of this species, O-methylleucotamine (10); the structures were confirmed by hydrolysis of O-methylleucotamine to galanthamine (11).[11] A new lactone alkaloid of <u>Clivia miniata</u> was assigned structure (12) as a result of ^{1}H n.m.r. and mass spectral studies and by comparison of the c.d. spectrum with that of related alkaloids.[12]

A full account has now been published of the conversion of tazettine into pretazettine (<u>cf</u>. Vol. 11, p.133).[13]

2 Synthesis

The total synthesis of Amaryllidaceae alkaloids has been reviewed (in Japanese).[14] Interest in this area continues and ingenious syntheses of the lycorine, crinine, and lycoramine ring systems have been reported.

The synthesis of compound (14), which has already been converted into lycorine, is described by Martin and Tu (Scheme 2); a key step is [4 + 2] cycloaddition of dienamide (13)[15] (<u>cf</u>. Stork's synthesis of 7-oxo-α-lycorane; Vol. 11, p.134). The readily accessible nitro derivatives (15) and (17) have been used by Seebach to prepare a possible trans-dihydrolycoricidine precursor (16) and 1-desoxy-2-lycorinone (18), respectively (Scheme 3).[16] A full account of the synthesis of ungminorine from acetyl-lycorine (cf. Vol. 10, p.137) is now available.[17]

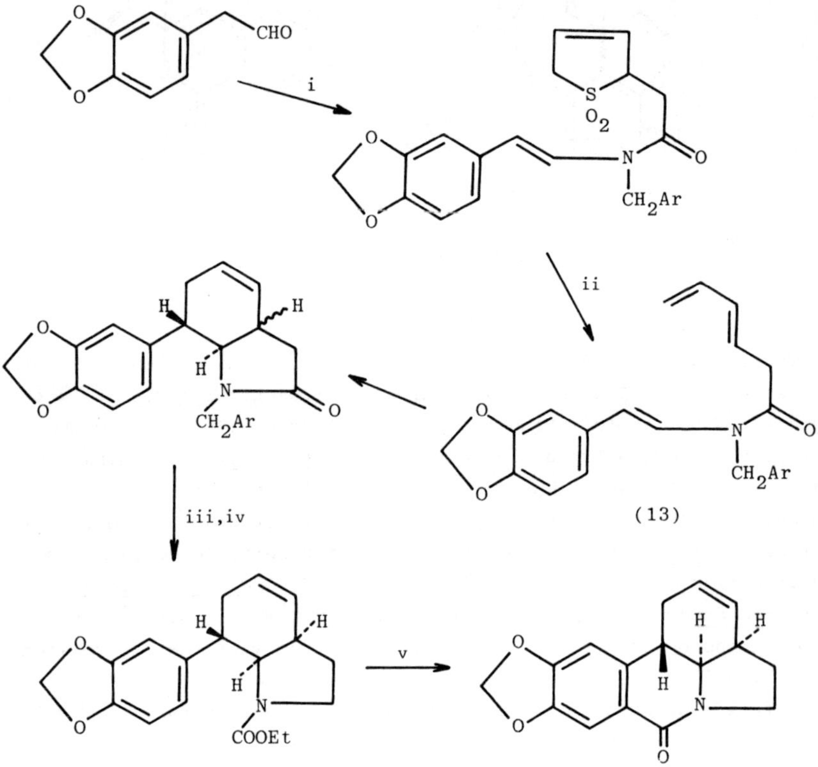

(Ar = p-MeOC$_6$H$_4$)

Reagents: i, p-MeOC$_6$H$_4$CH$_2$NH$_2$, PhMe, MgSO$_4$, at 0°C, then [sulfolene]CH$_2$COCl, PhNEt$_2$;
ii, xylene, heat; iii, LiAlH$_4$, Et$_2$O; iv, ClCOOEt, NaHCO$_3$, PhMe, heat;
v, POCl$_3$, at 90°C

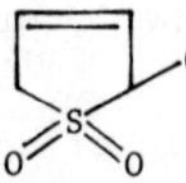

Scheme 2

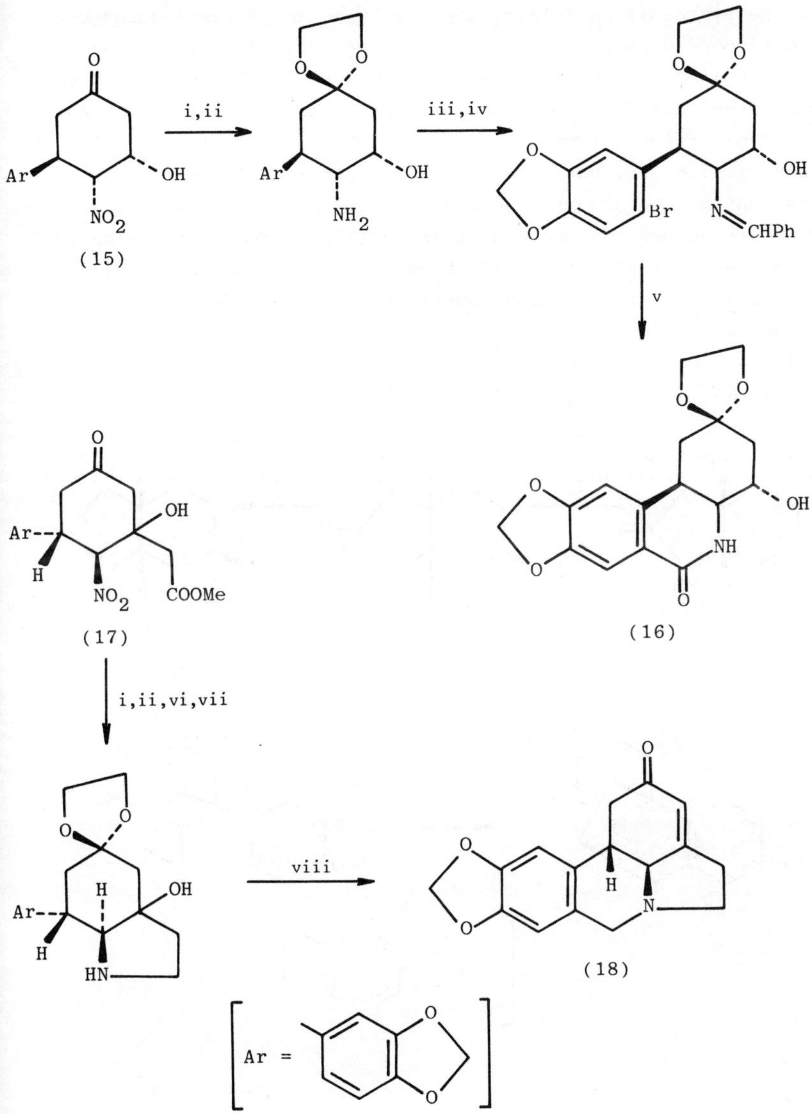

Reagents: i, HOCH$_2$CH$_2$OH, TsOH, PhH, reflux; ii, Raney nickel, H$_2$, EtOH, at 50°C; iii, Br$_2$, aq. (HOOC)$_2$, at 0°C; iv, PhCHO, I$_2$, PhH; v, CO, Pd(OAc)$_2$, PPh$_3$, NEt$_3$, MeOH, reflux; vi, xylene, reflux; vii, LiAlH$_4$, THF, at 50°C; viii, HCl, at 50°C, then 40% formalin, HCl, MeOH, at 50°C

Scheme 3

The synthesis of (±)-crinane involving an intramolecular ene reaction of an acylnitroso-enophile (cf. Vol. 12, p.154) has now been applied to the preparation of compound (19), which was then converted into dihydromaritidine (20) (Scheme 4).[18] A synthesis of dihydromaritidine from the allyl cyclohexanone derivative (21) has been carried out, but details are not yet readily available.[19] A new approach to the crinine ring system involves a 'directed' aza-Cope arrangement of the hydroxyamine (22) (Scheme 5); since the amino-ketone (24) has already been converted into crinine (23), this sequence of reactions constitutes a formal synthesis of the alkaloid.[20]

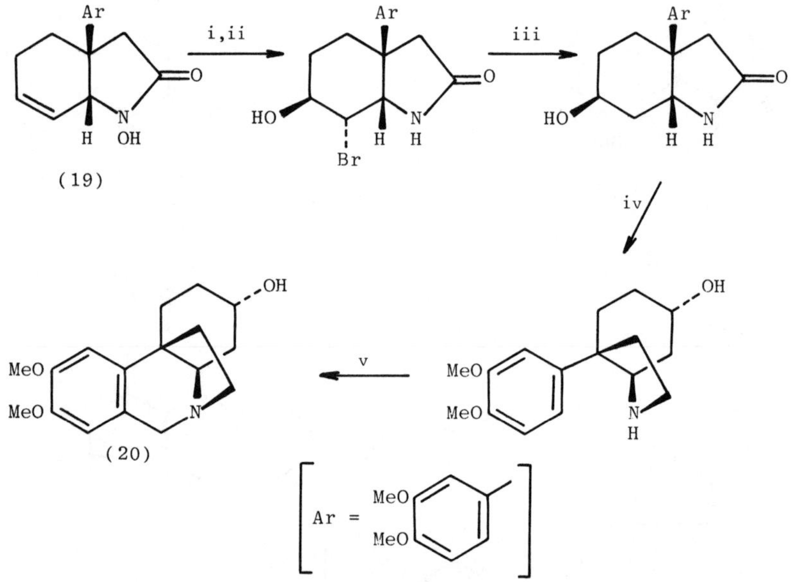

Reagents: i, TiCl$_3$, H$_2$O, MeOH, Na$_2$CO$_3$; ii, N-bromosuccinimide, DME, H$_2$O, at 0°C; iii, AIBN, Bu$^n_3$SnH, PhMe, reflux; iv, LiAlH$_4$, THF, reflux; v, aq. HCHO, conc. HCl or H$_2$C=$\overset{+}{N}$Me$_2$ I$^-$, THF, reflux

Scheme 4

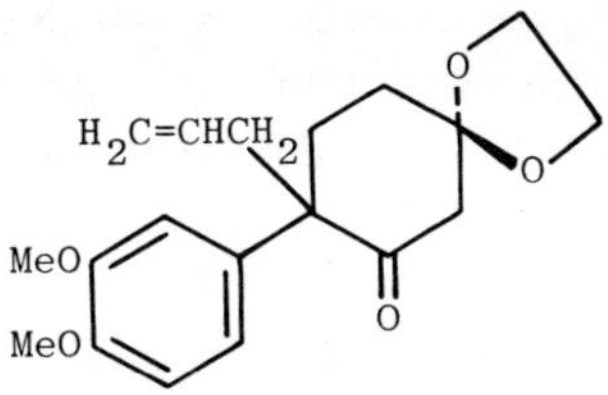

(21)

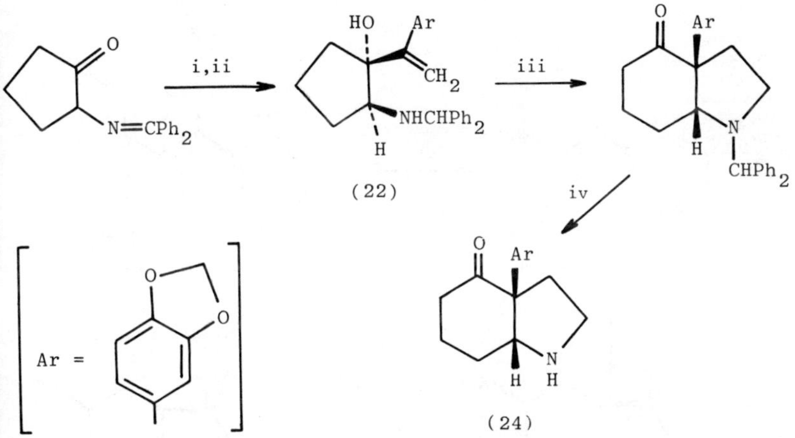

(22)

Ar = [1,3-benzodioxol-5-yl]

(24)

Reagents: i, ArC(Li)=CH$_2$; ii, NaCNBH$_3$; iii, HCHO, TsOH, Me$_2$SO; iv, cyclohexene, Pd/C, HCl, EtOH

Scheme 5

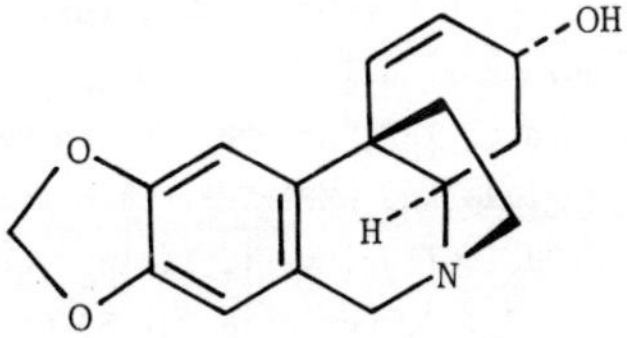

(23)

Martin and Garrison have described a new efficient stereoselective synthesis of dl-lycoramine (26) (cf. Vol. 9, p.141), based on the preparation and subsequent cyclisation of the 4,4-disubstituted cyclohexenone (25) (Scheme 6).[21]

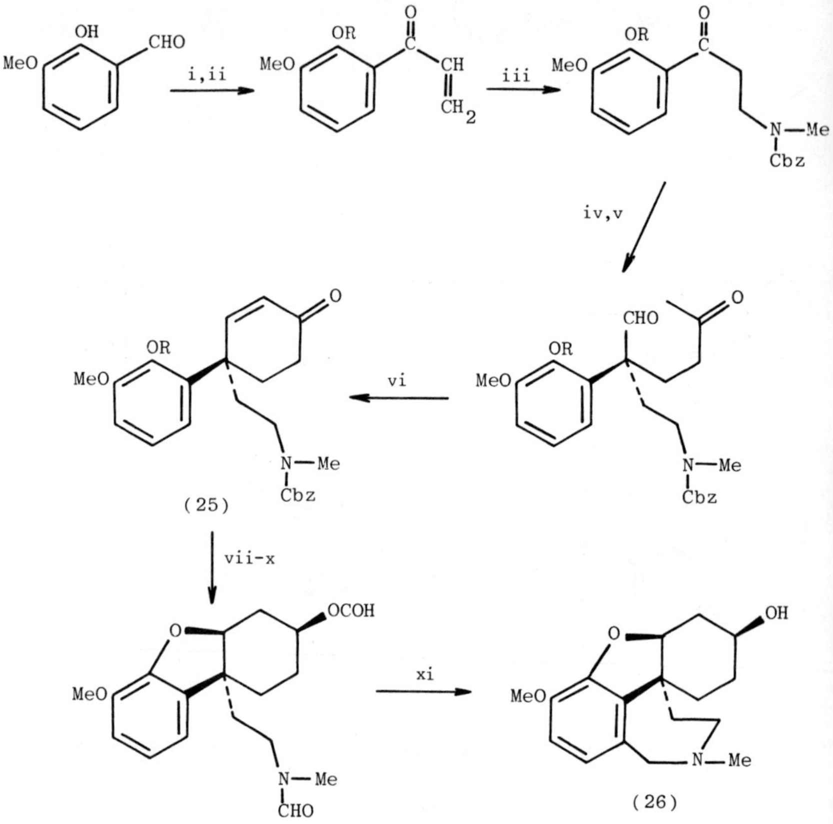

($R = -CH_2CH=CH_2$; Cbz = $PhCH_2OCO-$)

Reagents: i, $H_2C=CHCH_2Br$, DMF; ii, $H_2C=CHMgBr$, THF, then Jones oxidation; iii, MeNHCOOBz, camphorsulphonic acid; iv, diethyl [(N-benzylideneamino)-lithiomethyl]phosphonate, THF, at $-78°C$, then Bu^nLi; v, 2-(2-bromoethyl)-2-methyl-1,3-dioxolan, HMPA-THF, then aq. HCl; vi, 0.5M-KOH, aq. MeOH; vii, $RhCl_3$, EtOH, reflux; viii, $LiAlH_4$, glyme, at $-78°C$; ix, H_2, Pd/C, HCl, EtOH; x, $(HCO)_2O$, pyridine, at $80°C$; xi, $POCl_3$, at $85°C$, then $NaBH_4$, MeOH

Scheme 6

References

1. A.H.M.Alam and D.A.Murav'eva, Khim. Prir. Soedin., 1982, 401 (Chem. Abstr., 1982, 97, 69 321).
2. B.A.Sepulveda, P. del C.Pacheco, M.J.Silva, and R.Zemelman, Bol. Soc. Chil. Quim., 1982, 27, 178 (Chem. Abstr., 1982, 96, 214 307).
3. P. del C.Pacheco, M.J.Silva, P.G.Sammes, and W.H.Watson, Bol. Soc. Chil. Quim., 1982, 27, 289 (Chem. Abstr., 1982, 96, 214 311).
4. S.Tani, N.Kobayashi, H.Fuziwara, T.Shingu, and A.Kato, Chem. Pharm. Bull., 1981, 29, 3381.
5. S.Hung, G.Ma, and G.Sung, Huaxue Xuebao, 1981, 39, 529 (Chem. Abstr., 1982, 96, 139 653).
6. G.Sung, Fenxi Huaxue, 1981, 9, 520 (Chem. Abstr., 1982, 96, 162 994).
7. W.Doepke and E.Sewerin, Z. Chem., 1981, 21, 71.
8. S.Ghosal, P.H.Rao, D.K.Jaiswal, Y.Kumar, and A.W.Frahm, Phytochemistry, 1981, 20, 2003.
9. A.A.Ali, M.K.Mesbah, and A.W.Frahm, Planta Med., 1981, 43, 407.
10. S.Kobayashi, K.Yuasa, Y.Imakura, M.Kihara, and T.Shingu, Chem. Pharm. Bull., 1980, 28, 3433.
11. S.Kobayashi, K.Yuasa, K.Sato, and T.Shingu, Heterocycles, 1982, 19, 1219.
12. W.Doepke and S.A.Roshan, Z. Chem., 1981, 21, 223.
13. S.Kobayashi, M.Kihara, T.Shingu, and K.Shingu, Chem. Pharm. Bull., 1980, 28, 2924.
14. Y.Tsuda, Yakugaku Zasshi, 1981, 101, 295.
15. S.F.Martin and C.-Y.Tu, J. Org. Chem., 1981, 46, 3763.
16. T.Weller and D. Seebach, Tetrahedron Lett., 1982, 23, 935.
17. J.Toda, T.Sano, Y.Tsuda, and Y.Itatani, Chem. Pharm. Bull., 1982, 30, 1322; J.Toda, T.Sano, and Y.Tsuda, Heterocycles, 1982, 17, 247.
18. G.E.Keck and R.H.Webb, J. Org. Chem., 1982, 47, 1302.
19. O.Hoshino, S.Sawaki, N.Shimamura, A.Onodera, M.Yamazaki, M.Yuasa, and B.Umezawa, Tennen Yuki Kagobutsu Toronkai Koen Yoshishu, 24th, 1981, 445.
20. L.E.Overman and L.T.Mendelson, J. Am. Chem. Soc., 1981, 103, 5579.
21. S.F.Martin and P.J.Garrison, J. Org. Chem., 1981, 46, 3567; 1982, 47, 1513.

11
Erythrina and Related Alkaloids

BY A. S. CHAWLA AND A. H. JACKSON

1 Isolation and Structure Determination

A review[1] on Erythrina and related alkaloids covering the literature from November 1966 to the end of May 1979 has appeared; it summarizes the work published on isolation, structure determination, biosynthesis, synthesis and pharmacology of Erythrina, Homoerythrina, and Cephalotaxus alkaloids, and the abnormal Erythrina alkaloids from Cocculus species were also included.

Preliminary studies[2] of the alkaloid content of the leaves of four Erythrina species revealed the presence of α- and β-erythroidines (1,2) in E. berteroana, E. poeppigiana, and E. salviiflora but not in E. macrophylla. The presence of 8-oxoerythroidines was detected in E. berteroana and E. salviiflora. The most commonly occurring alkaloids, also usually present in greatest abundance, e.g., erysodine (3a) and erysopine (3b), were found in small amounts only in E. poeppigiana. The presence of erythratidine (4a) was noted in E. macrophylla, E. poeppigiana, and E. salviiflora, but not in E. berteroana. Erythraline (3c) was present only in E. macrophylla. Erybidine (5), erythratidinone (4b), and isoboldine (6) were found only in E. poeppigiana. Some of the 11-hydroxylated derivatives of Erythrina alkaloids were also tentatively identified from their mass spectra. Thus erythrinine (3d), 11-hydroxyerythratine (4c), and 11-hydroxyerysotinone (4d) were detected in E. macrophylla, whilst E. poeppigiana contained 11-hydroxyerythratidine (4e) and its C-2 epimer (4f). Similar studies[3] of the seeds of six Erythrina species (E. acanthocarpa, E. lanata, E. melanacantha, E. rubrinervia, E. subumbrans, and E. variegata) showed that the most abundant alkaloids in these species were erysodine (3a), erysopine (3b), and erysovine (3e); these were also present in substantial quantities as glycosides, as well as 'free' bases. In none of these six species was the presence of α- or β-erythroidine (1,2) observed. Erysoline (3f) was detected in E. lanata, E. rubrinervia, and E. subumbrans, while E. lanata also contained erysonine (3g). Erythratidine (4a) and, either erysotine (4g) or erysosalvine (4h), were present in E. melanacantha. Erythratine (4i) and small amounts of 11-hydroxy derivatives of erythratine and epierythratine, and erysoflorinone (4j) were detected in E. subumbrans. Aguilar and co-workers[4] isolated α- and β-erythroidines (1,2) from the flowers of E. americana and commented that these were the active principles responsible

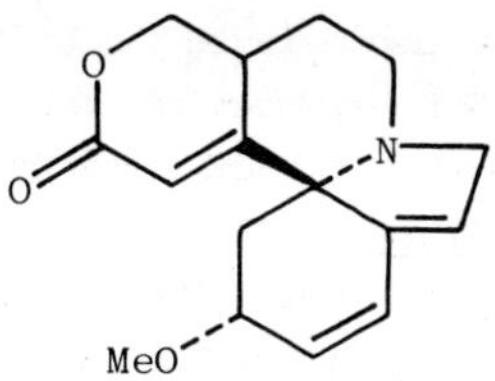

(1)

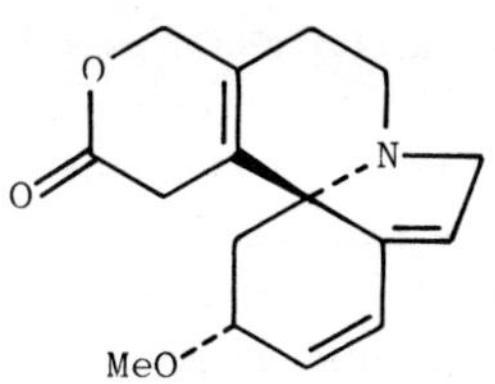

(2)

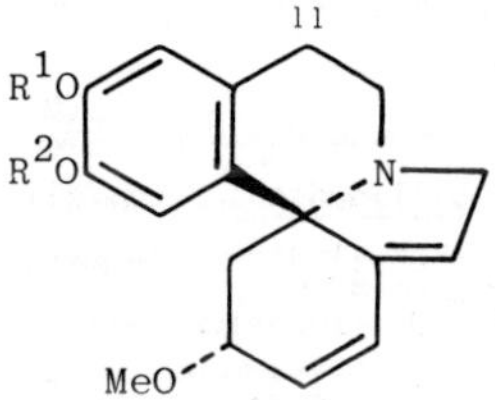

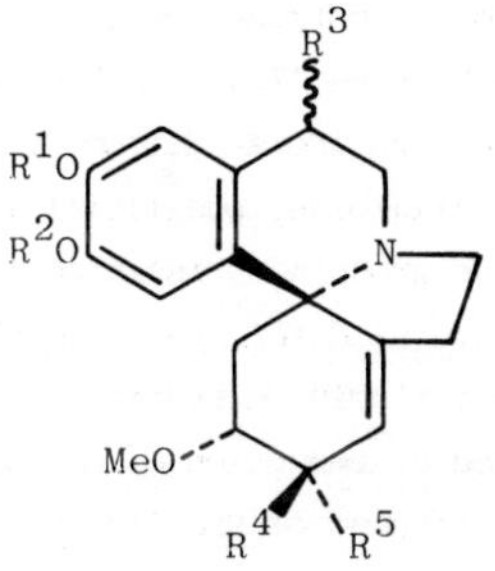

(3) a; $R^1 = H$, $R^2 = R^3 = Me$
   b; $R^1 = R^2 = H$, $R^3 = Me$
   c; $R^1R^2 = -CH_2-$, $R^3 = Me$
   d; 11-OH deriv. of (3c)
   e; $R^1 = R^3 = Me$, $R^2 = H$
   f; $R^1 = Me$, $R^2 = R^3 = H$
   g; $R^1 = R^3 = H$, $R^2 = Me$
   h; $R^1 = R^2 = R^3 = Me$
   i; 11-OH deriv. of (3h)

(4) a; $R^1 = R^2 = OMe$, $R^3 = R^4 = H$, $R^5 = OH$
   b; $R^1 = R^2 = OMe$, $R^3 = H$, $R^4R^5 = O$
   c; $R^1R^2 = -CH_2-$, $R^3 = R^4 = OH$, $R^5 = H$
   d; $R^1 = H$, $R^2 = Me$, $R^3 = OH$, $R^4R^5 = O$
   e; $R^1 = R^2 = Me$, $R^3 = R^5 = OH$, $R^4 = H$
   f; $R^1 = R^2 = Me$, $R^3 = R^4 = OH$, $R^5 = H$
   g; $R^1 = R^3 = H$; $R^2 = Me$; $R^4, R^5 = H, OH$
   h; $R^1 = Me$; $R^2 = R^3 = H$; $R^4, R^5 = H, OH$
   i; $R^1R^2 = -CH_2-$, $R^3 = R^5 = H$, $R^4 = OH$
   j; $R^1 = R^2 = R^3 = H$, $R^4R^5 = O$

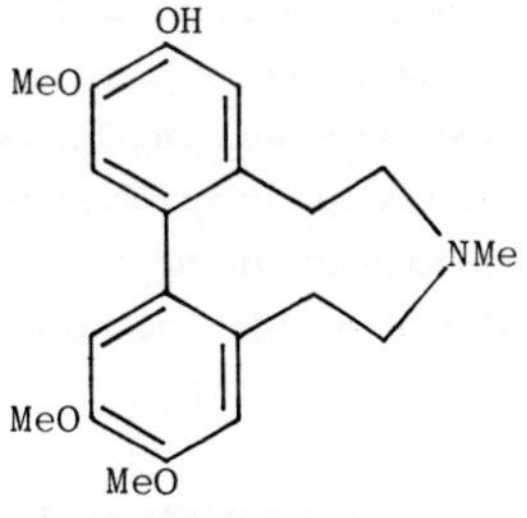

(5)

for the hypnotic activity of flower extracts; they also reported the ^{1}H NMR (60 MHz) spectra of these compounds. Erysotrine (3h), erythrartine (3i), hypaphorine, and two novel alkaloids, erysotrine N-oxide (7a) and erythrartine N-oxide (7b), were isolated from the flowers of E. mulungu.[5] This is the first report of naturally occurring N-oxides in the Erythrina genus. The N-oxide structures were assigned by spectroscopic analyses and confirmed by comparing the spectral data (ir, ^{1}H NMR) with the synthetic N-oxides obtained by treating erysotrine or erythrartine with m-chloroperbenzoic acid. The ^{13}C NMR (25.2 MHz) spectral data of erysotrine, erythrartine and their corresponding N-oxides, and erythristemine N-oxide (7c) have been reported.[5] Erysodine (3a), erysovine (3e), and hypaphorine were isolated from E. stricta bark.[6]

Two novel alkaloids, phellibilidine (8)[7] and isophellibilidine (9),[8] of the homoerythroidine group were isolated from Phelline billiardieri (Fam. Ilicaceae) and their structures determined from chemical and spectral data. Reduction of (8) with lithium aluminium hydride afforded a diol (10), identical with the product obtained by reduction of O-methylisophellibiline (11), the major alkaloidal constituent of Phelline billiardieri (cf. Vol. 2, p. 201), with the same reagent.[7] The proposed structure (9) for isophellibilidine was also confirmed by partial synthesis from (11).[8] On treatment with 1M KOH/MeOH at room temperature, (11) yielded the dienic salt (12) and then the acid (13) (85%). Treatment of the acid with mercuric acetate (THF) afforded the mercuric derivative (14), which on reaction with $NaBH_4/O_2$ [9] in DMF solution gave isophellibilidine (53%) and (15), (18%). Three new homoerythrina alkaloids, namely homoerythratine (16a), phellinine (17a), and O-methylphellinine (17b), were isolated from the leaves of Phelline brachyphylla and their structures established by spectral analyses[10] (especially ^{1}H NMR at 400 MHz). The structure of homoerythratine was also confirmed by its chemical correlation with epihomoerythratine (16b), previously isolated from Phelline comosa (cf. Vol. 2, p. 201).

The isolation of isoharringtonine, homoharringtonine, hydroxycephalotaxine, cephalotaxine, drupacine, epi-phelline, alkaloid 6, and wilsonine from Cephalotaxus sinensis has been reported.[11] Wilsonine had a weak antileukemic activity in mice. Cephalotaxinone, (+)-acetylcephalotaxine, cephalotaxinamide, demethylneodrupacine, deoxyharringtonic acid, and isoharringtonic acid were also isolated from Cephalotaxus hainanensis in addition to eleven alkaloids reported previously.[12]

2 Synthesis

Synthetic work on Erythrina and related alkaloids has again moved at a rather slow pace during the past year. However, Diels-Alder reactions of Δ^2-pyrroline-4,5-diones with activated butadienes yielded single hydroindole adducts, suggesting that the reaction proceeds in a regio- and stereo-selective

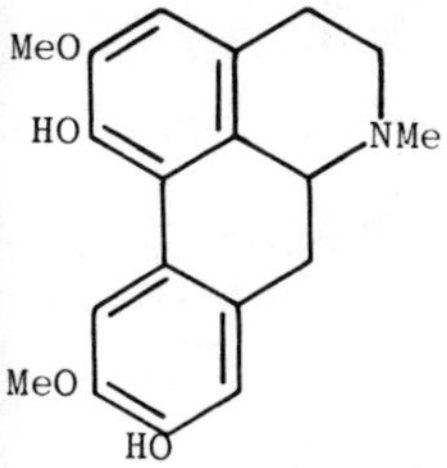

Erythrina and Related Alkaloids

(6)

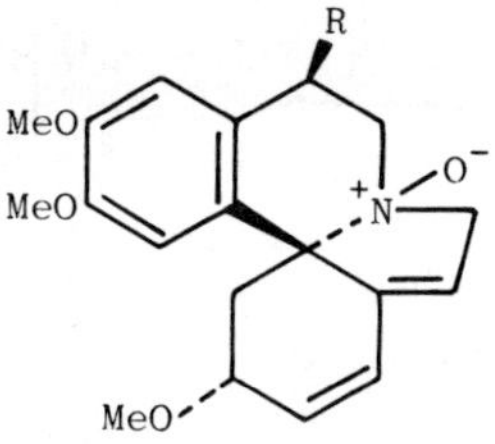

(7) a; R = H
    b; R = OH
    c; R = OMe

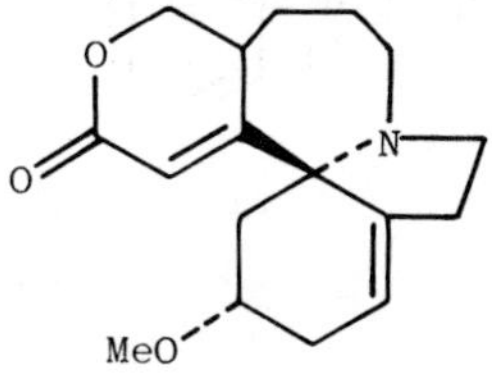

(8)

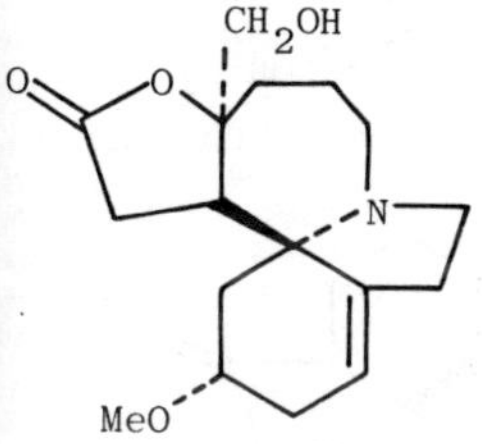

(9)

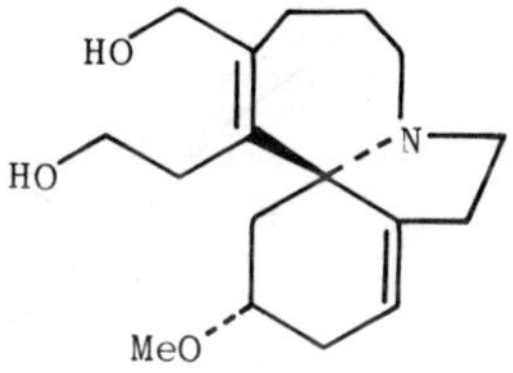

(10)

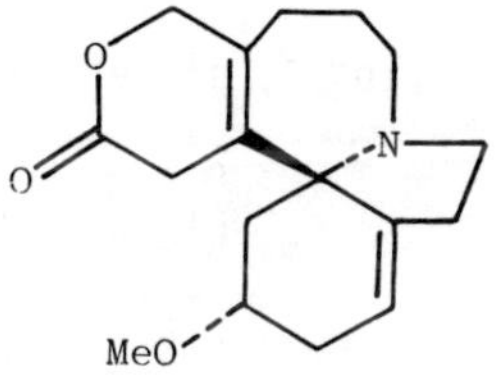

(11)

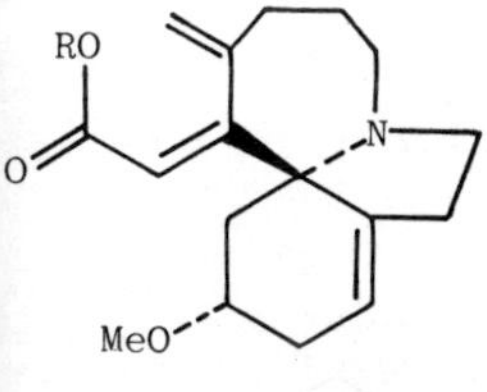

(12) R = K
(13) R = H

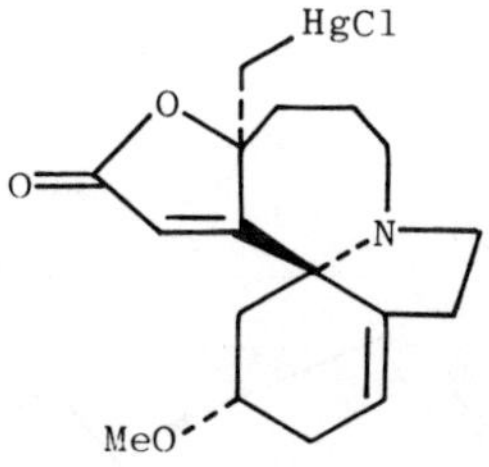

(14)

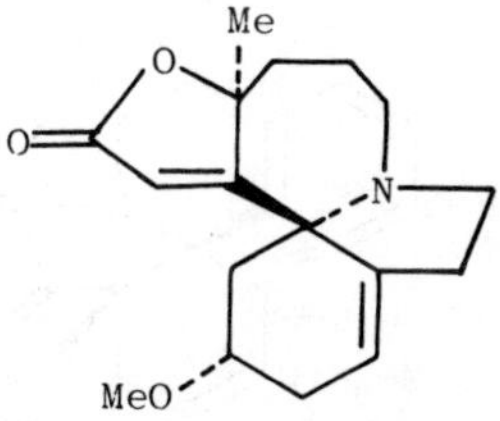

(15)

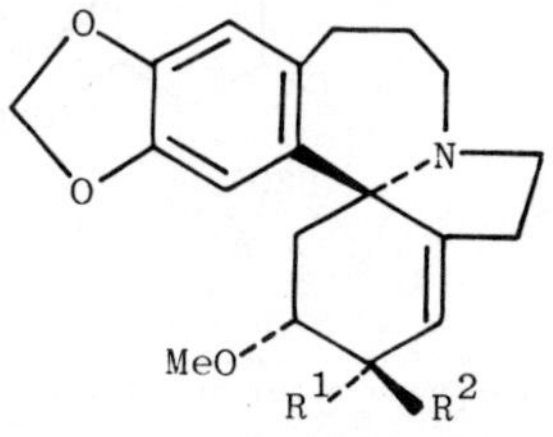

(16) a; R^1 = H, R^2 = OH
     b; R^1 = OH, R^2 = H

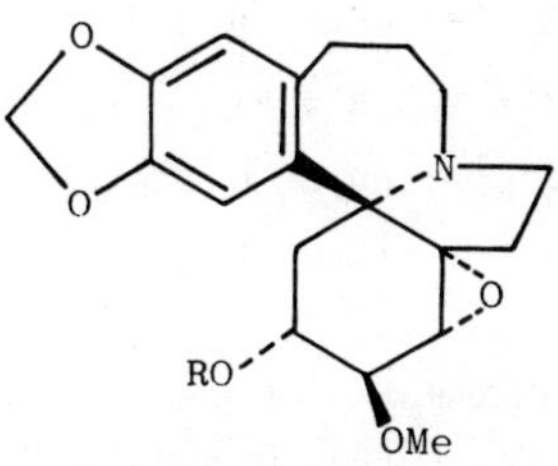

(17) a; R = H
     b; R = Me

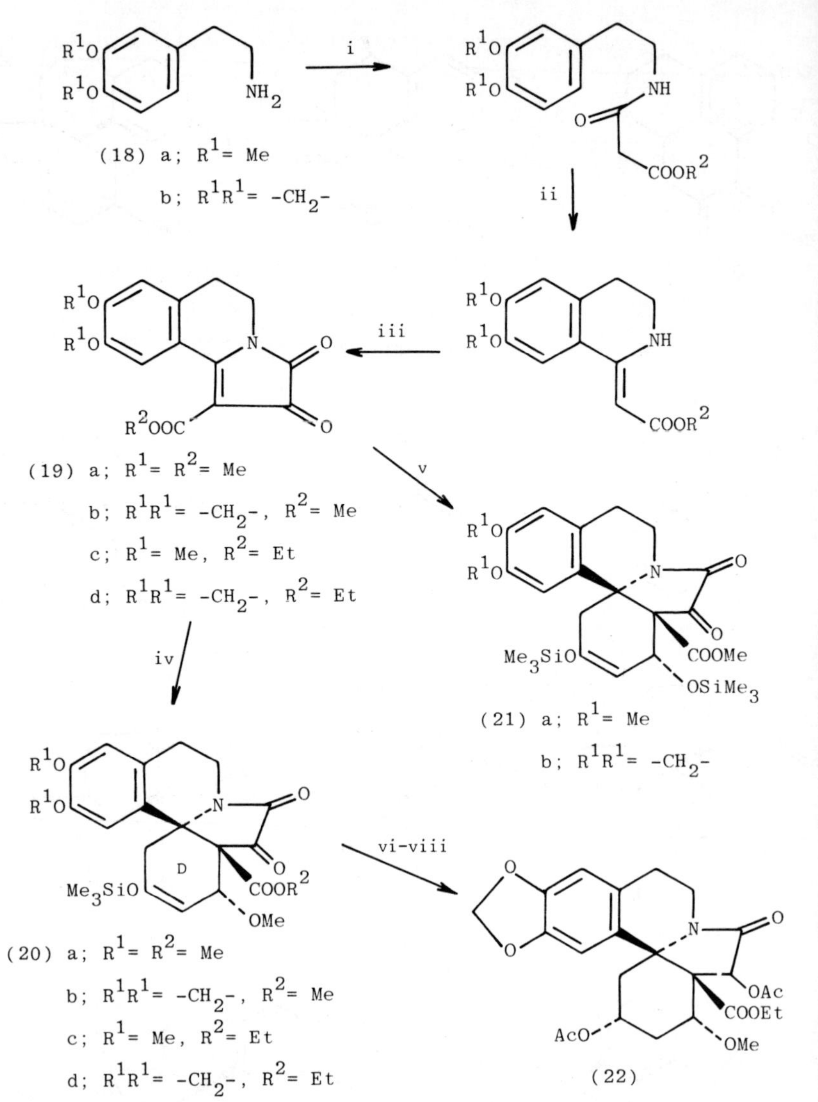

(18) a; R^1 = Me
b; R^1R^1 = $-CH_2-$

(19) a; $R^1 = R^2$ = Me
b; R^1R^1 = $-CH_2-$, R^2 = Me
c; R^1 = Me, R^2 = Et
d; R^1R^1 = $-CH_2-$, R^2 = Et

(20) a; $R^1 = R^2$ = Me
b; R^1R^1 = $-CH_2-$, R^2 = Me
c; R^1 = Me, R^2 = Et
d; R^1R^1 = $-CH_2-$, R^2 = Et

(21) a; R^1 = Me
b; R^1R^1 = $-CH_2-$

(22)

Reagents: i, $ClCOCH_2COOR$ (R = Me or Et); ii, polyphosphate ester in $CHCl_3$;
iii, $(COCl)_2$ in Et_2O, at $0°C$; iv, $MeOCH=CHC(OSiMe_3)=CH_2$;
v, $Me_3SiOCH=CH-C(OSiMe_3)=CH_2$; vi, KF in THF, at $20°C$; vii, $NaBH_4$;
viii, Ac_2O, pyridine

Scheme 1

manner.[13] This reaction was applied to the synthesis of ring-D-functionalized erythrinan derivatives and gave acceptable yields (Scheme 1). Isoquinolino-pyrrolinediones (19) prepared from β-arylethylamines (18), on condensation with 1-methoxy-3-trimethylsilyloxybutadiene, gave the corresponding 1,4-cycloadducts (20), whereas the diones (19a) and (19b) on reaction with 1,3-bis(trimethyl-silyloxy)butadiene gave the adducts (21a and 21b), respectively. The adduct (20d), on treatment with KF in THF followed by reduction with $NaBH_4$ and acetylation, yielded the diacetate (22). The structure and stereochemistry of the adducts were established by chemical and spectroscopic means, and confirmed by X-ray analysis of the derived acetate. This method may provide a simple route to the synthesis of Erythrina alkaloids, since the 6-alkoxycarbonyl group is easily removable from such 6-alkoxycarbonyl-7,8-dioxoerythrinans (cf. Vol. 12, p. 157). The same group of workers has also described the stereocontrolled transformations of the cycloadducts (20a) and (20b), and (21a) and (21b), into the alkaloids erysotrine (27a) and erythraline (27b)[14] respectively (Scheme 2). Partial reduction of (21a) and (21b) with a stoichiometric amount of $LiBH_4$, followed by acid treatment, afforded the unsaturated ketones (23a) and (23b), respectively, in ca. 80% yields. Similar treatment of (20a) and (20b) also gave the same ketones in comparable yields. Demethoxycarbonylation of (23a) and (23b) with $MgCl_2$-DMSO (cf. Vol. 12, p.157), with concomitant dehydration afforded the dienones (24a and 24b); Meerwein-Pondorff reduction of the latter gave the corresponding alcohols (25a) and (25b) in good yield, and which had the same configuration as that of natural alkaloids. Methylation of the alcohols with CH_3I catalysed by $KOH-Et_4NBr$[15] furnished (±)-erysotramidine (26a) (84%) and (±)-8-oxoerythraline (26b) (86%), respectively. The 8-oxoalkaloids, on reduction with $LiAlH_4$-$AlCl_3$,[16] afforded (±)-erysotrine (27a) (62%) and (±)-erythraline (27b) (71%). The identities of these compounds were confirmed by spectral comparisons with the natural alkaloids.

Sano and co-workers[17] also showed that photocycloaddition of 2-trimethyl-silyloxybutadiene to (19c) proceeded in a regio- and stereo-selective manner to give a single 1,2-adduct (28) (90%), which underwent a thermal [1,3]-shift in boiling toluene to give the erythrinan derivative (29) (64%) (Scheme 3). On hydrolysis this afforded the ketone (30) (90%). 1-Methoxy-3-trimethyl-silyloxybutadiene underwent similar cycloaddition to (19c), giving the product (31) (72%), which was then rearranged into the erythrinan derivative (32); on hydrolysis, the latter afforded the conjugated enone (33). This method for the preparation of functionalized erythrinan derivatives has considerable synthetic potential as it shows a reverse regio-selectivity to the normal Diels-Alder reaction, and the silyl enol ether produced is capable of further manipulation.

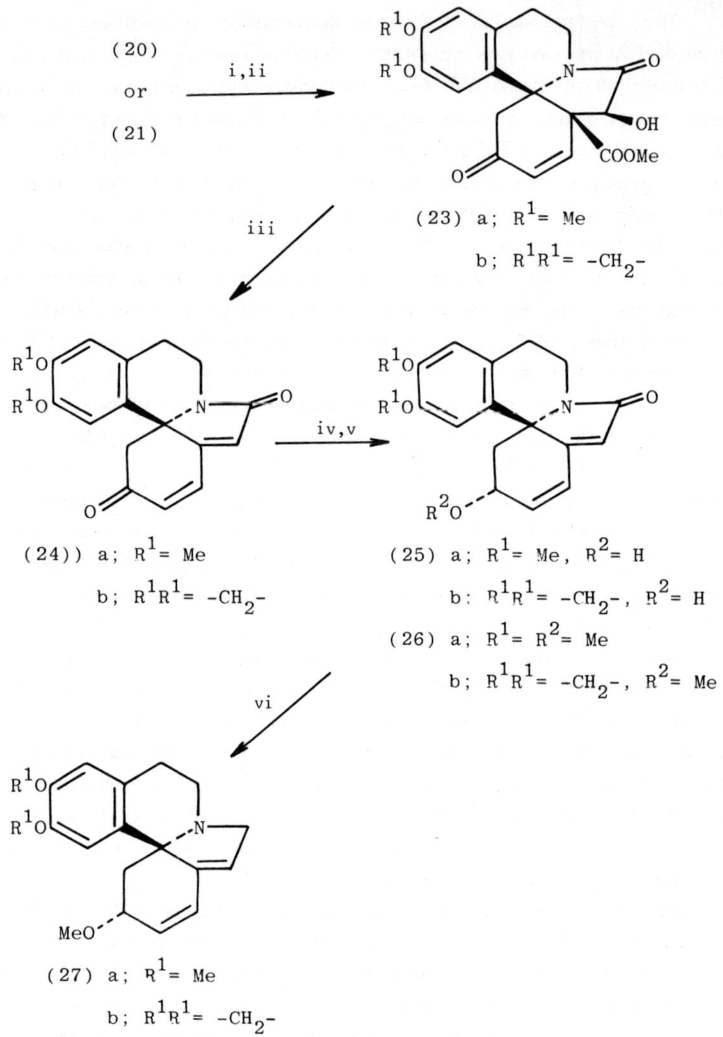

(20) or (21) →(i, ii)→ (23) a; R^1 = Me
b; R^1R^1 = $-CH_2-$

↓ iii

(24) a; R^1 = Me
b; R^1R^1 = $-CH_2-$

→(iv, v)→

(25) a; R^1 = Me, R^2 = H
b; R^1R^1 = $-CH_2-$, R^2 = H
(26) a; R^1 = R^2 = Me
b; R^1R^1 = $-CH_2-$, R^2 = Me

↓ vi

(27) a; R^1 = Me
b; R^1R^1 = $-CH_2-$

Reagents: i, LiBH$_4$ in THF, at -60°C, for 20 minutes; ii, 5% HCl-THF (1:1), reflux for 1 hour; iii, MgCl$_2$ in DMSO, at 140°C, for 1 hour; iv, Al(OPr^i)$_3$, Pr^{i}OH; v, MeI, KOH, Et$_4$N$^+$ Br$^-$; vi, LiAlH$_4$, AlCl$_3$

Scheme 2

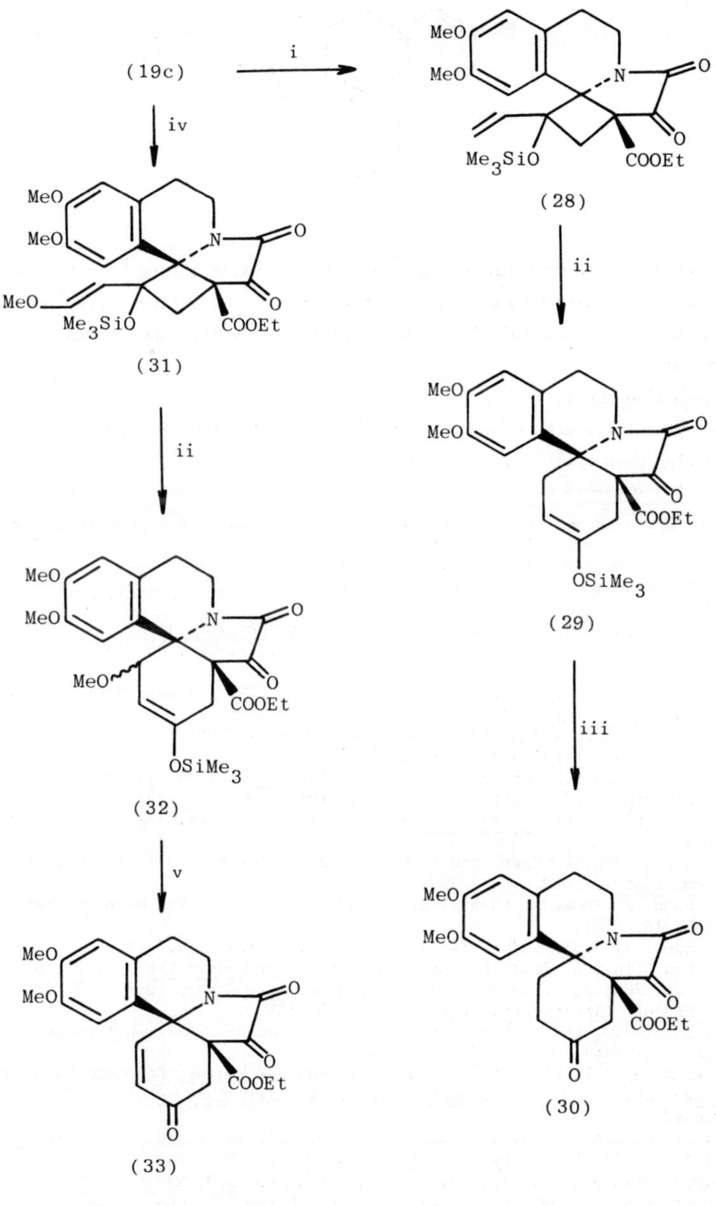

Erythrina and Related Alkaloids

Reagents: i, $H_2C=C(OSiMe_3)CH=CH_2$, $h\nu$; ii, heat; iii, 5% HCl–THF (1:1) or KF in THF; iv, $MeOCH=CHC(OSiMe_3)=CH_2$, $h\nu$; v, 5% HCl–THF (1:1), reflux 1 hour

Scheme 3

(34)

Syntheses of homoharringtonine[18] and cephalotaxine esters[19] have been reported and 1-azaerythrinan[20] and 2-azaerythrinan[21] dervatives have also been synthesized; the compound (34) showed analgesic activity superior to that of codeine phosphate.[20a]

Acknowledgement:

We gratefully acknowledge support from the F.B. and B.A. Krukoff Memorial fund during preparation of this review.

References

1   S.F. Dyke and S.N. Quessy, in 'The Alkaloids', ed. R.G.A. Rodrigo, Academic Press, New York, 1981, Vol. 18, p. 1.
2   A.H. Jackson and A.S. Chawla, Allertonia, 1982, 3, 39.
3   A.H. Jackson, P. Ludgate, V. Mavraganis and F. Redha, Allertonia, 1982, 3, 47.
4   M.I. Aguilar, F. Giral, and O. Espejo, Phytochemistry, 1981, 20, 2061.
5   M.H. Sarragiotto, H.L. Filho, and A.J. Marsaioli, Can. J. Chem., 1981, 59, 2771.
6   H. Singh, A.S. Chwla, V.K. Kapoor, N. Kumar, D.M. Piatak, and W. Nowicki, J. Nat. Prod., 1981, 44, 526.
7   M.-F. Seguineau and N. Langlois, Phytochemistry, 1980, 19, 1279.
8   N. Langlois, Tetrahedron Letters, 1981, 22, 2263.
9   C.L. Hill and G.M. Whitesides, J. Am. Chem. Soc., 1974, 96, 870.
10 D. Debourges and N. Langlois, J. Nat. Prod., 1982, 45, 163.
11 L. Ren and Z. Xue, Zhongcaoyao, 1981, 12(6), 1; Chem. Abstr., 1982, 96, 3672.
12 Z. Xue, L. Xu, D. Chen, and L. Huang, Yaoxue Xuebao, 1981, 16(10), 752; Chem. Abstr., 1982, 96, 82690.
13 T. Sano, J. Toda, N. Kashiwaba, Y. Tsuda, and Y. Iitaka, Heterocycles, 1981, 16, 1151.
14 T. Sano, J. Toda, and Y. Tsuda, Heterocycles, 1982, 18, 229.
15 D. Reushling, H. Piatsch, and A. Linkies, Tetrahedron Lett., 1978, 615.
16 cf. A. Mondon, J. Zander, and H.-U. Menz, Ann. Chem., 1963, 667, 126; A. Mondon, Liebigs Ann. Chem., 1959, 628, 123.
17 T. Sano, J. Toda, Y. Horiguchi, K. Imafuku, and Y. Tsuda, Heterocycles, 1981, 16, 1463.
18 Y-K. Wang, Y-L. Li, S-F. Pan, S-B. Li, and W-K. Huang, Lan-Chou Ta Hsueh Hsueh Pao, Tzu Jan Kò Hsueh Pan, 1980, (1), 71; Chem. Abstr., 1981, 95, 220199.
19 S. Li, H. Sun, X. Lu, S. Zhang, F. Lu, J. Dai, and Y. Xu, Yaoxue Xuebao, 1981, 16(11), 821; Chem. Abstr., 1982, 96, 143126.
20 Nippon Shinyaku Co. Ltd., Jpn. Kokai Tokkyo Koho JP 81 100785 (Chem. Abstr., 1982, 96, 6927) and JP 81 100786 (Chem. Abstr., 1982, 96, 35616).
21 Nippon Shinyaku Co. Ltd., Jpn. Kokai Tokkyo Koho JP 81 108789, JP 81 110686, JP 81 113784, and JP 81 113785 (Chem. Abstr., 1982, 96, 104577 - 104580).

12
Indole Alkaloids

BY J. E. SAXTON

1 Introduction

The most recent volume in the Manske series is entirely devoted to indole alkaloids, and contains chapters on the bisindole alkaloids and the eburnamine-vincamine group.[1] An outstanding new general text on alkaloids, recently published, also deserves mention.[2] The 2-acylindole alkaloids[3a] and the synthesis of bisindole alkaloids and their derivatives[3b] have been reviewed.

2 Simple Alkaloids

2.1 Non-tryptamines. — A further synthesis of mupamine has been reported.[4] Details of the synthesis of hyellazole (1), together with a synthesis of 6-chlorohyellazole by the same route, have also been published.[5] Takano et al. have reported a new synthesis of hyellazole (1), the critical stage in which involves the cyclization of the enamine (2) in acetic anhydride—acetic acid (Scheme 1).[6]

Gramodendrine (3) is a new, dipiperidylindole alkaloid which occurs in the aerial parts of Lupinus arbustus ssp. calcaratus. Its structure was deduced from spectral data, and confirmed by a simple Mannich synthesis from indole, formaldehyde, and ammodendrine.[7]

The antiepileptic alkaloid pimprinine (4), which was initially isolated from Streptomyces pimprina, has now been shown to be a metabolite of Streptoverticillium olivoreticuli (Arai, Nakada et Suzuki) Baldacci, Farina et Locci, ISP 5105, together with pimprinethine (5) and pimprinaphine (6).[8a] Pimprinethine also occurs in the lipophilic extracts from the culture filtrate of Streptomyces cinnamomeus.[8b] All three metabolites have been synthesised,[8a] and pimprinine has also been synthesised by an independent group.[8c]

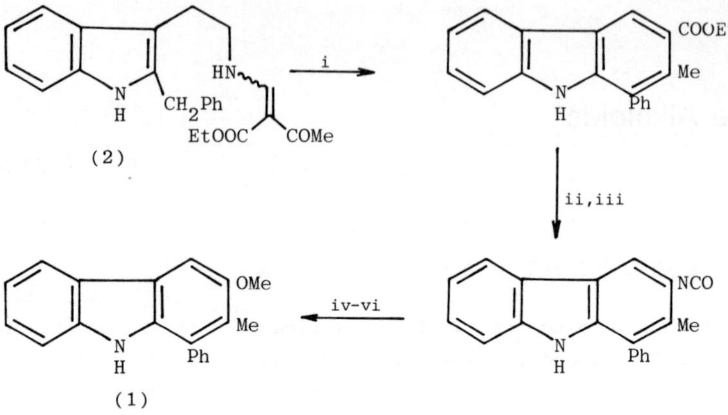

Reagents: i, Ac$_2$O, AcOH, reflux; ii, diphenylphosphoryl azide, MeCN, heat; iii, NaOH, H$_2$O, reflux; iv, H$_2$O, heat; v, NaOH, (CH$_2$OH)$_2$, reflux; vi, NaNO$_2$, H$_2$SO$_4$, MeOH, -15°C, then reflux.

Scheme 1

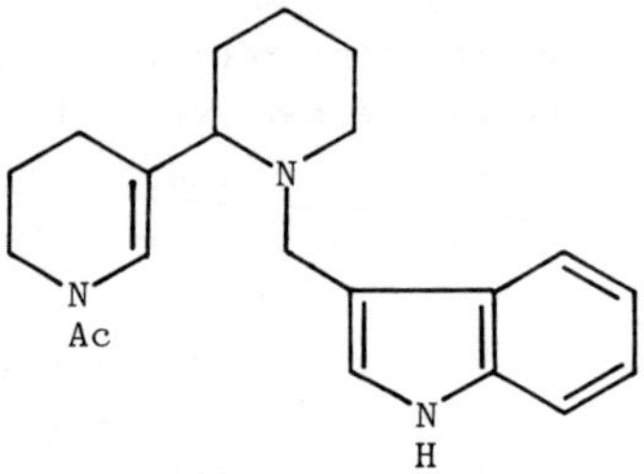

Gramodendrine (3)

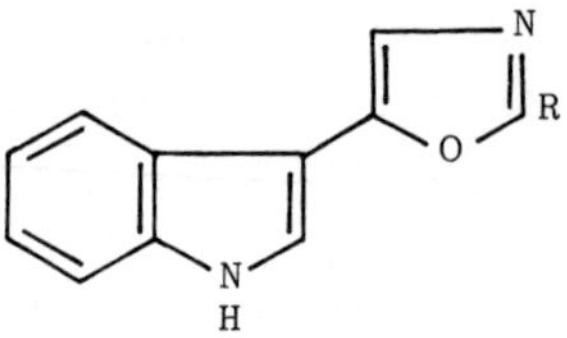

Pimprinine (4) R = Me
Pimprinethine (5) R = Et
Pimprinaphine (6) R = CH$_2$Ph

Hinnuliquinone, a pigment isolated from the fungus <u>Nodulisporium hinnuleum</u>, proves to be a bis-indolylquinone of structure (7).[9]

The occurrence of the mycotoxins penitrems D and E in <u>Penicillium crustosum</u> has again been noted.[10a] Chromatographic methods (h.p.l.c. and t.l.c.) for the separation of penitrems A-F have been investigated, and an h.p.l.c. method for their separation in one run has been developed.[10b]

A new indolosesquiterpene alkaloid, polyavolinamide, (8), has been isolated from the stems and stem bark of Nigerian Polyathia suaveolens,[11] in which it occurs with polyavolensin, polyavolensinol, and polyavolensinone.[12] A closely related compound, polyveoline (9), has been isolated from the trunk bark of the same species, growing in the Congo region.[13] The structure and stereochemistry of polyveoline were established by the X-ray method.[13] It should be noted that N-acetylpolyveoline has the same molecular formula as polyavolinamide, whose structure is less securely established. However, it is stated categorically[11] that deacetylpolyavolinamide is not identical with polyveoline, although the spectra of these two compounds show considerable similarity.

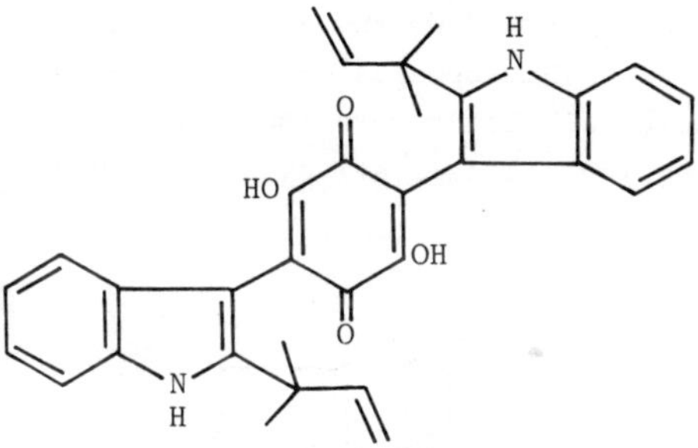

Hinnuliquinone (7)

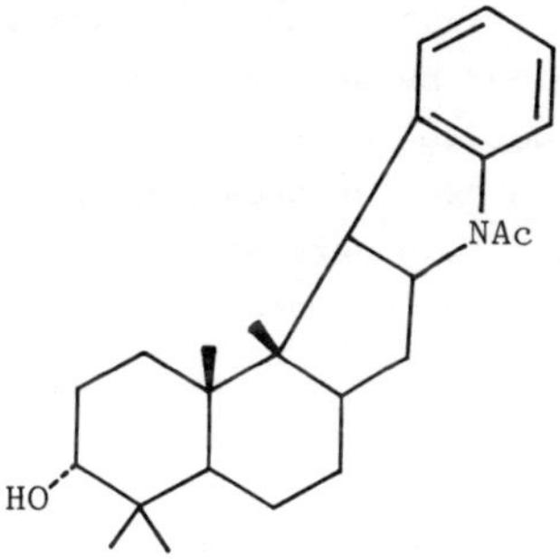

Polyavolinamide (8)

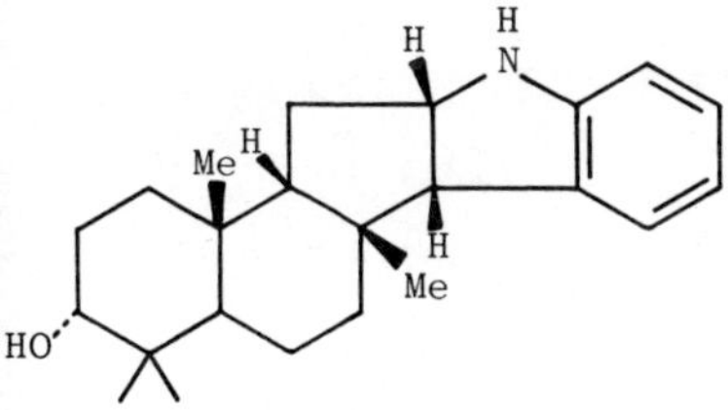

Polyveoline (9)

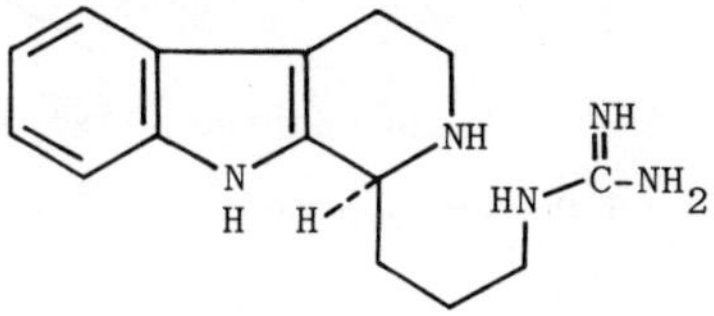

(−)-Trypargine (10)

2.2 Non-isoprenoid Tryptamines.

L-Hypaphorine is a major constituent of the seeds of Pterocarpus officinalis, and appears to be responsible for the rejection of these abundant seeds as a food source by at least one species of tropical, seed-eating rodent.[14a]

One of two new alkaloids isolated from the marine bryozoan Flustra foliacea proves to be 6-bromo-N_b-formyl-N_b-methyl-tryptamine.[14b]

(-)-Trypargine (10) is a guanidinopropyl-tetrahydro-β-carboline of potential pharmacological interest, which has recently been isolated from the skin of an African frog, Kassina senegalensis.[15a] Its structure was confirmed by a conventional Pictet-Spengler synthesis, and its absolute configuration deduced from its o.r.d. spectrum.[15b]

Strychnocarpine has been found to occur in the stem bark of a second Strychnos species, namely S. floribunda.[15c]

Annomontine (11a) and 6-methoxyannomontine (11b) are the first members of a new class of aminopyrimidino-β-carboline alkaloids, which occur in the trunk and root bark of Annona montana Macf. (fam. Annonaceae).[16] The structures of these alkaloids were deduced by spectral analysis and confirmed by the X-ray crystal structure determination of (11b). This species was investigated on an earlier occasion, when seven alkaloids were isolated, of which five were shown to be isoquinoline derivatives; the other two were not identified.

Annomontine (11a) R = H   (12a) R = H
6-Methoxyannomontine (11b) R = OMe   (12b) R = OH

The alkaloid content of two New Caledonian species of Dutaillyea (fam. Rutaceae) has been investigated.[17] The leaves of D. oreophila contain N,N-dimethyl-5-methoxytryptamine, 2-methyl-6-methoxy-1,2,3,4-tetrahydro-β-carboline, together with hordenine and kokusaginine, and the stem bark contains kokusaginine and three other furanoquinoline bases. The leaves of D. drupacea yielded N,N-dimethyl-5-methoxytryptamine; again the stem bark contained only quinoline alkaloids.

The root bark of Brucea antidysenterica Mill., used in folk medicine in various parts of Africa for the treatment of dysentery, tumorous growths, and asthma, contains canthin-6-one (12a);[18a] its 1-hydroxy derivative (12b) has been isolated from the heartwood of Ailanthus giraldii Dode,[18b] and from the root bark of A. altissima, together with 1-(1,2-dihydroxyethyl)-4-methoxy-β-carboline.[18c]

The most recent synthesis of rutaecarpine (14) simply involves coupling of the tetrahydropyridoquinazoline (13) (readily obtained by condensation of 2-piperidone with anthranilic acid) with benzene diazonium chloride, followed by Fischer indole ring-closure (Scheme 2).[19a] The regiochemical and stereochemical course of the reduction of cis- and trans-hexahydrorutaecarpine (Ring E saturated) has been studied.[19b]

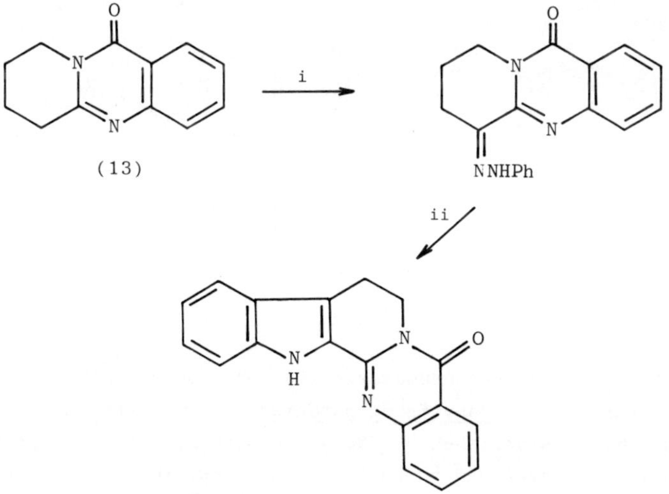

Reagents: i, PhN_2^+ Cl^-; ii, PPA, 180°C

Scheme 2

(±)-Elaeocarpidine (15) has been neatly synthesised by condensation of tryptamine with the amine bisacetal (16) (Scheme 3).[20] Presumably the preferential 5-exo-trig cyclization of the dialdehyde released from the bisacetal (16) to the pyrrolinium aldehyde (17) prior to Pictet-Spengler cyclization with tryptamine accounts for the regioselectivity of this synthesis. The stereoselectivity is explained by the fact that the unknown trans isomer (15), if formed, would undergo rapid acid-catalyzed equilibration of the aminal function to afford the more stable cis isomer.

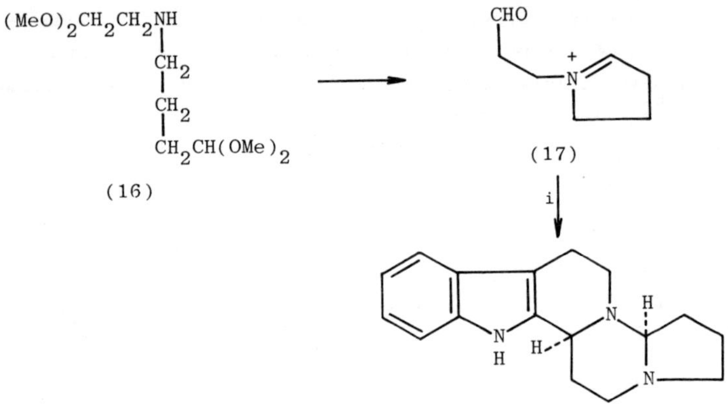

Elaeocarpidine (15)

Reagents: i, tryptamine, pH 5.5, MeOH, reflux 3 h, then pH 1.5, reflux 42 h

Scheme 3

A second toxin, neosurugatoxin (18), has been isolated from the Japanese Ivory Shell, Babylonica japonica.[21] The structure of neosurugatoxin, established by the X-ray method, differs from that of surugatoxin, the first toxin to be isolated, in having a five-membered ring C and a seven-membered ring D in place of the two six-membered rings in surugatoxin; in addition, neosurugatoxin is a disaccharide, composed of xylose and myo-inositol units in place of the myo-inositol unit present in surugatoxin. Neosurugatoxin has been estimated to have 100 times the antinicotinic activity of surugatoxin.[21]

Two new metabolites isolated from cultures of Chaetomium globosum have been shown[22] to be 19-O-acetylchaetoglobosin D (19) and 19-O-acetylchaetoglobosin B, which differs from (19) only in having the allylic double bond in position 5,6. The ^{13}C n.m.r.

Neosurugatoxin (18)

spectra of several chaetoglobosins, and of cytochalasin G, a metabolite of Pseudeurotium zonatum, have also been recorded and interpreted.[22]

19-O-Acetylchaetoglobosin D (19)

An independent synthesis[23] of the spirocyclic oxindole-lactone (20) constitutes a second formal synthesis of (±)-tryptoquivaline G, since (20) has already been converted into tryptoquivaline G by Büchi et al.[24a]

3 Isoprenoid Tryptamine and Tryptophan Derivatives

3.1 Non-terpenoid Alkaloids. — Flustrabromine (21) is a new isoprenoid 6-bromotryptamine derivative, isolated from the marine bryozoan Flustra foliacea L.[25] Four new 6-bromoeserine derivatives isolated from this same source are flustramide A (22)[14b] (cf. flustramine A),[24b] flustramine C (23), flustraminol A (24), and flustraminol B (25).[26]

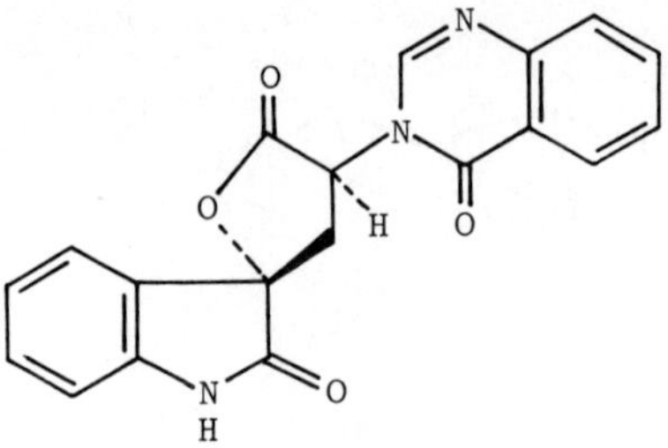

(20)

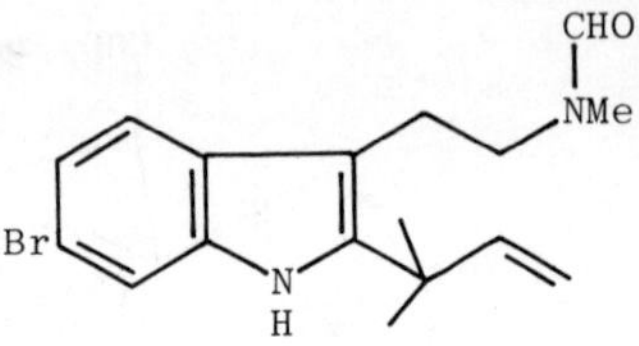

Flustrabromine (21)

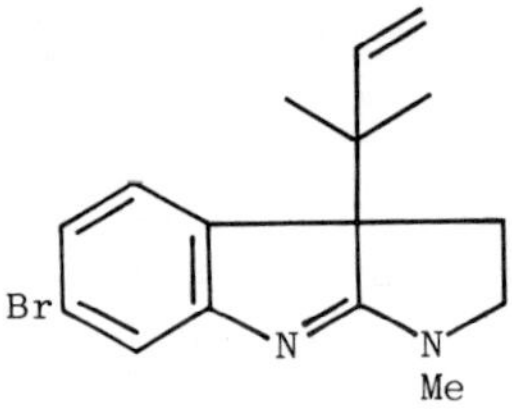

Flustramine C (23)

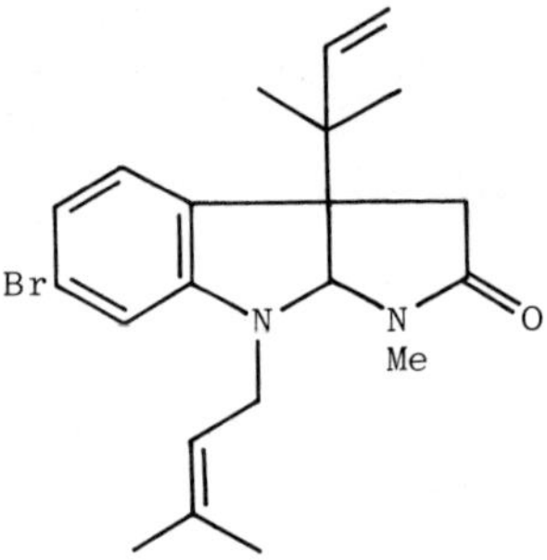

Flustramide A (22)

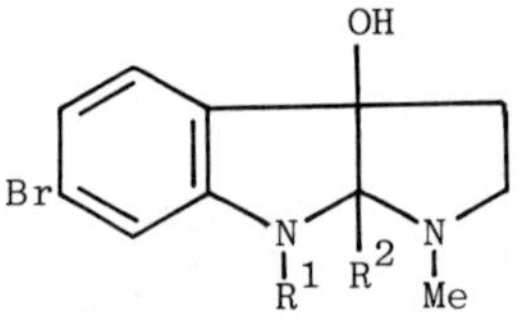

Flustraminol A (24) R^1= H, R^2= -CMe$_2$CH=CH$_2$
Flustraminol B (25) R^1= -CH$_2$CH=CMe$_2$, R^2= H

Lyngbyatoxin A is a highly inflammatory, vesicatory substance which has been isolated from the lipid extract of Lyngbya majuscula Gomont, a blue-green alga found in Hawaiian coastal waters which is responsible for swimmer's itch, and is also toxic to fish.[27a] Lyngbyatoxin appears to be a primary irritant, not an allergen, since all subjects coming into contact with it are affected. The structure (26) of lyngbyatoxin is reminiscent of that of teleocidin (27) (a constituent of several strains of Streptomyces), which is also a potent skin irritant.[27b,c] Lyngbyatoxin, teleocidin, and dihydroteleocidin B exhibit a potent tumour-producing activity when painted on mouse skin, and the possibility that they exhibit a similar activity on humans can not be overlooked.[27c]

A 7-isoprenyltryptophan unit also appears to be the basis for

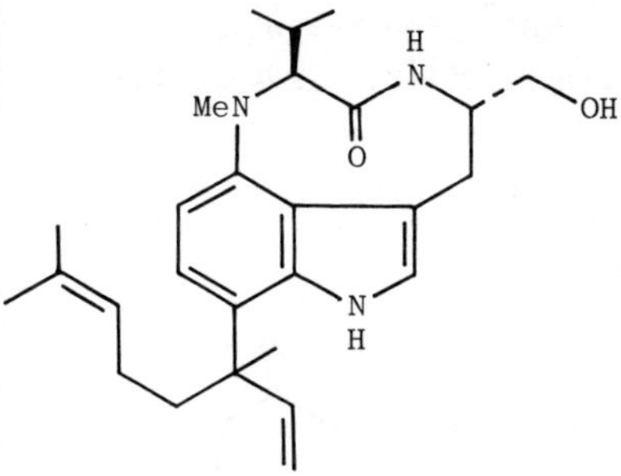

Lyngbyatoxin (26) 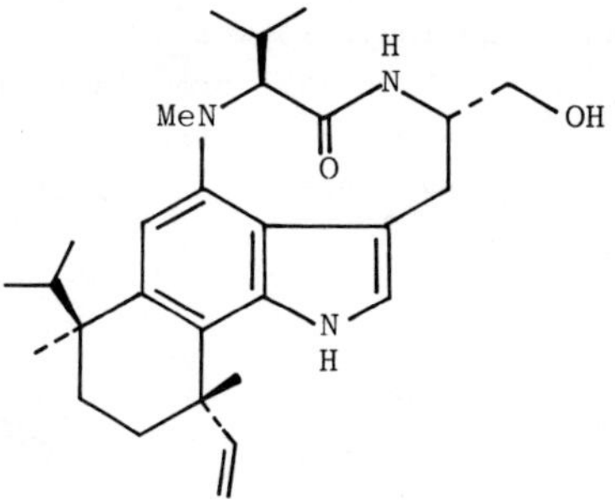   Teleocidin (27)

the structure of astechrome (28), an iron-containing pigment isolated from Aspergillus terreus.[28] The structure of astechrome was elucidated by studying the compounds obtained, e.g. (29) and (30), following the removal of the iron atom (Scheme 4).

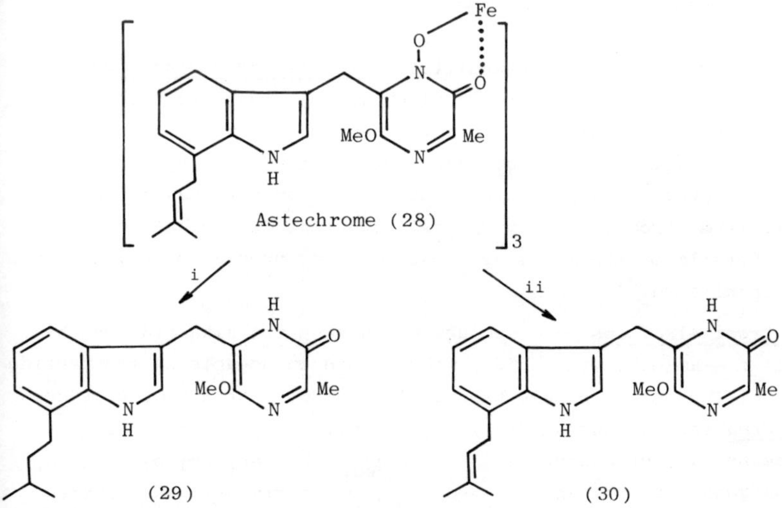

Astechrome (28)

(29)   (30)

Reagents: i, H_2, PtO_2; ii, PBr_3

Scheme 4

Other metabolites derived from isoprenyltryptophan and amino-acids which have been encountered recently include roquefortine, which is the principal polar metabolite of Penicillium crustosum,[10] and aszonalenin (31), which was isolated[29] together with its N-acetyl derivative (32) from Aspergillus 1FO8817. N-Acetylaszonalenin (32) proves to be identical with metabolite LL-S490β, isolated in 1973 from an unidentified Aspergillus species by Ellestad et al.[30]

Aszonalenin (31) R = H
N-Acetylaszonalenin (32) R = Ac

A new tremorgenic metabolite of Penicillium verruculosum, which causes severe tremors in mice, is 8-acetoxyverruculogen (33).[31] The structure of (33) was deduced mainly from spectral comparison with verruculogen and the isolation of 4-hydroxyproline from its hydrolysis products; the stereochemistry was finally established from the X-ray crystal structure determination.

Details of the first synthesis of deoxybrevianamide E have now been published.[32]

3.2 Ergot Alkaloids.— Two new pairs of ergot alkaloids which contain α-aminobutyric acid as the second amino-acid in the peptide half of the molecule have been isolated from a strain of Claviceps purpurea that produces mainly ergotoxine.[33] The alkaloids concerned are ergobutyrine (34) and its 8-epimer, ergobutyrinine, and ergobutine (35) and its 8-epimer, ergobutinine. These are the first ergot alkaloids to be encountered in which α-aminobutyric acid occurs as second amino-acid in the peptide chain.

Axenic cultures of Sphacelia sorghi, an ergot fungus parasitic on sorghum, produce dihydroergosine as major alkaloid, together with two other minor alkaloids, one of which, 9'-hydroxydihydroergosine (36), was produced in considerably greater amounts following the addition of allo-4-hydroxyproline to the culture medium.[34]

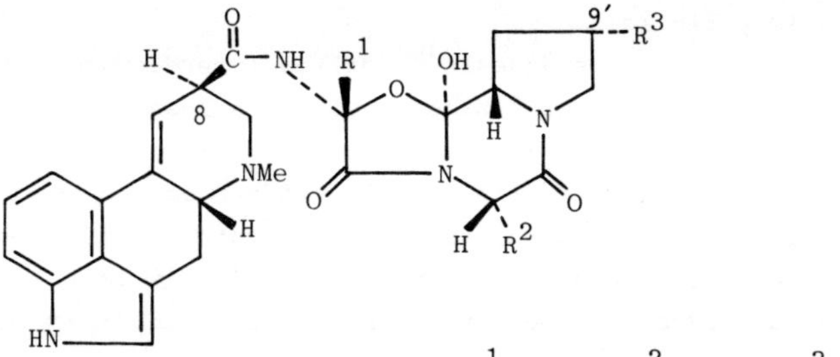

	R^1	R^2	R^3
Ergobutyrine (34)	CHMe$_2$	Et	H
Ergobutine (35)	Et	Et	H
9'-Hydroxydihydroergosine (36)	Me	CH$_2$CHMe$_2$	OH

Synthetic work reported during the past year has been exclusively concerned with tricyclic alkaloids and simple ergoline derivatives. Kozikowski and Greco[35] have developed the synthesis of a tricyclic intermediate (38), the crucial stage in which involves the intramolecular [3 + 2] dipolar cycloaddition of an azide group to an alkene with concomitant loss of nitrogen [(37) → (38)] (Scheme 5); this approach affords promise as a possible route

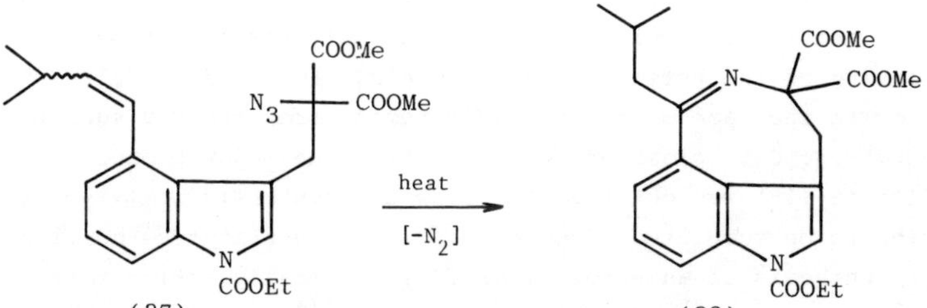

Scheme 5

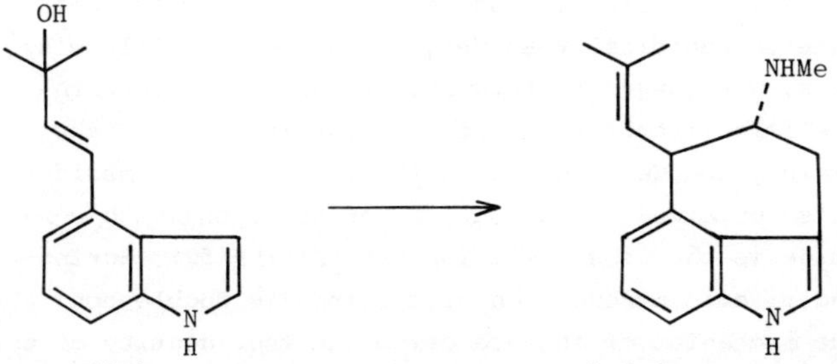

6,7-Secoagroclavine (40)

to clavicipitic acid.

An alternative synthesis[36a] of the intermediate (39) constitutes yet another synthesis of (±)-6,7-secoagroclavine (40).[36b]

Kozikowski's versatile nitrile oxide route to the ergot alkaloids has afforded the first total synthesis of (+)-paliclavine (41), and therefore of (+)-paspaclavine (42) as well.[37] Asymmetry was introduced at the outset by the reaction of N-tosylindole-4-aldehyde with the optically active Wittig reagent (43). Unfortunately the intramolecular nitrile oxide cycloaddition stage proved not to be stereospecific and the product, after re-acetylation and removal of the tetrahydropyranyl protecting group, was a mixture of (44) and the diastereoisomer epimeric at both positions 9 and 10. Fortunately, the mixture proved to be separable following mesylation of the primary hydroxyl group, and a selenoxide elimination then gave the isoxazoline (45). Introduction of the final asymmetric centre was achieved by methylation followed by hydride reduction. Initial experiments in which sodium borohydride was used as reducing agent gave one product almost exclusively, which was shown to be 5-epipaliclavine. However, the use of lithium aluminium hydride as reducing agent gave a mixture of (46) and its C-5 epimer; this presumably resulted from complexing of the heterocycle with $LiAlH_4$, followed by delivery of hydride to the face of the molecule remote from the C-9 substituent. Although the C-5 epimer of (46) was still the major product, sufficient (46) was obtained to allow the synthesis of (+)-paliclavine to be completed (Scheme 6). This constitutes the first total synthesis of an ergot alkaloid in optically active form.

A new synthesis of (±)-lysergine (47)[38a] starts essentially from the tricyclic ketone (48), which has previously[38b] been converted in one stage into the spirocyclic lactone (49) (Scheme 7). Reduction of the lactone and conjugated double bond proceeded essentially stereospecifically at C-8. The product, following re-benzoylation, was converted into its primary mesylate with simultaneous elimination of the tertiary hydroxyl group; the product was mainly the desired alkene (50), but some isomeric enamide was also obtained. The only other stage on which comment needs to be made is the debenzoylation stage, (51)→(52), during which a serendipitous introduction of the indolic double bond occurred; the mechanism of this reaction and the identity of the

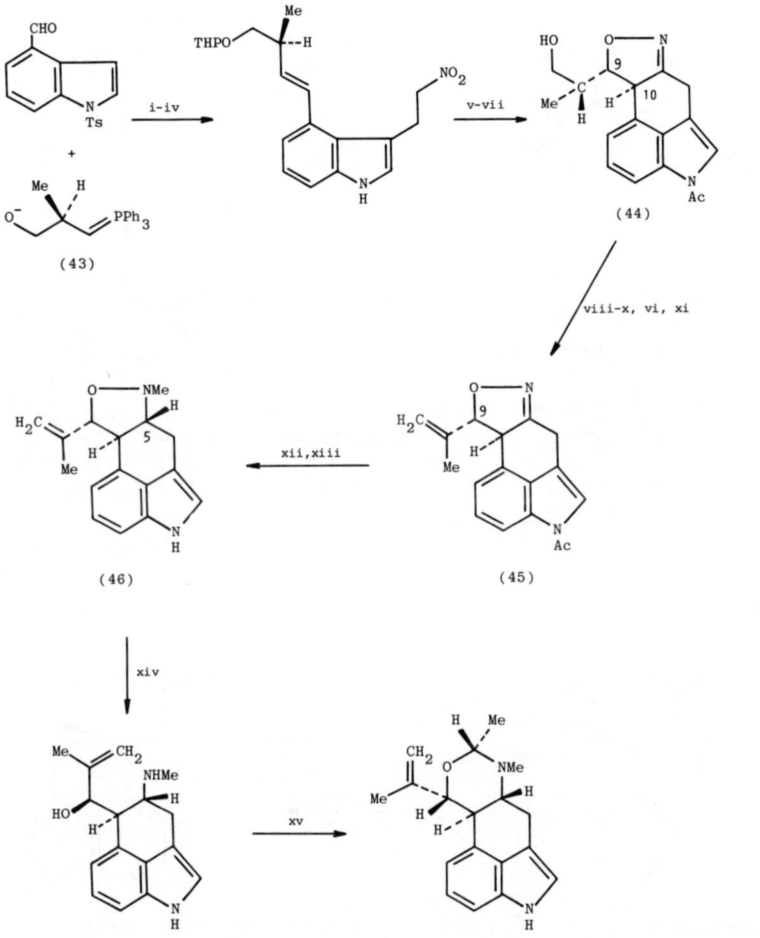

(43)

(44)

(46)

(45)

(+)-Paliclavine (41)

(+)-Paspaclavine (42)

Reagents: i, heat; ii, dihydropyran, pyridine, TsOH, CH_2Cl_2; iii, KOH, MeOH; iv, $H_2C{=}CHNO_2$; v, PhNCO, NEt_3; vi, Ac_2O, DMAP; vii, Dowex 50W-X8; viii, MesCl, NEt_3; ix, separation of diastereoisomers; x, PhSeNa; xi, $NaIO_4$; xii, $Me_3O^+\ BF_4^-$; xiii, $LiAlH_4$; xiv, Hg/Al; xv, MeCHO.

Scheme 6

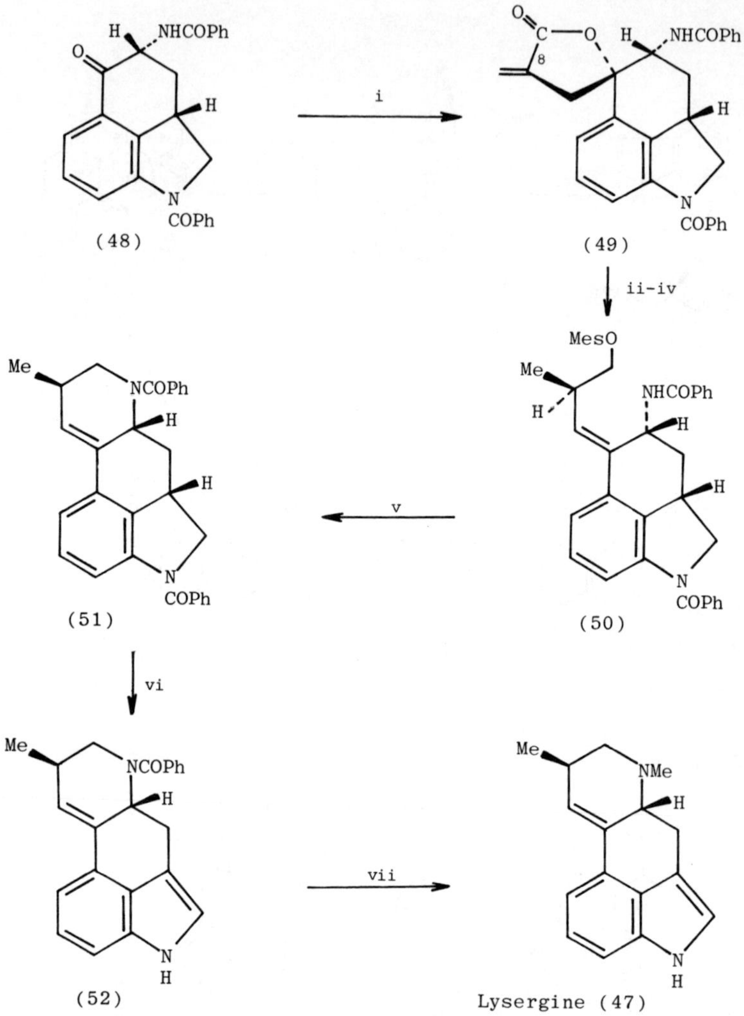

Reagents: i, BrCH$_2$C(=CH$_2$)CO$_2$Et, Zn; ii, NaBH$_4$, MeOH; iii, PhCOCl, MeOH; iv, MesCl, NEt$_3$; v, NaH, DMF; vi, Bu^{t}OK, THF, H$_2$O; vii, DIBAL, THF, NaBH$_3$CN, CH$_2$O.

Scheme 7

specific oxidant involved have not been elucidated.[38a] An alternative synthesis of dihydrosetoclavine (54), from the previously prepared[12b] intermediate (53), has been reported (Scheme 8).[39] Protection of the indolic nitrogen in (53), followed by release of the ketone group, gave a ketone (55) which on bromination by means of copper(II) bromide in the presence of ethyl orthoformate gave a bromoacetal (56). Subsequent stages are unexceptional, the final product being a mixture of (±)-dihydrosetoclavine (54) and its C-8 epimer.

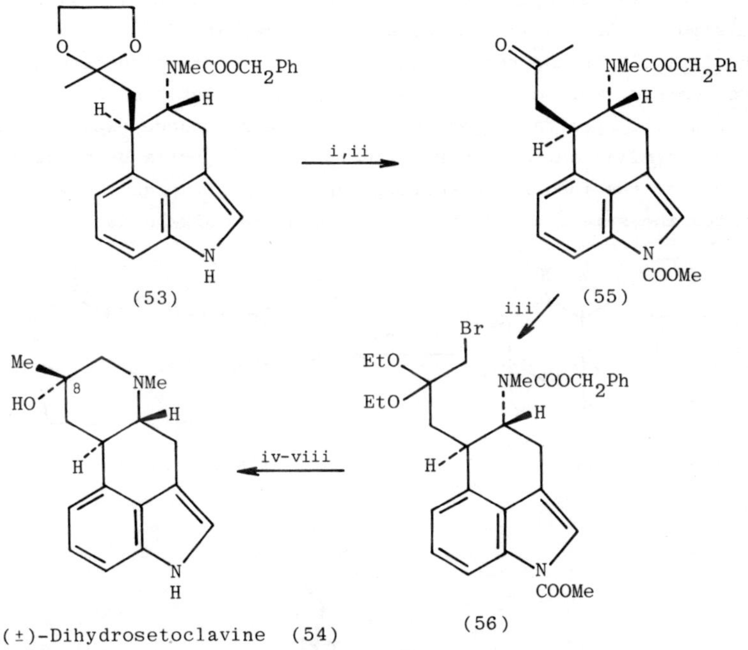

Reagents: i, $ClCO_2Me$, NAH; ii, TsOH, Me_2CO; iii, $CuBr_2$, $CH(OEt)_3$, EtOH; iv, H_2, Pd/C; v, NEt_3, Δ; vi, 1.5% HCl, DME, H_2O; vii, MeLi, THF, Et_2O, 0°C; viii, 3% KOH, MeOH, H_2O.

Scheme 8

A number of new ergoline derivatives have been prepared for pharmacological evaluation. These include the methyl esters of several 6-alkyl-6-nordihydrolysergic acids,[40a] various ind. -N-alkylergolines,[40b] and several ergoline derivatives containing a substituent at position 2 of the indole ring system.[40c]

4 Monoterpenoid Alkaloids

4.1 Alkaloids of *Aristotelia* and *Borreria* Species. — Four new *Aristotelia* alkaloids have been reported during the past year. Makomakine and makonine, two minor alkaloids of A. serrata, derive their names from the Maori name, makomako, for this plant.[41] Makomakine is a β-substituted indole derivative which, since it can be converted by hydrobromic acid into aristoteline (57), is formulated as (58). Tasmanine, a minor alkaloid of A. peduncularis,[42] is the related oxindole (59). Its structure and stereochemistry were deduced from a complete analysis of its ^{1}H and ^{13}C n.m.r. spectra, together with double-resonance experiments. A closely similar set of n.m.r. signals in the spectra of aristoserratenine, the indolenine analogue of tasmanine, presumably indicates the same stereochemistry in these two alkaloids.

Makomakine (58)   47% HBr →   Aristoteline (57)

Tasmanine (59)

Makonine shares many of the characteristics of aristotelinone (60), but has one additional olefinic carbon atom, and the non-indolic nitrogen appears to be tertiary. Its structure (61) becomes clear from its formation by the oxidation of aristotelinone

(60) with mercuric acetate.[41]

Aristomakine, the third new alkaloid from A. serrata, is isomeric with aristoteline (62), but has an olefinic double bond not conjugated with the indole nucleus.[43] Since the substitution pattern in the region of the indole ring is the same in aristomakine and aristoteline (n.m.r. spectra), and since the non-indolic nitrogen atom is secondary, and carries an isopropyl group, aristomakine is formulated as the product (63) of fission of the 13,14 bond in aristoteline, with introduction of a 14,15 double bond. This conclusion was confirmed by a thorough analysis of its n.m.r. spectra.[43]

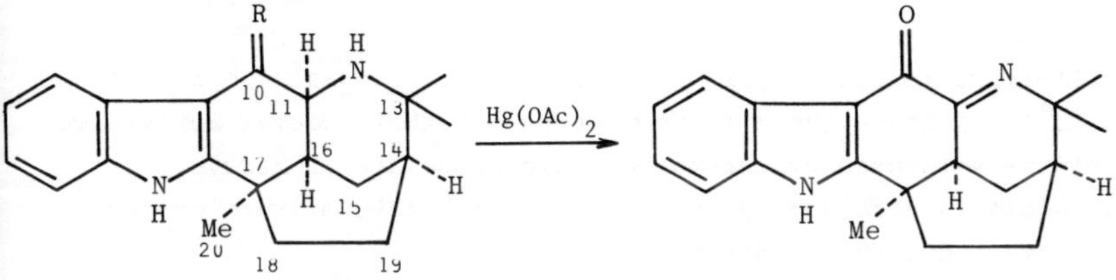

Aristotelinone (60) R = O
Aristoteline (62) R = H_2

Makonine (61)

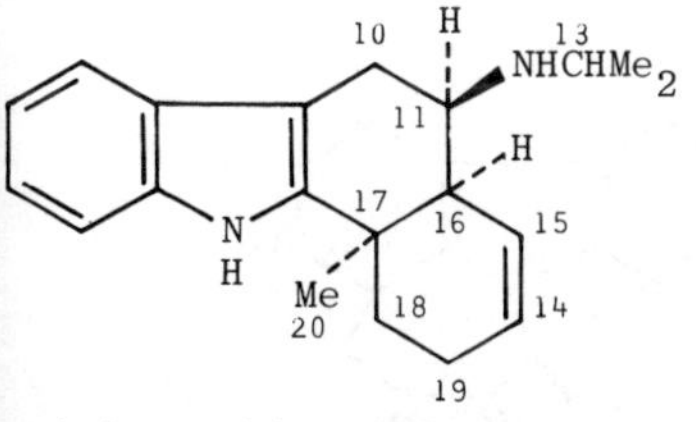

Aristomakine (63)

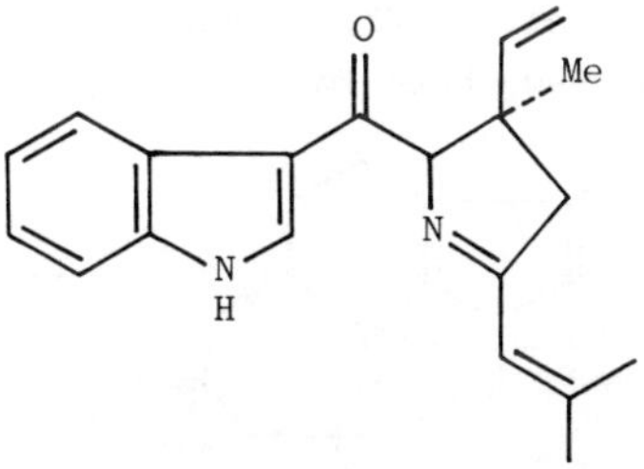

Dehydroborrecapine (64)

Two new alkaloids have also been isolated from Borreria capitata.[44] One of these is dehydroborrecapine (64); the other, borrecoxine, is simply the epoxide of dehydroborrecapine at the isopropylidene group.

4.2 Corynantheine-Heteroyohimbine-Yohimbine Group, and Related Oxindoles.— The chemotaxonomic significance of the pattern of indole alkaloid content in plants of the tribe Naucleeae (fam. Rubiaceae) has been discussed in some detail.[45]

In a search for new biosynthetic intermediates among the minor constituents of the leaves of Pauridiantha lyalli (Baker) Brem, Levesque et al. have isolated two new glycosides, which prove to

6'-*trans*-Feruloyl-lyaloside (65) R = H
6'-*trans*-Sinapoyl-lyaloside (66) R = OMe

be 6'-trans-feruloyl-lyaloside (65) and 6'-trans-sinapoyl-lyaloside (66).[46] Neither these two compounds nor their acetyl derivatives could be separated by chromatographic procedures; however, methanolysis afforded lyaloside and a separable mixture of methyl ferulate and methyl sinapate.

New extractions of the leaves of Nauclea latifolia Sm. have resulted in the isolation of two further alkaloids, of which naucleofoline (67) is new.[47] The second alkaloid, nauclechine, has already been found in N. diderrichii.

Naucleofoline (67)

3,14-Dihydrodecussine (68) R = H
10-Hydroxy-3,14-dihydro-
          decussine (69) R = OH

Strychnos decussata (Pappe) Gilg. is another plant which has been subjected to re-examination.[48a] Four new alkaloids have been isolated from the stem bark, in addition to those found earlier. Two of these new bases were shown to be 3,14-dihydrodecussine (68) and its 10-hydroxy derivative (69); the third, rouhamine, is obviously closely related to decussine, since it is rapidly formed by decomposition of decussine, but its structure is still unknown. The fourth alkaloid isolated is bisnordihydrotoxiferine. Decussine and dihydrodecussine were also found in the stem bark of S. dale and S. elaeocarpa.[48a] Decussine and rouhamine also occur, together with akagerine, in the stem bark of S. floribunda.[15c]

Normelinonine B accompanies eighteen alkaloids of the strychnine group in the root bark of Sri Lankan S. nux vomica.[48b]

Sixteen alkaloids have been isolated from the leaves, stem bark, and root bark of Anartia meyeri (G. Don) Miers (sub-tribe Tabernaemontanineae), a shrub found in Guyana, which has not previously been studied. Of these, angustine and 16-epi-pleiocarpamine belong to this group.[49] Peschiera echinata (Aublet) A. DC., another Guyanese member of the sub-tribe Tabernaemontanineae, also appears to be a prolific source of indole alkaloids, and has so far yielded 22 alkaloids, which include angustine, 16-epi-isositsirikine, and pleiocarpamine.[50]

The presence of vallesiachotamine in the leaves and roots of Rhazya stricta has again been noted.[51] The leaves of Alstonia lanceolifera S. Moore have so far yielded four alkaloids, including 10-methoxydeplancheine.[52] This is the first reported extraction of this species; the plant material extracted in 1975, which was claimed to be A. lanceolifera, was subsequently[24c] identified as A. boulindaensis Boiteau.

Strychnohirsutine and tetradehydrostrychnohirsutine, two new alkaloids isolated from the stem and root bark of Brazilian Strychnos hirsuta Spruce ex Bentham, have been shown to be the tetrahydro-β-carboline derivative (70) and the related β-carboline, respectively.[53] The ring system present in (70) has been encountered previously in talbotine (71), although in the latter, ring E has a different biogenesis, as the numbering in (70) and (71) indicates.

Rotundifoleine N_b-oxide has been found in Mitragyna rubrostipulata[54] and mavacurine and fluorocurine in the root bark of Strychnos variabilis.[55]

The leaves and stems of Rauwolfia volkensii Stapf , from the Kilimanjaro region, having been studied earlier, Akinloye and Court have now turned their attention to the roots. Twenty-six alkaloids were identified, including (in this group) tetrahydroalstonine, melinonine-A, isoreserpiline, reserpiline, aricine,

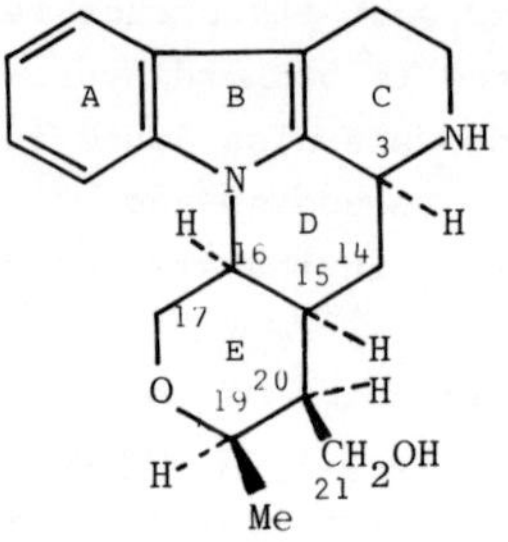

Strychnohirsutine (70)

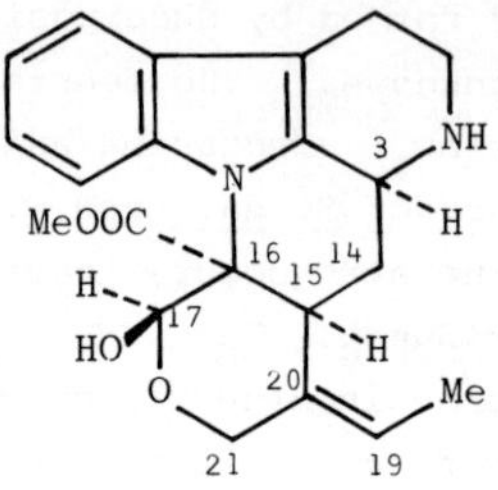

Talbotine (71)

ajmalicine, yohimbine, α-yohimbine, carapanaubine, isoreserpiline pseudoindoxyl, reserpine, rescinnamine, methyl deserpidate, 18-hydroxyyohimbine, alstonine, and a new alkaloid, rauvolcinine, which is tentatively formulated as 17,20-epicabucinine (72), on the basis of its spectrographic properties.[56] Rauvolcinine would thus appear to be a new member of the small group of heteroyohimbine alkaloids in which ring E is a hydroxydihydro derivative of that found in the vast majority of alkaloids of this group.

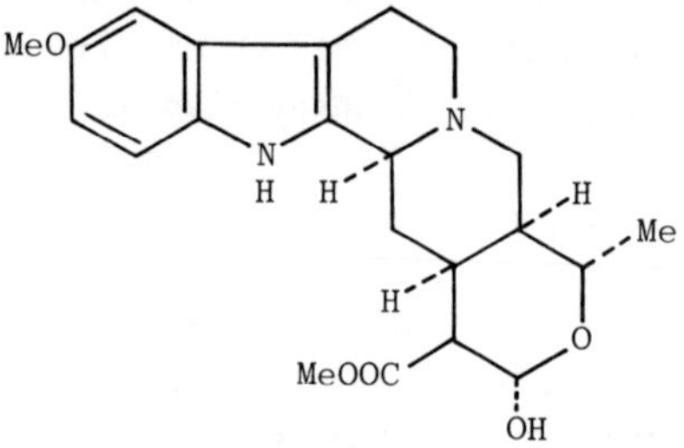

Rauvolcinine (72)

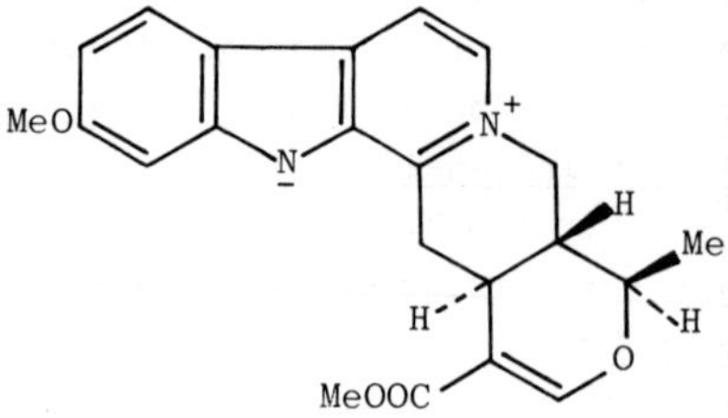

Serpenticine (73)

New extractions of <u>Uncaria elliptica</u> have demonstrated the variability of this species.[57] Whereas alkaloids of the roxburghine group appear to be confined to <u>U</u>. <u>elliptica</u>, four specimens of this plant, collected in S. Thailand, contained no roxburghines, and the major alkaloid proved to be different in each of the four samples. These four alkaloids were identified as rauniticine, isorauniticine (the first time this alkaloid has been encountered in this genus), 19-epiajmalicine, and 19-epi-3-isoajmalicine. None of these alkaloids has previously been isolated from <u>U</u>. <u>elliptica</u>.[57]

Serpenticine is a new anhydronium base isolated from <u>Rauwolfia</u> <u>omitoria</u> Afz.[58] Since, on the basis of its spectroscopic properties, tetrahydroserpenticine is formulated as 19-epitetra-hylline, serpenticine must have the constitution expressed in 73).

The alkaloid content[59] of Kenyan <u>Hunteria zeylanica</u> presumably <u>H</u>. <u>zeylanica</u> var. <u>africana</u>) shows some striking differences from that reported[12c] for Sri Lankan <u>H</u>. <u>zeylanica</u> (variety unknown). The Kenyan plants yielded twenty alkaloids, including isositsirikine, geissoschizol, pleiocarpamine, and yohimbol; dihydrocorynantheol, epiyohimbol, and 17-hydroxy-16-desmethoxycarbonyl-16,17-dihydro-epiajmalicine, isolated[12c] from the Sri Lankan variety, were apparently not found.[59]

<u>Rauwolfia nitida</u> Jacq. is a large shrub or small tree indigenous to the West Indies, extracts of the roots of which have been used in native medicine as an emetic and cathartic. Three previous investigations resulted in the isolation of eleven alkaloids from the leaves and roots. A more thorough examination of the roots[60] has revealed the presence of 33 alkaloids, of which alstonine, serpentine, ajmalicine, tetraphylline, 3-isoajmalicine, reserpine, isoreserpiline, isoreserpinine, reserpiline, yohimbine, 11-methoxy-yohimbine, α-yohimbine, alloyohimbine, 17-$\underline{O}$-acetyl-alloyohimbine, 18-hydroxy-yohimbine, deserpidine, pseudoreserpine, and reserpinine belong to this group.[60] No new alkaloids were detected.

16-Epivenenatine (74a) and 16-epialstovenine (74b) are two new alkaloids found in the root bark of <u>Alstonia venenata</u>.[61]

Tissue[62] or cell[63] cultures of <u>Catharanthus roseus</u> continue to be investigated, in the search for improved methods of production of pharmacologically important alkaloids. Three such investigations have recently been reported, and although several alkaloids were obtained, including some heteroyohimbine and yohimbine alkaloids, no new ones were encountered, and none of the valuable bisindole alkaloids was obtained. However, pleiocarpamine, one of the alkaloids isolated, had not previously been found in <u>C</u>. <u>roseus</u>.[62]

The ^{13}C n.m.r. spectra of 3-isoajmalicine, 19-epiajmalicine, and 3-isorauniticine,[64a] as well as those of pteropodine, isopteropodine, and five related oxindole bases,[64b] have been reported and analyzed.

16-Epivenenatine (74a) 3β-H
16-Epialstovenine (74b) 3α-H

(75)

The C-3, N_b bond in tetrahydro-β-carboline alkaloids can be reductively cleaved by means of chloroformic ester and sodium cyanoborohydride; tetrahydroalstonine, for example, affords the tetracyclic products (75).[65] This is a potentially useful reaction which has already been applied with success in the synthesis of mavacurine.[12d]

It has been proposed[24d] that the biogenesis of the less abundant 19R heteroyohimbine alkaloids involves 1,4-addition of an enol to a Z-alkene, e.g. (76). A similar intermediate is presumably involved in the conversion of cathenamine (77a) into 19-epicathenamine (77b) by means of alumina in chloroform. However, only the E-alkene (78) has been isolated from Guettarda eximia, and the related alkaloids geissoschizine and the isositsirikines all have the E configuration. Evidence for the formation of (76) has now been obtained[66] by the reduction of 4,21-dehydrogeissoschizine chloride (78) by means of sodium borohydride in methanol, which gave the Z-isositsirikines (79) in addition to the expected E-isositsirikines, tetrahydroalstonine, and 19-epiajmalicine. Evidently 4,21-dehydrogeissoschizine (78) equilibrated with its Z-isomer (76), presumably via the dienamine (80); reductive trapping of (76) then gave, as minor products, the Z-isomers (79).

There has again been no dearth of synthetic activity in this area during the period under review. Three new syntheses of (±)-deplancheine (81) include one[67] in which the alkaloid is directly obtained by reduction of the vinylogous amide (82); no trace of the Z-isomer of (81) was found (Scheme 9). A second group of workers has developed[68] two routes to deplancheine, of which the preferred route is also outlined in Scheme 9. The

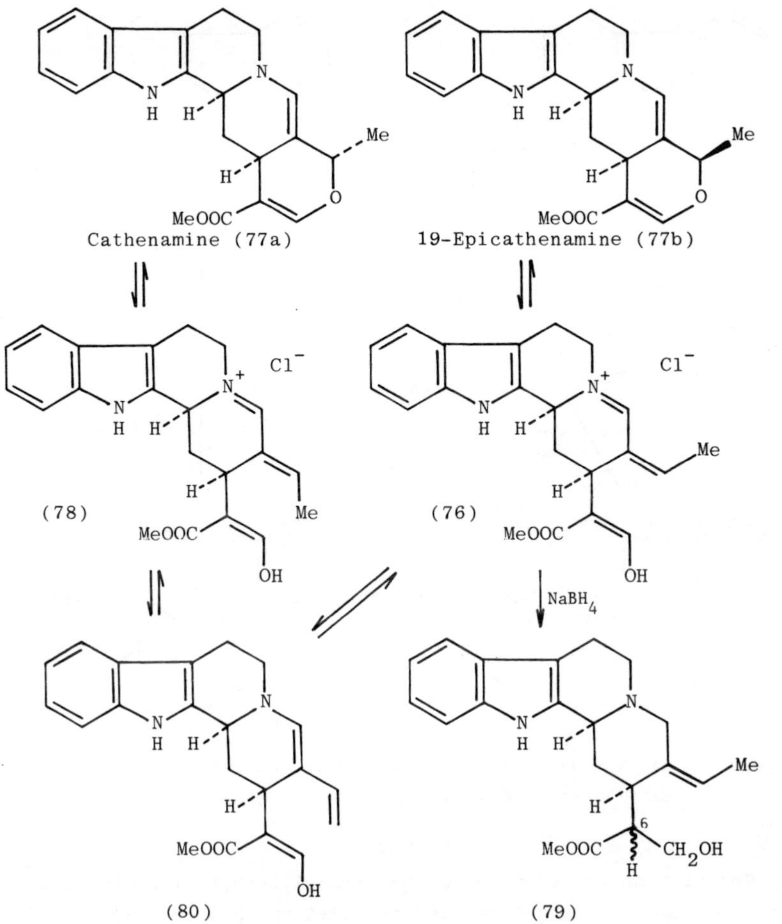

Indole Alkaloids

Cathenamine (77a)

19-Epicathenamine (77b)

(78)

(76)

(80)

(79)

enamine derived from 1-methyl-3,4-dihydro-β-carboline (83) was added to methyl 2-(diethylphosphono)acrylate (84) to give, after reduction of the adduct with sodium borohydride, the lactam (85). A Wittig-Horner reaction of the anion from (85) with acetaldehyde gave the enamide (86), careful reduction of which gave (±)-deplancheine. In the penultimate stage [(85)→(86)], again no trace of Z-isomer was obtained.[68]

A new synthesis of indoloquinolizidines substituted at positions 15 and 20 (indole alkaloid numbering) may prove useful in the synthesis of alkaloids of the corynantheine group.[69]

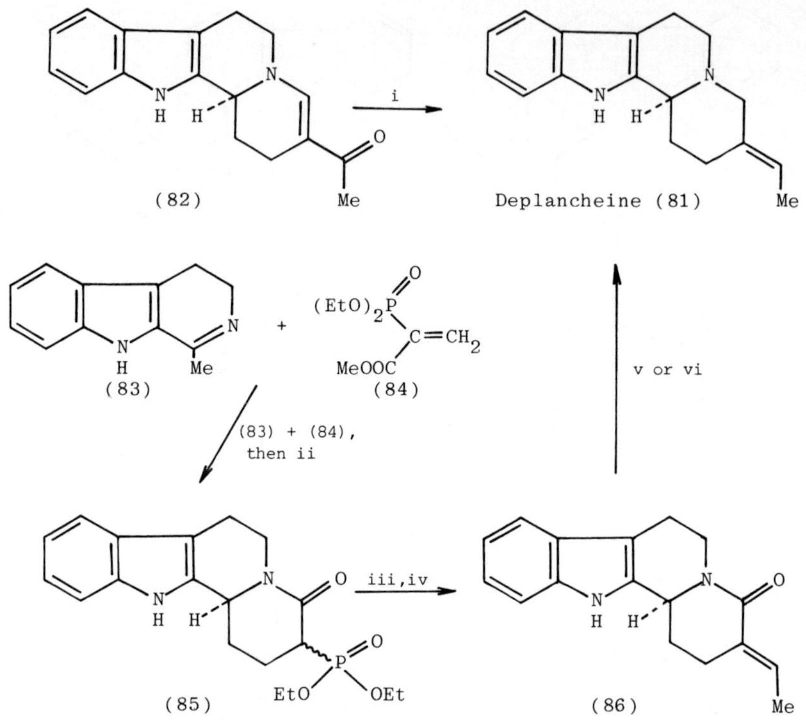

i
(82)
Deplancheine (81)
(EtO)$_2$P
C=CH$_2$
MeOOC
(83)
(84)
v or vi
(83) + (84), then ii
iii,iv
EtO
OEt
(85)
(86)

Reagents: i, $LiAlH_4$; ii, $NaBH_4$, MeOH, $0^{o}C$; iii, NaH, DME; iv, MeCHO; v, $LiAlH_4$, DME, $-78^{o} \rightarrow 0^{o}C$; vi, AlH_3, Et_2O, DME.

Scheme 9

The dithionite reduction of N-(β-indolylethyl)-isoquinolinium salts, followed by acid-catalyzed cyclization, has been applied[70] to the synthesis of dihydrogambirtannine; obvious stages then led to ourouparine and gambirtannine. Other gambirtannine derivatives have been prepared by the enamine photocyclization route.[71]

The dithionite reduction of isoquinolinium salts has also been applied to the synthesis of the (±)-desmethylhexahydrovallesiachotamine lactones (87a) and (87b); these were also prepared from vallesiachotamine (88) (Scheme 10).[72] This appears to be the first correlation between synthetic and natural vallesiachotamine derivatives.

Some progress towards the synthesis of vinoxine has been made by the synthesis of the desired ring system (89a, b) by mercuric

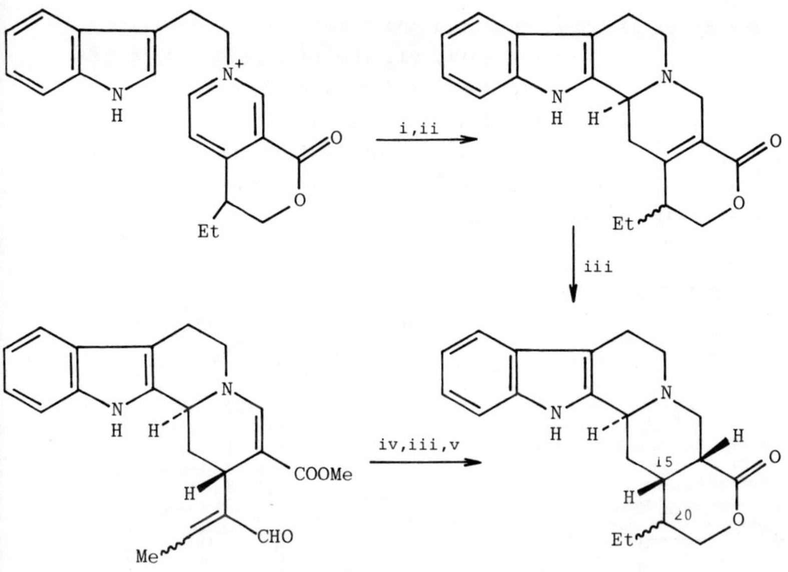

Scheme 10

Vallesiachotamine (88)

(87a) 20S
(87b) 20R

Reagents: i, $Na_2S_2O_4$; ii, MeOH, HCl; iii, H_2, PtO_2; iv, $NaBH_4$; v, $NaBH_4$, AcOH.

Scheme 10

acetate oxidation of the N-(4-pyridylmethyl)-indole derivatives (90a, b), followed by cyclization (Scheme 11).[73]

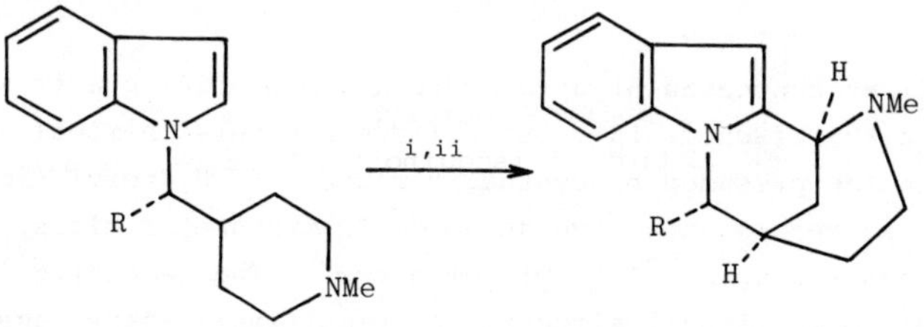

(90a) R = H
(90b) R = COOMe

(89a) R = H
(89b) R = COOMe

Reagents: i, $Hg(OAc)_2$, H_2O, EDTA (disodium salt); ii, $NaBH_4$.

Scheme 11

Takano et al. have reported the first enantioselective synthesis of (-)-antirhine (91), via the lactone-aldehyde (93) (Scheme 12).[74] This intermediate (93) affords promise as a useful synthon for indole alkaloid synthesis; although it was prepared by a tortuous 16-stage synthesis from the chiral lactone (92), itself prepared[75] from L-glutamic acid, the overall yield of (93) from (92) is claimed[74] to be 14%.

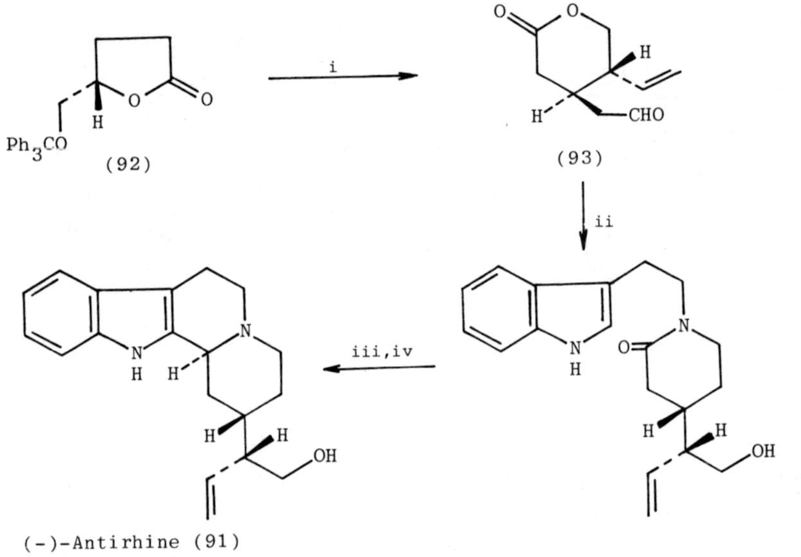

Reagents: i, 13 steps; ii, tryptamine, $NaBH_3CN$, H_2O, MeOH, pH 6; iii, DIBAL, $-78°C$; iv, H^+, H_2O

Scheme 12

The first synthesis of 3α-dihydrocadambine (94) starts with secologanin (95) (Scheme 13), but follows a purely chemical course, rather than the presumed biosynthetic route.[76] Hydroxylation of the protected secologanin gave an epimeric mixture of diols, in which the desired epimer (96) predominated. The remaining stages are unexceptional, although the penultimate stage (acetal hydrolysis followed by cyclization) has so far been achieved only in low yield through the agency of formic acid; a variety of other acidic conditions failed to give the desired product.

Kametani's enamide (97)[12e] has been converted into

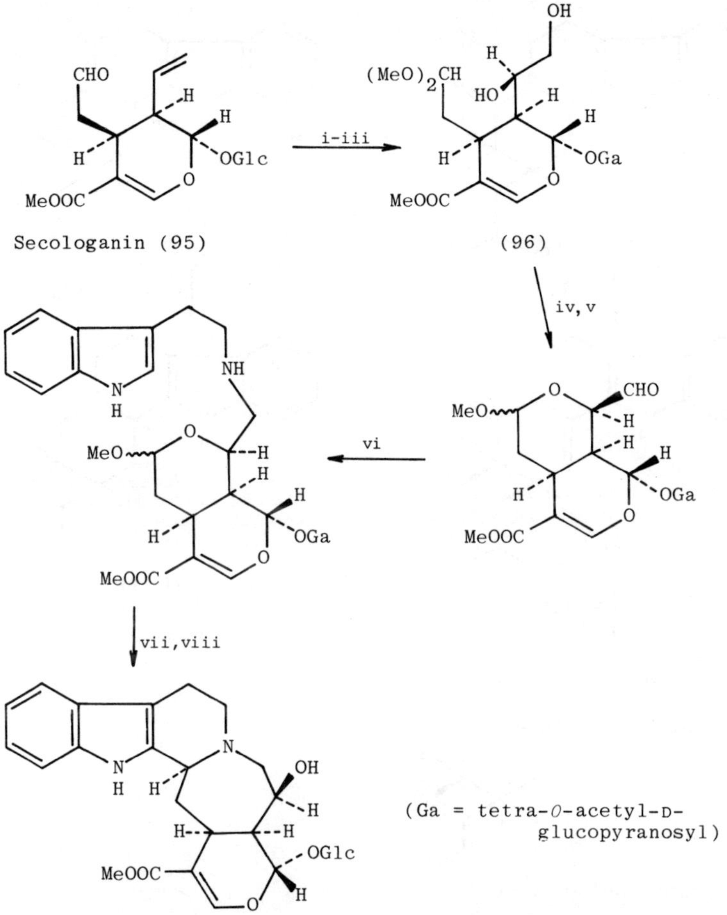

Indole Alkaloids

Secologanin (95) → (96)

3α-Dihydrocadambine (94)

(Ga = tetra-*O*-acetyl-D-glucopyranosyl)

Reagents: i, Acetylation; ii, acetal formation; iii, hydroxylation; iv, Amberlite IR 120, MeOH; v, PCC, CH_2Cl_2, NaOAc, Celite; vi, tryptamine, $NaBH_3CN$; vii, 90% HCO_2H, Δ; viii, K_2CO_3, MeOH.

Scheme 13

(±)-corynantheal (98) (Scheme 14),[77] thereby completing another formal synthesis of corynantheine and ajmalicine. A point of stereochemical interest in this synthesis follows from the hydrogenation of (97), which gives a mixture of C-3 epimeric products; however, it was found that the alcohols (99a, b) resulting from reduction of this mixture by means of $LiAlH_4$ could

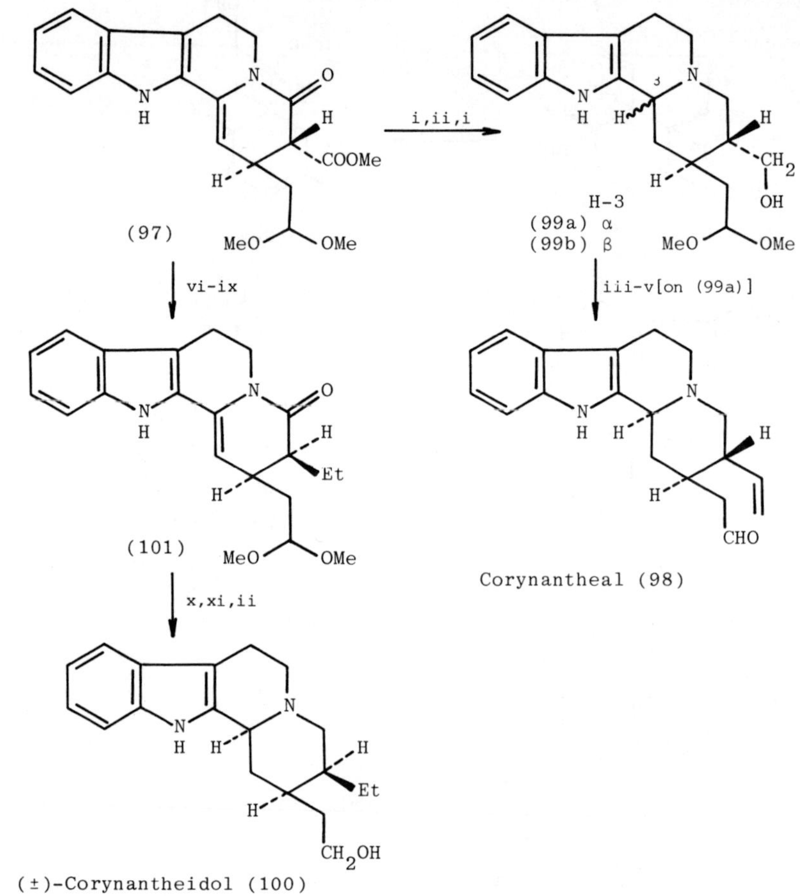

Reagents: i, H_2, PtO_2; ii, LiAlH$_4$; iii, DMSO, DCC, TFA, py; iv, $H_2C=PPh_3$;
v, TsOH, acetone; vi, NaH, DMF, EtI; vii, KOH, MeOH; viii, DMF, PhH,
at 100°C; ix, separation of diastereoisomers; x, H_2, Pd/C; xi, TsOH.

Scheme 14

be quantitatively converted into the desired <u>normal</u> isomer (99a) by
prolonged treatment with Adams' catalyst in methanol under hydrogen.

In variants on their earlier synthesis[12e] of dihydrocorynantheol, the same group of workers have used the enamide (97) in syntheses of (±)-corynantheidol (100) (Scheme 14) and (±)-hirsutinol [3,20-epimer of (100)].[78] In the former sequence, hydrogen-

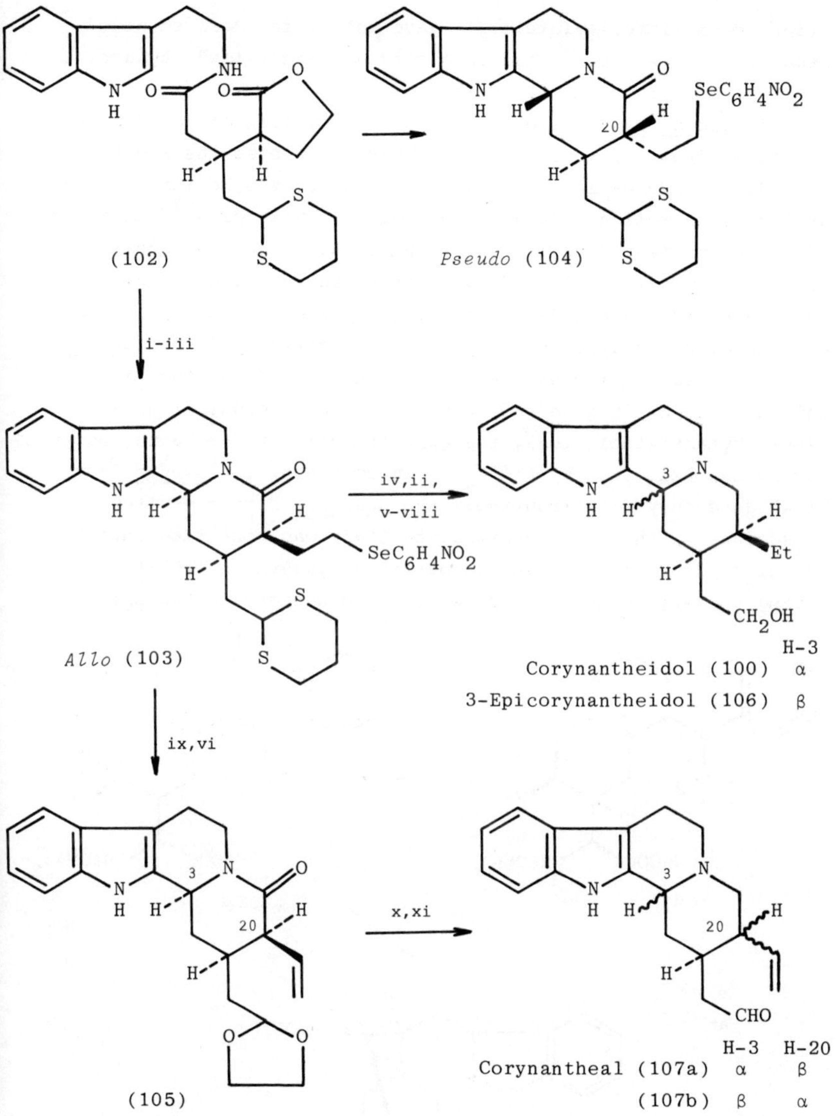

Indole Alkaloids

(102)

Pseudo (104)

Allo (103)

Corynantheidol (100) H-3 α
3-Epicorynantheidol (106) β

(105)

Corynantheal (107a) H-3 α H-20 β
         (107b)   β   α

Reagents: i, POCl$_3$, CH$_2$Cl$_2$, Δ; ii, NaBH$_4$, MeOH; iii, O$_2$N-C$_6$H$_4$SeCN, Bu$_3$P, THF; iv, MeI, H$_2$O, MeCN; v, Ac$_2$O, NaOAc; vi, m-Cl-C$_6$H$_4$CO$_3$H; vii, H$_2$, PtO$_2$; viii, LiAlH$_4$; ix, (CH$_2$OH)$_2$, TsOH, PhH; x, LiAlH$_4$—AlCl$_3$, THF, $-20°$ → $0°$C; xi, 60% AcOH, Δ.

Scheme 15

ation of the intermediate (101) gave solely the desired allo isomer, which was converted into corynantheidol (100) by acetal hydrolysis followed by reduction.

Takano et al. have extended their syntheses of corynantheine relatives from (±)-norcamphor, and have completed the synthesis of the 18,19-dihydro-alkaloids with all four stereochemical configurations, as well as three of the four Δ^{18} alkaloids.[79] Examples of these syntheses are given in Scheme 15. The previously prepared tryptamide (102)[12f] was converted into a mixture of the selenides (103) and (104), the latter (pseudo) isomer presumably arising by epimerization at C-20 of the initially formed epiallo isomer. Elaboration of the allo selenide (103) by familiar methods ultimately gave a mixture of corynantheidol (100) and 3-epicorynantheidol (106), the epimerization at C-3 having occurred during the reduction by LiAlH$_4$. In contrast, the pseudo isomer (104) gave only (±)-hirsutinol, the pseudo analogue of (100). Subsequently, the allo intermediate (103) was converted into the Δ^{18}-aldehydes, which proved to be (±)-corynantheal (107a) (C-20 having epimerized during the reaction sequence) and the epi-allo

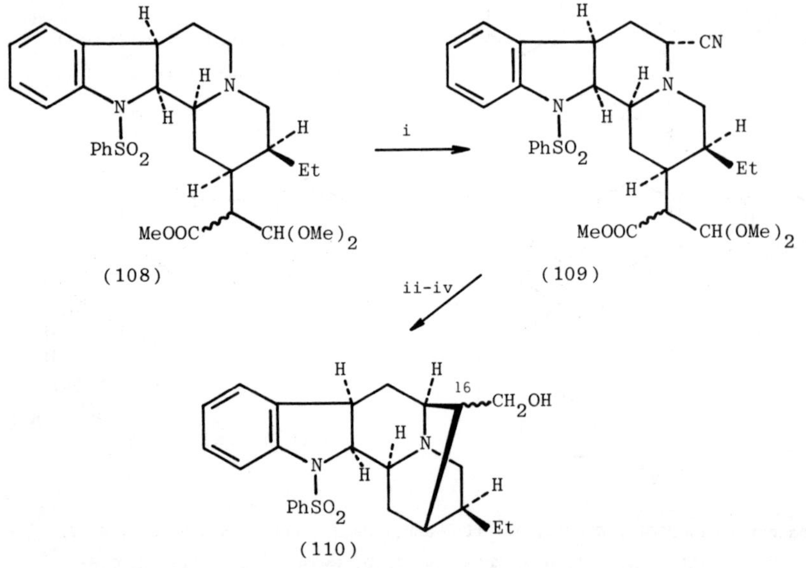

Reagents: i, O$_2$, hν, MeOH, KCN, MeCOCO$_2$Na, Rose Bengal; ii, NaOH, H$_2$O, dioxan; iii, AcOH; iv, NaBH$_4$, MeOH.

Scheme 16

isomer (107b). In confirmation, corynantheal (107a) was reduced (NaBH$_4$, then H$_2$-PtO$_2$) to dihydrocorynantheol, and (107b) was reduced to 3-epicorynantheidol (106).[79]

Some photochemical reactions of corynantheidine derivatives have been reported[80] that could find application in the partial synthesis of alkaloids of the sarpagine group. Thus, photosensitized oxidation of the corynantheidine derivative (108) gave an aminonitrile (109), which was transformed into the pentacyclic compound (110) (Scheme 16); unfortunately, the low yield obtained prevented its conversion into a known sarpagine derivative.

Details of the syntheses[81a] of ajmalicine and 19-epiajmalicine by Uskoković and his collaborators have now been published.[81b]

Sakai's partial synthesis of the heteroyohimbine alkaloids from the bicyclic intermediate (111) has been extended to the

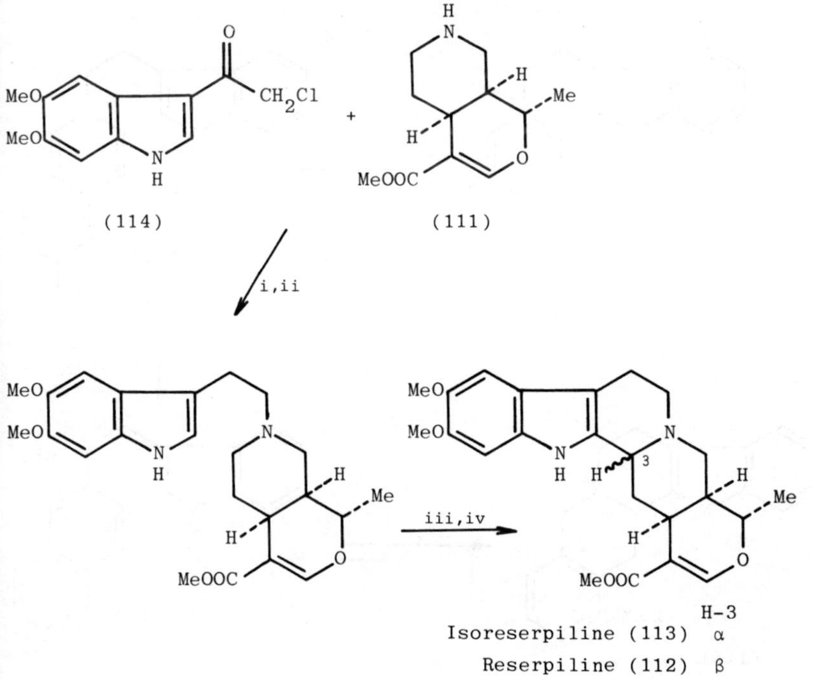

Isoreserpiline (113) H-3 α
Reserpiline (112) β

Reagents: i, NaHCO$_3$, DMF, 60°C; ii, NaBH$_4$ (excess), AcOH (excess), dioxan, Δ; iii, Hg(OAc)$_2$, AcOH; iv, NaBH$_4$.

Scheme 17

synthesis of reserpiline (112) and 3-isoreserpiline (113).[82] Owing to the sensitivity of 5,6-dimethoxytryptophyl bromide, it could not be used in condensation with (111); hence the chloroketone (114) was used instead, and the carbonyl group in the product was removed by reduction with a large excess of sodium borohydride in acetic acid. Otherwise the synthesis follows the pattern of that reported earlier (Scheme 17).

In the yohimbine series, a new synthesis of yohimbane, alloyohimbane, and alloyohimbone by the reductive photocyclization of enamides has been reported,[83] and details of an earlier synthesis of yohimbane and alloyohimbane have been published.[84]

A neat four-stage synthesis (Scheme 18)[85] of the pentacyclic ester (115) constitutes a short, formal synthesis of yohimbine, which has earlier been prepared from (115) by two groups of workers. The critical stage in the brief synthesis of (115)

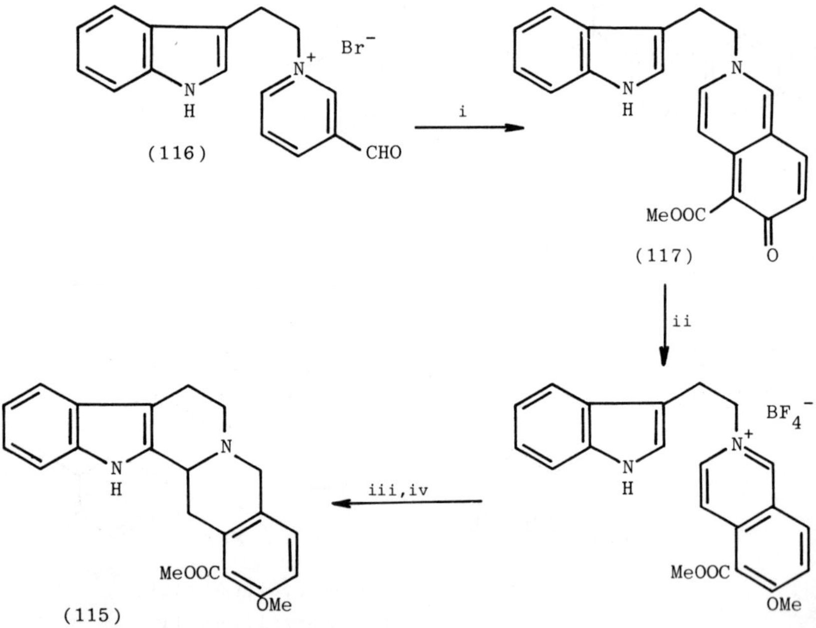

Reagents: i, MeCOCHNaCO$_2$Me; ii, Me$_3$O$^+$ BF$_4^-$, MeNO$_2$; iii, NaBH$_4$, THF; iv, TFA.

Scheme 18

involved 1,4-addition of sodioacetoacetic ester to the pyridinium salt (116). The resulting 1,4-dihydropyridine adduct cyclized by an internal aldol condensation; dehydration followed by dehydrogenation [with (116) as hydrogen acceptor?] then gave the tetracyclic product (117), which was converted into (115) by established methods.[85]

A novel approach to yohimbine alkaloid synthesis by Meyers et al.[86] may be illustrated by the synthesis of the pentacyclic lactam (118) (Scheme 19). The essential stage involves alkylation of an α-aminocarbanion derived from tetrahydro-β-carboline, the indole nitrogen atom having been first protected to avoid dianion formation. In spite of earlier reports, the methoxymethyl group was found to be a satisfactory protecting group, and could be cleanly removed by treatment with acid, then alkali.

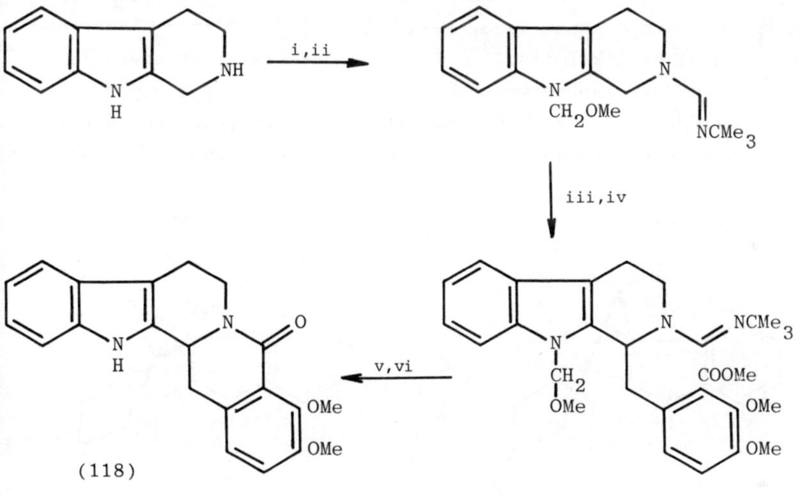

Reagents: i, $Me_3CN=CHNMe_2$, PhMe; ii, $KOBu^t$, Et_2O, $ClCH_2OMe$, 18-crown-6; iii, Bu^tLi, THF, $-25°$ C; iv, 2-ethoxycarbonyl-3,4-dimethoxybenzyl chloride; v, NH_2NH_2, AcOH, H_2O, MeOH; vi, 3M HCl, r.t., then NaOH, H_2O, r.t.

Scheme 19

4.3 Sarpagine-Ajmaline-Picraline Group. — New extractions have revealed the presence of normacusine B in the stem bark of

Strychnos dolichothyrsa[87] and, together with O-methylmacusine B and 16-epi-O-methylmacusine B, in the root bark of Sri Lankan S. nux vomica.[48b] Polyneuridine has been reported for the first time to occur in the leaves and roots of Rhazya stricta.[51] Rauwolfia volkensii roots contain[56] sarpagine, tetraphyllicine and its 17-O-acetate, perakine, peraksine, raucaffrinoline, ajmaline and its 17-O-acetate, and suaveoline.[56] R. nitida root bark contains a similar array of indole alkaloids, including normacusine B, vellosimine and its N_a-methyl derivative, lochnerine, sarpagine, nortetraphyllicine and its 17-O-acetate, tetraphyllicine, norajmaline, ajmaline, ajmalidine, peraksine, raucaffrinoline, and suaveoline.[60] Hunteria zeylanica leaves and stem bark have yielded the rather more elaborate members of this sub-group; thus, the leaves contain picrinine, picralinal, corymine, norisocorymine, 3-epidihydrocorymine and its 3-acetate and 17-acetate, and lanceomigine, and the roots contain a new base, 10-hydroxy-16-epiaffinine, together with nine other alkaloids.[59] It is of interest that there is little or no overlap between the alkaloids of the two organs.

Tabernulosine and 12-demethoxytabernulosine, two new alkaloids which occur with vincadiffine in the leaves and stems of W. African Tabernaemontana glandulosa Stapf, prove to be 10,12-dimethoxy-picrinine (119a) and 10-methoxypicrinine (119b), respectively.[88]

Tabernulosine (119a) R = OMe
12-Demethoxy-
  tabernulosine (119b) R = H

(120) R = H
(121) R = PhCO
(122) R = 3,4,5-(MeO)$_3$C$_6$H$_2$CO

In other extractions, vobasine has been isolated from the leaves of Peschiera echinata,[50] voachalotine, vincamajine, quaternine, and N_a-demethylquaternine occur in the trunk bark of New Caledonian Alstonia legouixiae van Heurck et Muell.,[89] and three new bases from this group, namely 10,11-dimethoxy-N_a-methyl-deacetylpicraline (120), its 17-O-benzoate (121), and its 17-O-(3,4,5-trimethoxybenzoate) (122), have been found in the leaves of Alstonia lanceolifera.[52]

A quaternary alkaloid, designated Alkaloid Q_3, isolated from Aspidosperma peroba F. Allem ex Sald., is tentatively formulated as the N_b-methyl-16-epi-pericyclivine ammonium ion.[90] Finally, raucaffricine and perakine have been isolated from Rauwolfia caffra Sond.,[91] vomilenine has been obtained from the tissue culture extracts of R. serpentina,[92] and akuammiline from the tissue culture of C. roseus.[62]

The configurations at C-2 and C-17 in ajmaline and several related alkaloids have been correlated with their n.m.r. spectra.[93] A stereospecific long-range coupling was observed between the β-proton at C-6 and an α-proton at C-17, while a β-proton at C-2 and an α-proton at C-17 gave rise to a n.O.e. of ∿ 10%. The ^{13}C n.m.r. spectra of ajmaline, isoajmaline, and their C-17 epimers sandwicine and isosandwicine,[94] as well as those of raucaffricine and perakine,[91] have been reported and analyzed.

It has been pointed out[95] that the ^{13}C n.m.r. data reported[12g] for ajmalinol, a new alkaloid of Rauwolfia vomitoria, are inconsistent with the proposed 11-hydroxyajmaline structure. Some of the data are very close to those reported for ajmaline itself, others are consistent neither with an ajmaline nor an 11-hydroxy-ajmaline structure. Hence the structure of ajmalinol is at present obscure.

A second group of workers has elucidated the structure of koumine (123) by X-ray analysis of its hydrobromide; n.m.r. data, and some degradation experiments, were also reported.[96]

Koumine (123)

The reactions of sandwicine and several of its derivatives, and notably their behaviour in the von Braun degradation, have been investigated in a study of structure - anti-arrhythmic activity relationships in the ajmaline-sandwicine group of alkaloids.[97]

The laboratory conversion of dregamine (124)[98] into 20-epi-ervatamine (125) suggests that alkaloids of the vobasine group may be involved in the biosynthesis of the ervatamine group of alkaloids. Support for this view comes from the report that the same conversion (Scheme 20) can be enzymically achieved by means of rat liver microsomes in the presence of NADPH and oxygen.[99]

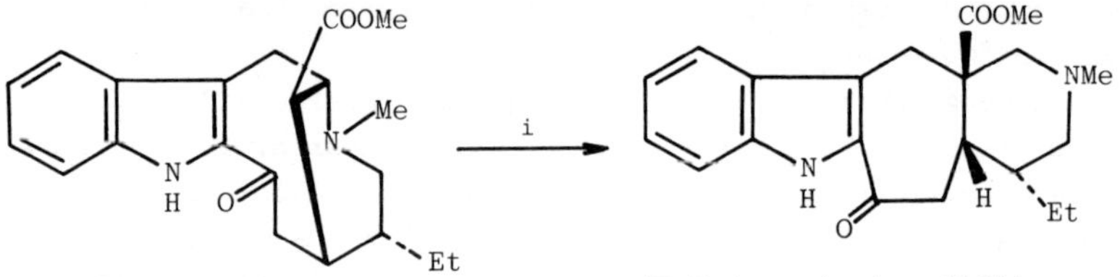

Dregamine (124)　　　　　　　　　20-Epiervatamine (125)

Reagent: i, Rat liver microsomes, NADPH, O_2.

Scheme 20

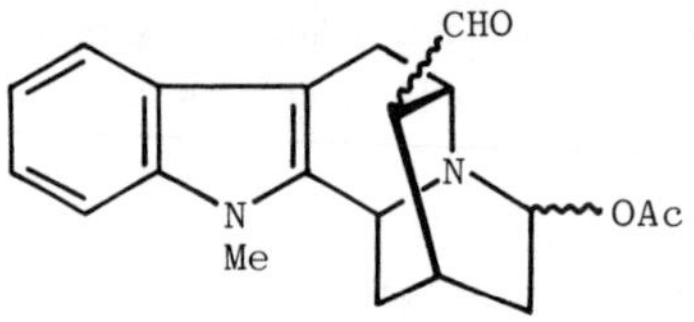

(126) 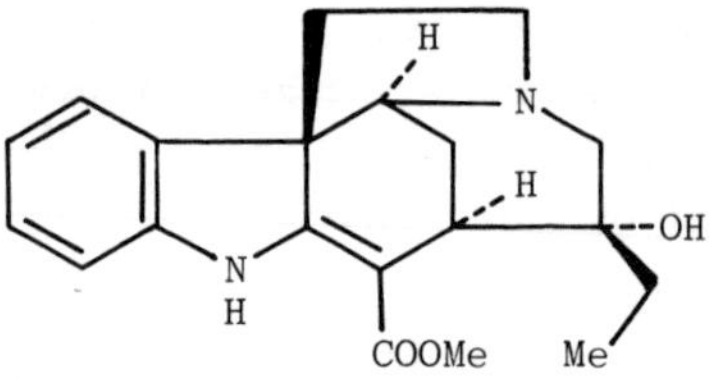　　　　　　　　(127)

A new approach to the synthesis of ajmaline has been described,[100] which to date has progressed as far as the epimeric aldehydes (126).

4.4 Strychnine-Akuammicine-Ellipticine Group. — Akuammicine has been found in the roots of Rauwolfia volkensii,[56] and the presence of its 10-hydroxy derivative, sewarine, in the leaves and roots of Rhazya stricta has again been noted.[51] Tubotaiwine has been isolated from the leaves of Anartia meyeri,[49] from the stem bark of Hunteria zeylanica,[59] and from the leaves of Peschiera echinata.

livacine occurs in the stem and root bark of this last species.⁵⁰
eacetylisoretuline and isorosibiline occur in the stem bark of
Strychnos floribunda,¹⁵ᶜ and tubotaiwine, condylocarpine, nor-C-
luorocurarine, the Wieland-Gumlich aldehyde and its 18-deoxy
erivative, and 11-methoxydiaboline in the stem bark of S. dolicho-
thyrsa.⁸⁷ This last alkaloid was also found in the stem bark of
S. urceolata.⁸⁷

Eighteen of the twenty-two alkaloids isolated from the root
ark and leaves of Sri Lankan S. nux vomica are strychnine
erivatives, and include several that have not previously been
obtained from this species.⁴⁸ᵇ The alkaloids identified were
strychnine, its 10- and 12-hydroxy derivatives, β-colubrine,
rucine, 12-hydroxy-11-methoxystrychnine, the N_b-oxides of
strychnine, brucine, 12-hydroxystrychnine and 12-hydroxy-11-methoxy-
strychnine, isostrychnine I and its 19,20-dihydro derivative,
rotostrychnine, vomicine, 3-hydroxystrychnine, 3,12-dihydroxy-
strychnine, 3,12-dihydroxy-11-methoxystrychnine, and 3-hydroxy-10,
11-dimethoxystrychnine.

The alkaloid content of these specimens of S. nux vomica
clearly differentiates them from Indonesian plants, which, for
example, contain no 12-hydroxy-11-methoxystrychnine and only traces
of 12-hydroxystrychnine. In view of this, it seems likely that the
Sri Lankan variety is a hybrid of the true S. nux vomica and the
jungle liane S. wallichiana Steud. ex DC. Certainly the alkaloid
content is that expected from such a hybrid, although morphologic-
ally S. nux vomica appears to be the dominant plant.⁴⁸ᵇ

Some further ¹³C n.m.r. data on pyridocarbazoles related to
ellipticine have been published.¹⁰¹ The hydroboration-oxidation
of akuammicine results unexpectedly in Markownikov hydration of the
19,20 double bond, to give 20α-hydroxy-19,20-dihydroakuammicine
(127).¹⁰² Presumably initial complexing of the borane with N_b
directs the regiochemical course of the reaction.

A second report on the Leuchs sulphonation of strychnine has
been published.¹⁰³

Brucine is recommended¹⁰⁴ as a convenient and effective
reagent for the resolution of tertiary acetylenic alcohols. The
evidence available to date indicates that the acetylenic alcohol
must have two aryl groups, or one aryl group and one bulky alkyl
group, attached to the hydroxyl-bearing carbon atom. The X-ray
crystal structure analysis of one such brucine complex was also
reported.

As in recent years, much of the new synthetic work in this area has been devoted to the ellipticine group of alkaloids. There are, however, two notable new contributions in the strychnine area. Takano's approach[105] involves sulphonium salt formation on the thiolactam (128) followed by proton abstraction and thio-Claisen rearrangement, which gives the intermediate (129). Conventional stages then lead to the tetracyclic base (130) (Scheme 21), which has already been converted into tubifoline, tubifolidine, and tubotaiwine by Schumann and Schmid.[106] Polonovski-Potier reaction on the N_b-oxide of (130) gave exclusively the product (131), having the Strychnos framework; none of the alternative cyclization product having the aspidospermatidine skeleton was obtained.

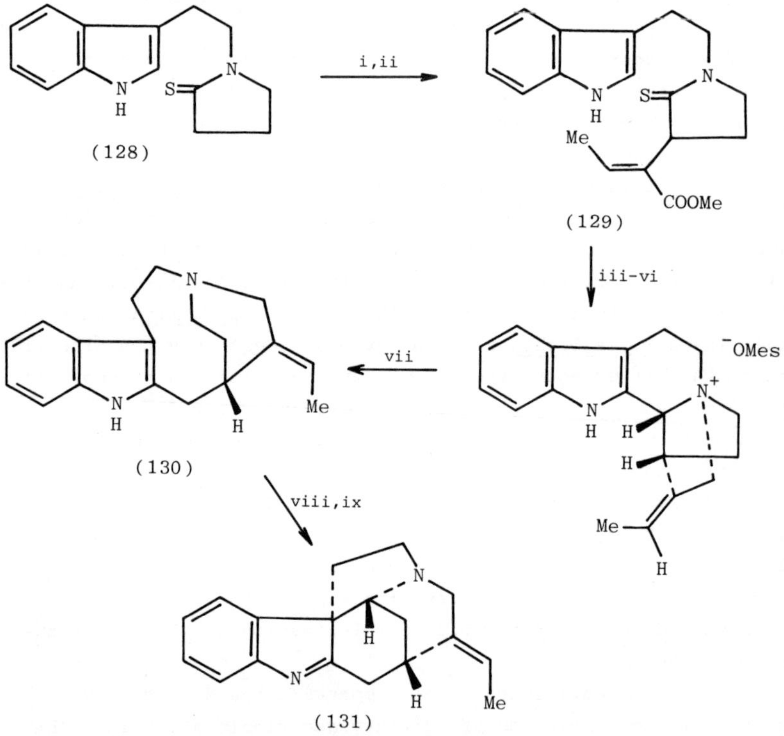

Reagents: i, $BrCH_2CH=CH-CO_2Me$, THF; ii, NaOMe, THF; iii, $POCl_3$; iv, $NaBH_4$; v, DIBAL, PhH; vi, MesCl, py; vii, Na, NH_3; viii, m-ClPBA; ix, $(CF_3CO)_2O$, CH_2Cl_2, $-78°C \rightarrow$ r.t.

Scheme 21

Indole Alkaloids

Ban's general entry[107] to the Strychnos and Aspidosperma alkaloids makes ingenious use of a novel photoisomerization of 1-acylindole derivatives to 3-acylindolenines. When applied to the tryptamine derivative (132), the derived indolenine (133) spontaneously rearranged, with formation of the indole-amide (134) in 80% yield (Scheme 22). Obvious stages then led to (135), which has already been converted into tubifoline and condyfoline by Harley-Mason et al.[108]

In the ellipticine series, Sainsbury et al.[109] have developed an improved synthesis of 3-[1-(3-ethylpyridyl)]-indoles, which were important intermediates in an earlier route to ellipticine. Several ellipticines, including 11-hydroxy- and 10-methoxy-ellipticine, were prepared by the new route.

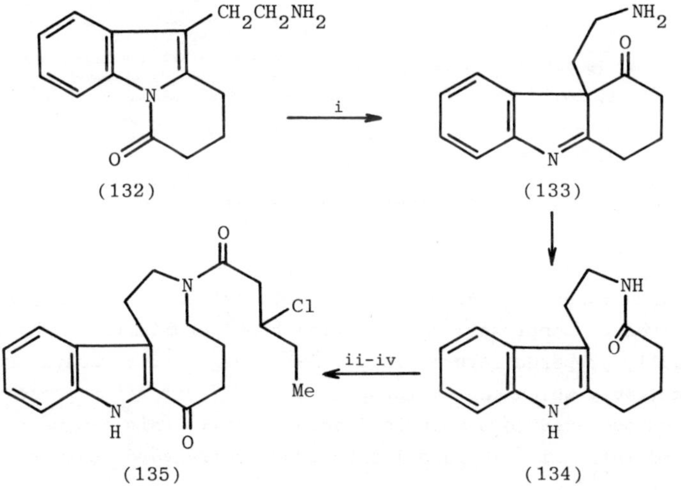

Reagents: i, MeOH, hν; ii, LiAlH$_4$; iii, EtCHClCOCl; iv, I$_2$O$_5$, THF, H$_2$O

Scheme 22

Other synthetic work includes a new synthesis[110] of ellipticine (137), in which the critical stage involves a 1,5-sigmatropic shift of hydrogen in a readily synthesised intermediate (136), followed by thermal cyclization (Scheme 23); and the synthesis of 9,10,11-trimethoxyellipticine by the modified Cranwell-Saxton route.[111] An efficient synthesis of 'ellipticine quinone' affords a useful improvement in syntheses of ellipticine and its derivatives via

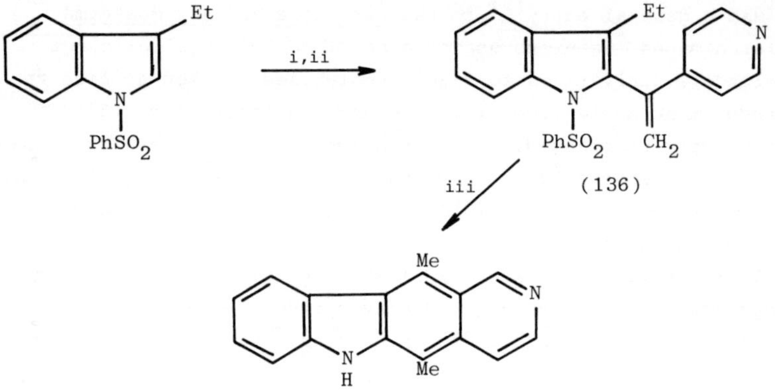

Ellipticine (137)

Reagents: i, Isonicotinic acid anhydride, $NaCH_2SOMe$; ii, $H_2C=PPh_3$; iii, 500° C.

Scheme 23

this intermediate.[112] A number of non-alkaloidal ellipticines have also been synthesised.[113]

A new synthesis of olivacine (138) (Scheme 24)[114] and guatambuine takes advantage of the stabilization of dihydropyridine derivatives by complexation with tricarbonylchromium(O). Thus, the readily prepared intermediate (139) gives a tricarbonylchromium complex (140), which can be formylated by the Vilsmeier procedure and dehydrogenated to demethylolivacine (141); this can be readily converted into olivacine and its N-methyl tetrahydro derivative, guatambuine.[114]

A new synthesis of desethyldasycarpidone has also been reported.[115]

4.5 Aspidospermine-Aspidofractine-Eburnamine Group. — Nine alkaloids have been extracted from the leaves and stems of Melodinus polyadenus Baill. Boit., six of which belong to this group; these are (−)-tabersonine, (+)-vincadifformine, (−)-11-methoxytabersonine, (−)-venalstonidine, (−)-venalstonine, and (+)-Δ^{14}-vincine.[116] (−)-12-Hydroxy-vincadifformine occurs in the leaves of Bonafousia tetrastachya (Humboldt, Bonpland, et Kunth.) Mgf.,[117] together with vincadifformine (previously noted), and kopsinine and eburnamine have been found in the stem bark of

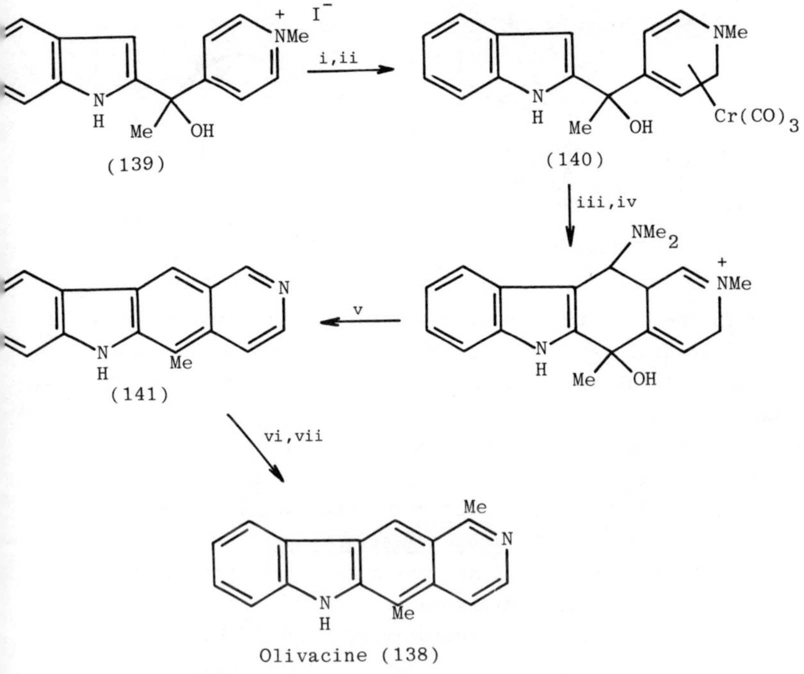

Olivacine (138)

Reagents: i, NaBH$_4$; ii, (MeCN)$_3$Cr(CO)$_3$; iii, DMF, POCl$_3$; iv, pyridine; v, Pd/C, 300° C; vi, MeLi, THF; vii, I$_2$.

Scheme 24

Hunteria zeylanica.[59]

Three dihydroindole alkaloids isolated from Kopsia officinalis Tsiang. are derived from fruticosine.[118] The structure of methyl 11,12-methylenedioxychanofruticosinate (Alkaloid C) (142) was determined by X-ray crystal structure analysis; it is thus only the third monoterpenoid indole alkaloid to possess a methylene-dioxy-group. Alkaloids A and B lack the methylenedioxy-group, and were deduced, on the basis of their ^{13}C n.m.r. spectra, to be methyl chanofruticosinate (143) and its de-N-methoxycarbonyl derivative (144).

The structure of 16βH-Δ^{14}-vincanol, a constituent of Melodinus celastroides, has been confirmed by X-ray crystal structure analysis.[119a]

The culture of several cell lines of Catharanthus roseus

(142)

(143) R = COOMe
(144) R = H

results in the formation of a complex mixture of indole alkaloids, including several Aspidosperma alkaloids, mainly of the anilino-acrylate or vindolinine sub-groups.[63] Tissue culture of C. roseus similarly produces several anilinoacrylate alkaloids, including 20R-hydroxytabersonine, which is now encountered for the first time in C. roseus.[62]

Dihydrovindoline suffers demethylation of its aromatic methoxy-group when incubated with Streptomyces griseus;[119b] this contrasts with the behaviour of vindoline, which is metabolised to products resulting from oxidation reactions in ring D.

Synthetic (±)-18-methylenevincadifformine[12h] has been resolved, and the dextrorotatory enantiomer shown, by comparison of c.d. spectra, to belong to the same stereochemical series as (+)-vinca-difformine and (+)-tabersonine.[120] Completion of the synthesis as before,[12h] using (+)-18-methylenevincadifformine, then established the absolute configuration of (−)-strempeliopine as shown in (145).

(−)-Strempeliopine (145)

The oxidative rearrangement of vincadifformine to vincamine has been much studied in recent years, and a number of procedures for this purpose have been developed. Perhaps the most efficient, however, is a one-pot process in which vincadifformine is ozonised at 60° C in dilute sulphuric acid-methanol; the product, obtained in 74% yield, is a 7:3 mixture of vincamine and 16-epivincamine.[121a] Similarly, tabersonine gives Δ^{14}-vincamine and its 16-epimer. The

Indole Alkaloids

same group of workers have also investigated the dye-sensitised photo-oxygenation of vincadifformine; after reduction of the reaction mixture with sodium thiosulphate, the related 16-hydroxy-indolenine derivative was obtained, which (without isolation) was transformed by rearrangement in acetic acid to vincamine in 46% yield.[121b] Tabersonine behaved similarly. The results are thus in broad agreement with those recently reported by Lévy and his collaborators.[12i]

This sub-group of alkaloids continues to be a popular area for synthesis, and several notable contributions have been published during the past year, including some which involve new methodology.

Wenkert et al.[122] have published details of their syntheses of quebrachamine, eburnamonine, vincadine, and epivincadine, and

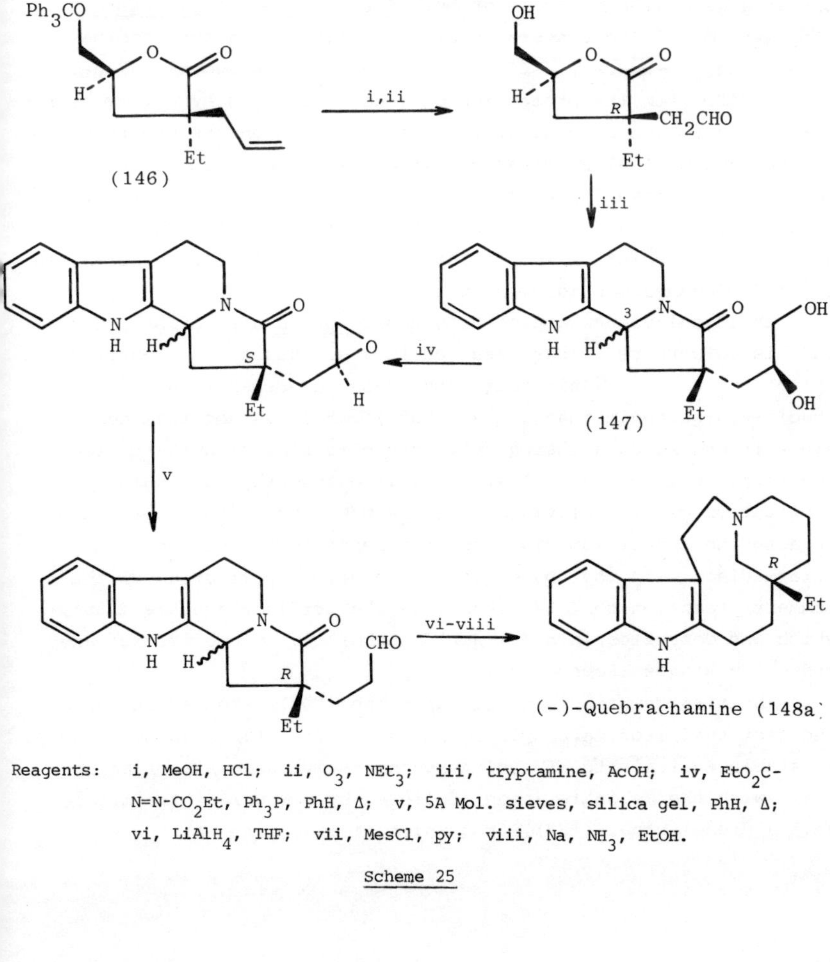

Reagents: i, MeOH, HCl; ii, O_3, NEt_3; iii, tryptamine, AcOH; iv, EtO_2C-N=N-CO_2Et, Ph_3P, PhH, Δ; v, 5A Mol. sieves, silica gel, PhH, Δ; vi, $LiAlH_4$, THF; vii, MesCl, py; viii, Na, NH_3, EtOH.

Scheme 25

Takano et al.[123] have extended their enantioselective synthesis to the preparation of both enantiomers of quebrachamine from a single chiral synthon, namely the lactone (146). (+)-Quebrachamine was synthesised earlier,[24e] and a second route from (146) is now reported.[123] The synthesis of (-)-quebrachamine was also achieved by two routes, one of which is outlined in Scheme 25. With the introduction of C-3, i.e. formation of (147), a pair of epimers was produced, and this isomerism persisted as far as the final stage of the synthesis. Separation of the epimers could be carried out, but was unnecessary since the asymmetry at C-3 was destroyed during the final stage, and both epimers gave (-)-quebrachamine (148a).[123]

Ban et al.[107] have used the intermediate lactam (134) (Scheme 22) as a general precursor for both Strychnos and Aspidosperma ring systems. The conversion of (134) into (±)-quebrachamine (148b) and (±)-N-acetylaspidospermidine (149) is shown in Scheme 26. Aside from the protection of indole nitrogen by a tetrahydropyranyl group, the only stage that requires comment is the reduction of the lactam (150) by means of $LiAlH_4$, which stopped at the carbinolamine stage, presumably owing to steric factors in the nine-membered ring, since removal of the indole nitrogen protecting group in acid solution was followed by cyclization, with formation of (±)-1,2-dehydroaspidospermidine (151).[107]

An ingenious new approach to Aspidosperma alkaloids by Magnus and his co-workers[124] has resulted in a synthesis of (±)-aspidospermidine (152). This route involves the preparation of an indole-2,3-quinodimethane, e.g. (153a), which is not isolated, since it undergoes a thermal electrocyclic ring-closure to give the tetracyclic product (154). It is noteworthy that this cyclization gives exclusively the cis C/D isomer (154), presumably because the transition state for the reaction is derived from an intermediate with the conformation (153b). Application of the Pummerer reaction to (154) then gave the trifluoroacetate (155), which was converted into (±)-aspidospermidine by cyclization and reduction stages (Scheme 27).

The first synthesis of limaspermine (156) takes advantage of the fact that iron carbonyl complexes of type (157) can be regarded as stable equivalents of cyclohexenone γ-cations. Reaction of (157) with the malonate anion, followed by removal of the metal, gave a dienol ether (158), from which limaspermine (156) was

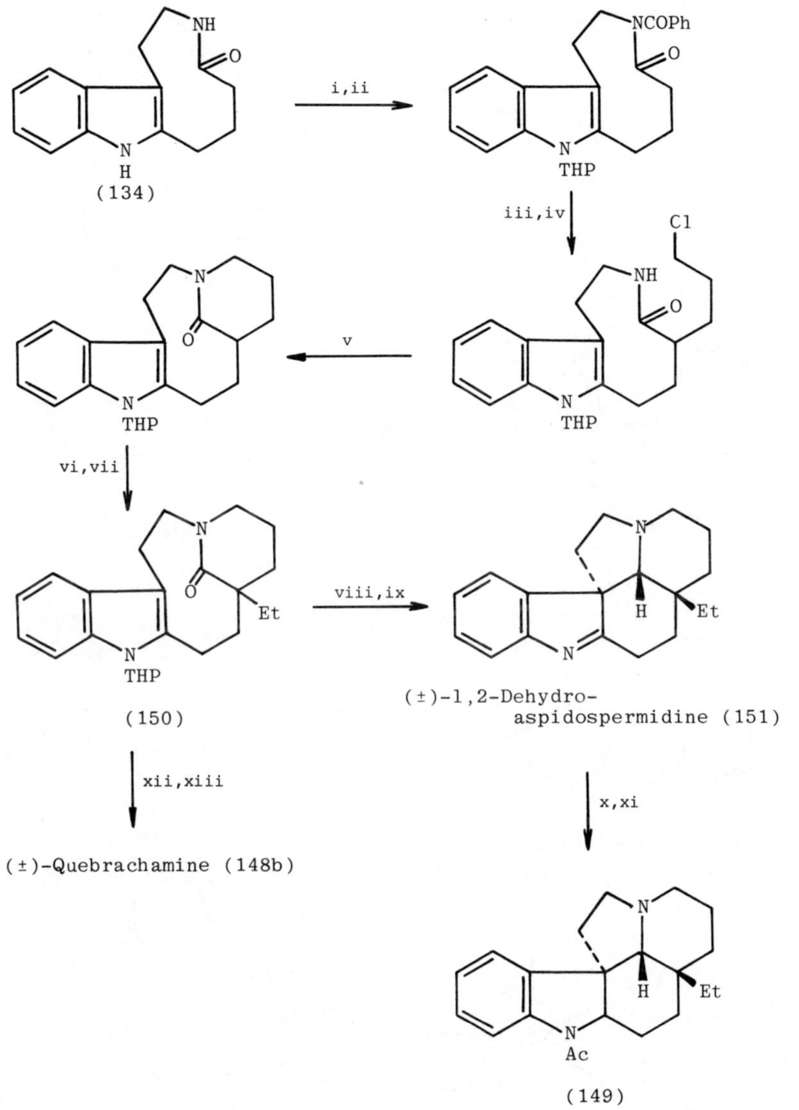

Indole Alkaloids

Scheme 26

Reagents: i, PhCOCl, NEt$_3$; ii, dihydropyran, camphor sulphonic acid; iii, LDA, ClCH$_2$CH$_2$CH$_2$I; iv, MeCH$_2$CH$_2$NH$_2$, CH$_2$Cl$_2$, r.t.; v, NaH, KI, 18-crown-6; vi, LDA, THF, HMPA; vii, EtI, $-60\,°$C; viii, LiAlH$_4$, THF; ix, 10% HCl, THF; x, LiAlH$_4$; xi, Ac$_2$O, py; xii, LiAlH$_4$ (excess), dioxan, Δ; xiii, H$^+$, H$_2$O.

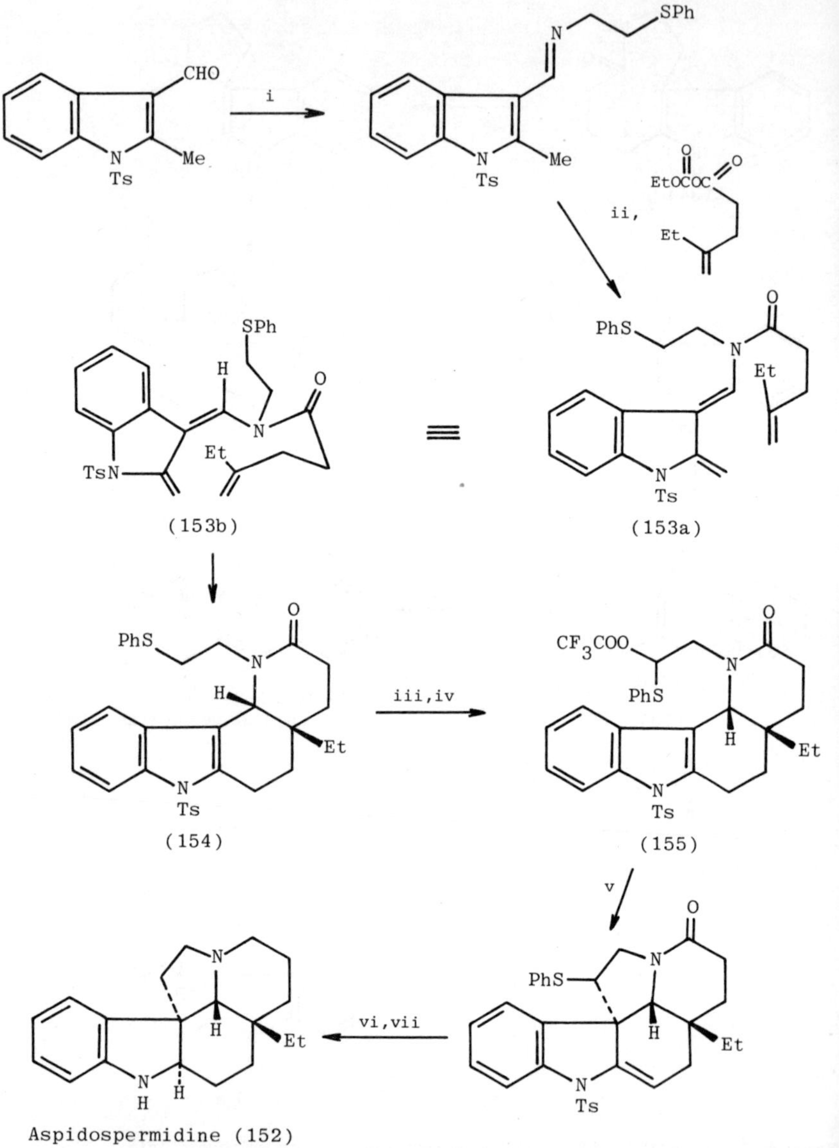

Aspidospermidine (152)

Reagents: i, PhSCH$_2$CH$_2$NH$_2$; ii, PhCl, 140° C; iii, m-ClPBA, NaHCO$_3$, CH$_2$Cl$_2$, 0° C; iv, TFAA, CH$_2$Cl$_2$, 0° C; v, PhCl, 130° C; vi, Raney Ni, EtOH; vii, LiAlH$_4$.

Scheme 27

obtained by conventional stages (Scheme 28). Unfortunately, removal of one of the malonate ester groups in (159) could not be

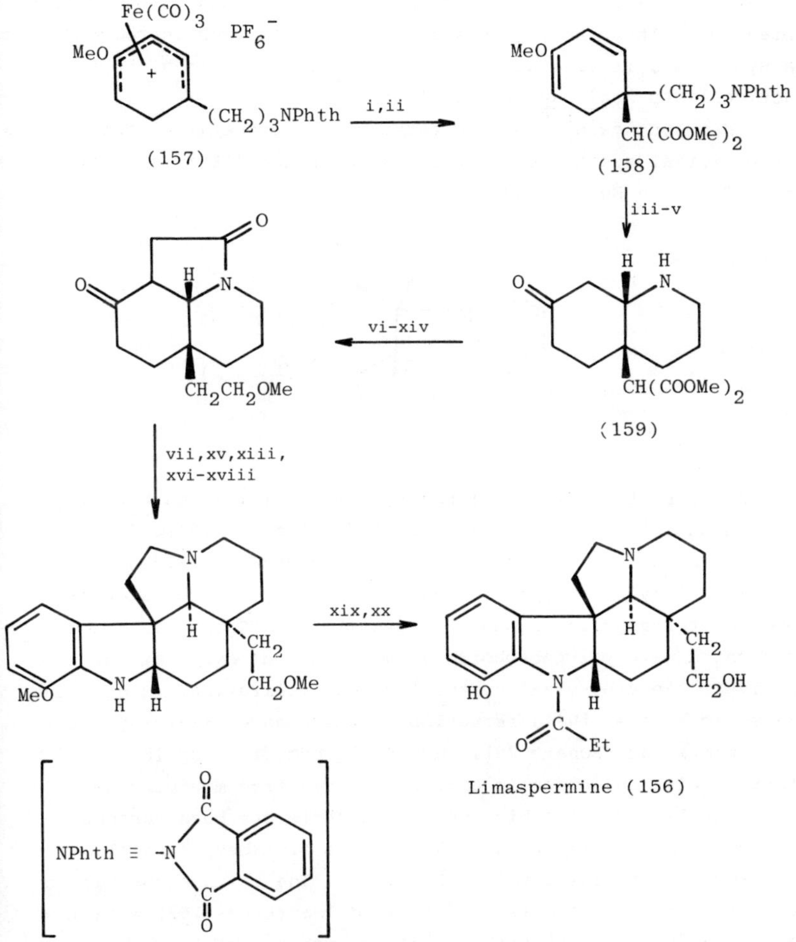

Limaspermine (156)

Reagents: i, KH(CO$_2$Me)$_2$; ii, Me$_3$NO, PhH, 50° C; iii, NH$_2$NH$_2$, MeOH; iv, (CO$_2$H)$_2$, MeOH, H$_2$O; v, NaHCO$_3$, MeOH, H$_2$O; vi, Ac$_2$O, py; vii, (CH$_2$OH)$_2$, TsOH, PhH; viii, NaCN, wet DMSO, 118° C; ix, LiBH$_4$, THF; x, NaH, MeI, THF; xi, Ca, NH$_3$, DME, EtOH; xii, ClCH$_2$COCl, py; xiii, EtOH, HCl; xiv, KOBu^t, Bu^{t}OH, PhH; xv, LiAlH$_4$, THF; xvi, o-MeOC$_6$H$_4$NHNH$_2$, HCl, EtOH; xvii, AcOH, 95° C; xviii, LiAlH$_4$, Et$_2$O; xix, EtCOCl, py; xx, Me$_3$SiI, CHCl$_3$, py.

Scheme 28

satisfactorily accomplished without prior protection of both amino and keto functions; subsequently, the primary alcohol function generated by reduction of the remaining ester group also had to be protected. In spite of the additional stages thus introduced into the synthesis, it remains a viable route to the more highly functionalised Aspidosperma alkaloids.[125]

A new synthesis[126] of the tetracyclic ketone (160) constitutes another formal synthesis of vindorosine, since (160) is a vital intermediate in Büchi's synthesis.

(160)

Lévy's route to the anilinoacrylate alkaloids has been applied in a synthesis of (±)-19-hydroxyaspidofractinine (161).[127] Condensation of the unstable, highly functionalised aldehydo-ester (162) with 2-hydroxytryptamine gave a mixture of three stereo-isomeric, tetracyclic oxindole esters (163). The major isomer, when heated with polyphosphoric acid, underwent cyclisation, hydrolysis, decarboxylation, and further cyclization, to give the hexacyclic ketone (164), reduction of which gave (±)-19-hydroxy-aspidofractinine (Scheme 29), identical (except in optical rotation) with an authentic sample prepared from minovincine.

Kuehne's versatile biomimetic synthesis has been further developed, and has led to a new synthesis of tabersonine (165), via 14-hydroxyvincadifformine (166).[128] The obvious approach to (166), i.e. the condensation of the indoloazepine (167) with an epoxyaldehyde, presents problems of regioselectivity at the epoxide ring-opening stage. This difficulty was neatly circumvented by use of the lactol-chloride (168), which under appropriate conditions, in a solvent-dependent reaction, gave a good yield of epimeric 14-hydroxyvincadifformines (166a, 166b), in which the β-hydroxy epimer (166a) predominated. Dehydration of (166a) then gave tabersonine (165) (Scheme 30). In contrast, attempts to dehydrate 14α-hydroxyvincadifformine (166b) resulted in contraction of ring D, with formation of hydroxymethyl-D-norvincadifformine

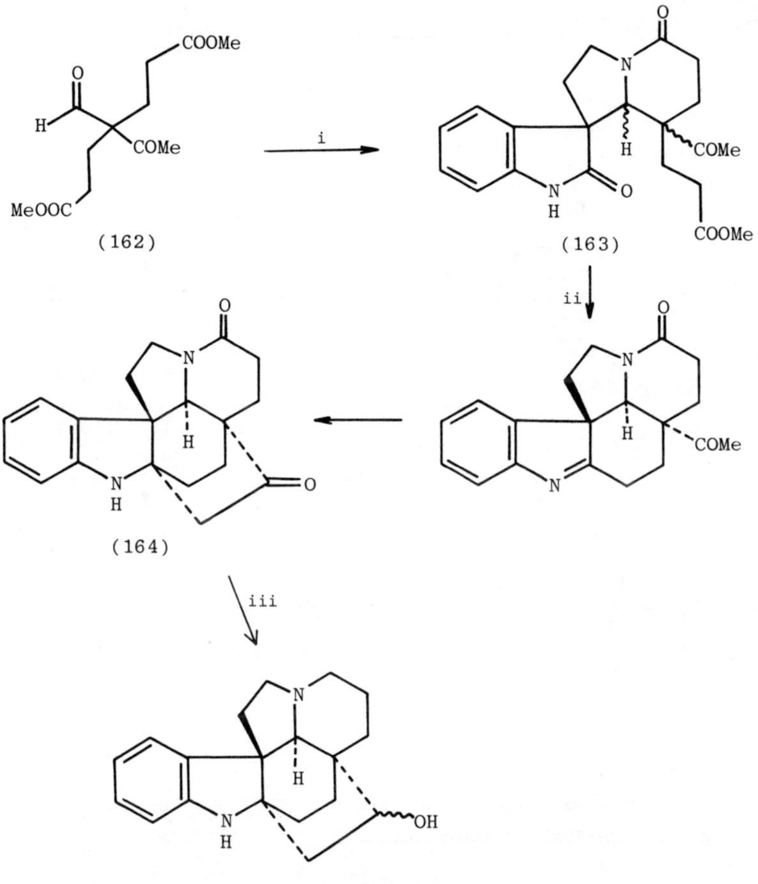

Indole Alkaloids

(162) → (163) → (164) → 19-Hydroxyaspidofractinine (161)

Reagents: i, 2-Hydroxytryptamine, Δ, 12 h, remove H_2O, then AcOH, Δ, 5h; ii, PPA, 120° C, 2 h; iii, $LiAlH_4$.

Scheme 29

(169); however, tabersonine could be obtained from (166b) by thiocarbamate pyrolysis.[128]

Other synthetic work in this area includes syntheses of desethylaspidospermidine,[129] eburnamonine (vincamone),[130,131] apovincamine,[132] and ethyl apovincaminate.[133] A new synthesis of (±)-homoeburnamonine constitutes another formal synthesis of (±)-vincamine.[134] Model studies in this area have resulted in the

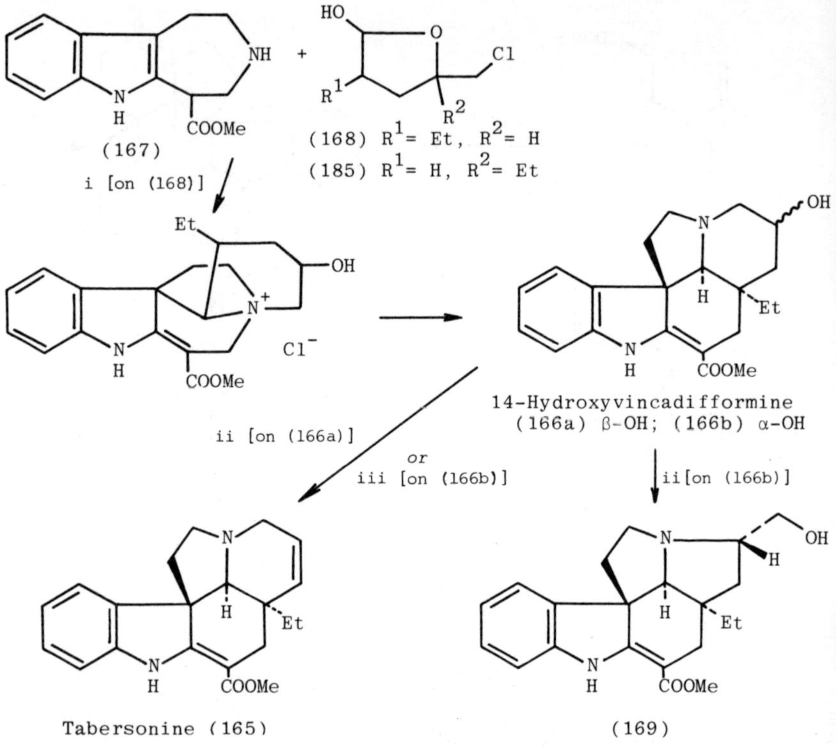

(167) + (168) R¹= Et, R²= H
(185) R¹= H, R²= Et

i [on (168)]

ii [on (166a)]

14-Hydroxyvincadifformine
(166a) β-OH; (166b) α-OH

or
iii [on (166b)]

ii [on (166b)]

Tabersonine (165)

(169)

Reagents: i, TsOH, MeOH, 65° C, 6 h; ii, PPh$_3$, CCl$_4$, NEt$_3$ [on (166a)]; iii, pyrolysis of thiocarbamate of (166b)

Scheme 30

construction of arylhydrolilolidines, which could be useful intermediates in <u>Aspidosperma</u> alkaloid synthesis,[135] and in the synthesis of a stabilised dehydrosecodine, in which the dihydropyridine ring is stabilised by methoxycarbonyl groups.[136]

4.6 Catharanthine-Ibogamine-Cleavamine Group.

— A total of twenty alkaloids has been isolated from various organs of <u>Anartia meyeri</u>; these include three new alkaloids, namely 11-hydroxycoronaridine (170), 11-hydroxyheyneanine (171), and 10-hydroxyheyneanine (172), which were found in the leaves.[49] The known alkaloids isolated included conopharyngine, jollyanine, voacangine, isovoacangine, ibophyllidine, coronaridine 7-hydroxyindolenine, coronaridine, heyneanine, 19-epiheyneanine, and eglandine.

	R^1	R^2	R^3
11-Hydroxy-coronaridine (170)	H	OH	H
11-Hydroxyheyneanine (171)	H	OH	OH
10-Hydroxyheyneanine (172)	OH	H	OH

10-Methoxyeglandine (173)

10-Hydroxyheyneanine (172) has also been found, together with another new alkaloid, 10-methoxyeglandine (173), in Peschiera echinata.[50] Twenty known alkaloids extracted from this plant included voacangine, voacangine 7-hydroxyindolenine, voacristine, 19-epivoacristine, 10-hydroxycoronaridine, ibogaine, coronaridine, voacangine pseudo-indoxyl, 3-oxovoacangine, and ibogaine 7-hydroxy-indolenine. (+)-20R-Pseudovincadifformine, (+)-20R-pandoline, and (+)-20S-pandoline have been found, together with six other alkaloids, in the leaves and stems of Melodinus polyadenus.[116]

Tissue cultures of C. roseus produce a number of alkaloids of the heteroyohimbine and Aspidosperma groups; the only iboga alkaloid found was catharanthine, which proved to be one of the two major alkaloids, and constituted 16.4% of the total alkaloid content.[62]

A new synthesis of ibogamine[137] and two syntheses of epi-ibogamine[137,138] have been reported recently. In the first of these syntheses (Scheme 31) the quinuclidine ring system was constructed via an internal Michael reaction on the unsaturated ester-ketone (174). The product consisted of a mixture of (175) (minor product) and its epimer at the future C-20. A sequence of unexceptional stages then led to (±)-ibogamine (176); similarly, the epimer of (175) ultimately gave (±)-epi-ibogamine.[137]

In the second synthesis of (±)-epi-ibogamine (177) the quinuclidine ring system (178) was prepared by Diels-Alder reaction of N-benzyl-2-pyridone with methyl acrylate. Conventional steps then led to the tertiary base (179), which was quaternised with

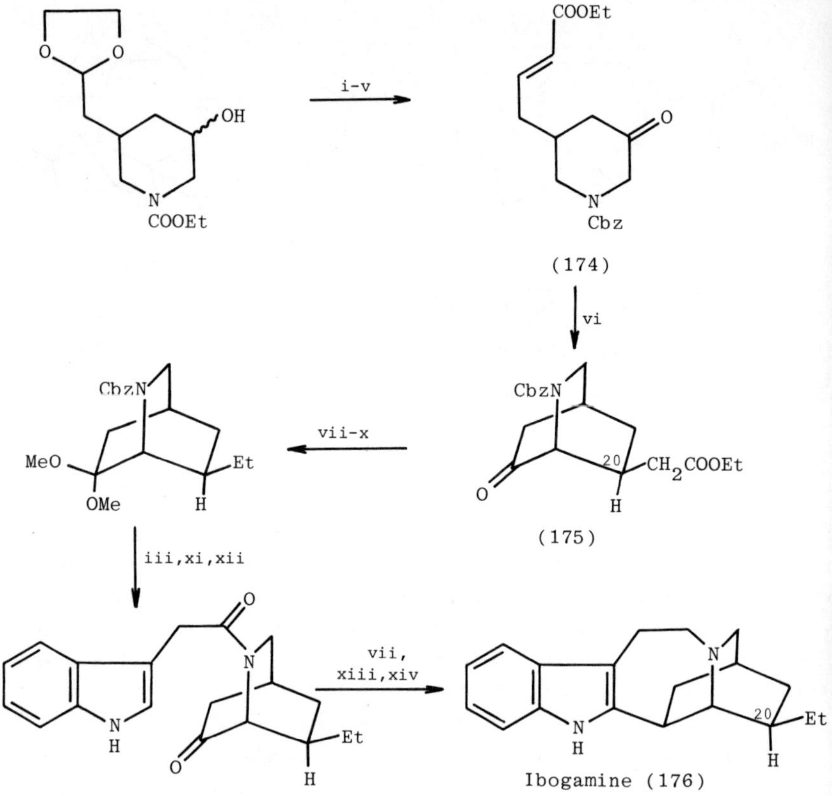

Scheme 31

Reagents: i, Alkaline hydrolysis; ii, $ClCO_2CH_2Ph$, base; iii, 1% HCl, acetone; iv, $(EtO)_2POCH_2CO_2Et$, PhH, base, $0°$ C; v, Jones' oxidation; vi, NaH, dioxan, Δ; vii, $HC(OMe)_3$, TsOH, MeOH, Δ; viii, $LiAlH_4$, Et_2O; ix, MesCl, NEt_3, PhH; x, Zn, NaI, DME, Δ; xi, H_2, Pd/C, MeOH; xii, β-indolylacetyl chloride; xiii, TsOH, PhH, Δ; xiv, $LiAlH_4$, $AlCl_3$, THF, r.t.

β-indolylethyl bromide and debenzylated to the tertiary base (180). Cyclization by Trost's method and reduction by $NaBH_4$ finally gave (±)-epi-ibogamine (177) (Scheme 32). This paper also records a new synthesis of (±)-desethylibogamine.[138]

Kuehne's biomimetic synthesis has now been applied to the synthesis of the C-20 epimeric ibophyllidines (Schemes 33, 34)[139]

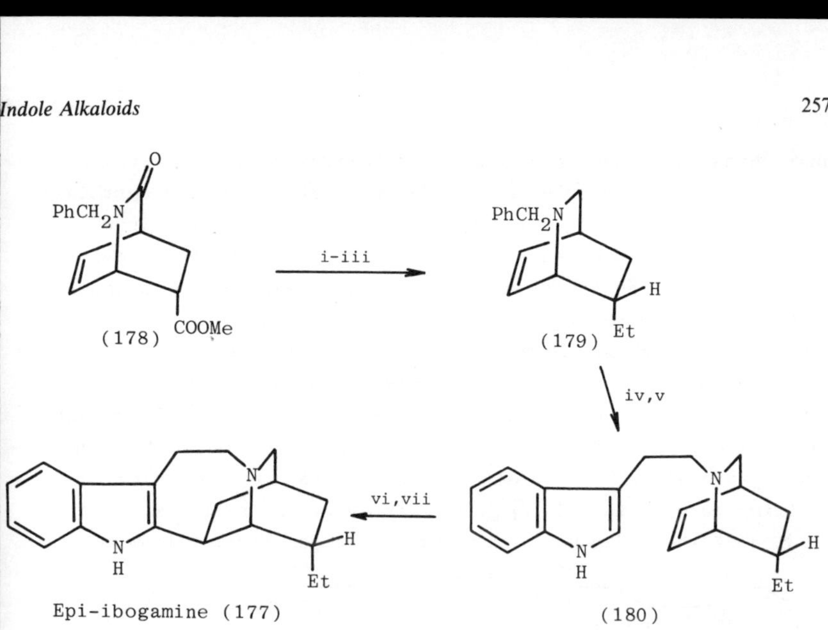

Epi-ibogamine (177)

(180)

Reagents: i, LiAlH$_4$, THF, 60° C; ii, TsCl, NEt$_3$; iii, MeMgI, Li$_2$CuCl$_4$, THF, Et$_2$O; iv, β-indolylethyl bromide; v, n-C$_3$H$_7$SLi, HMPA; vi, (MeCN)$_2$PdCl$_2$, AgBF$_4$, NEt$_3$; vii, NaBH$_4$.

Scheme 32

and the pandolines.[128] Condensation of the indoloazepine ester (167) with 4-bromohexanal gave a mixture of tertiary bases which on

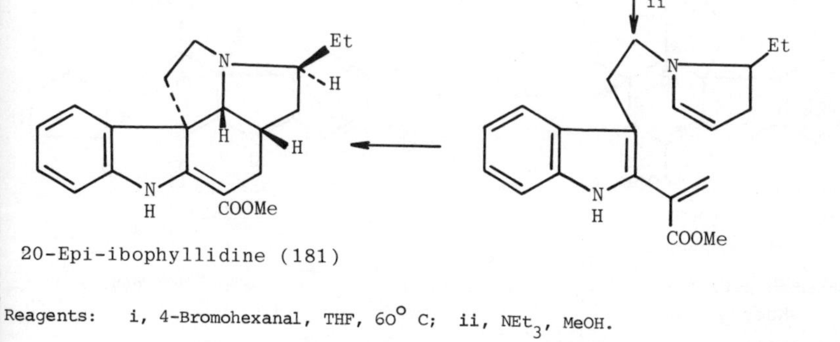

20-Epi-ibophyllidine (181)

Reagents: i, 4-Bromohexanal, THF, 60° C; ii, NEt$_3$, MeOH.

Scheme 33

quaternization, fragmentation, and recyclization gave 20-epi-ibophyllidine (181) as sole product (Scheme 33). For the synthesis of ibophyllidine (184), a modified strategy was adopted, in which the tetracyclic intermediate (182) was prepared by the normal Kuehne method, and the five-membered ring D was formed in the final stages of the synthesis. It is of interest that the epimeric mixture of quaternary bromides (183) fragmented and cyclized to give a <u>single</u> product of the stereochemistry shown in (182). The vital reversal of configuration at C-3 and C-7 presumably occurred during the final hydrogenolysis-cyclization-hydrogenation stage, since protonation at C-16 affords an indoleninium ion which can equilibrate with the 3,7-epimeric series by reversible Mannich fission of the 3,7-bond. The final hydrogenation, following

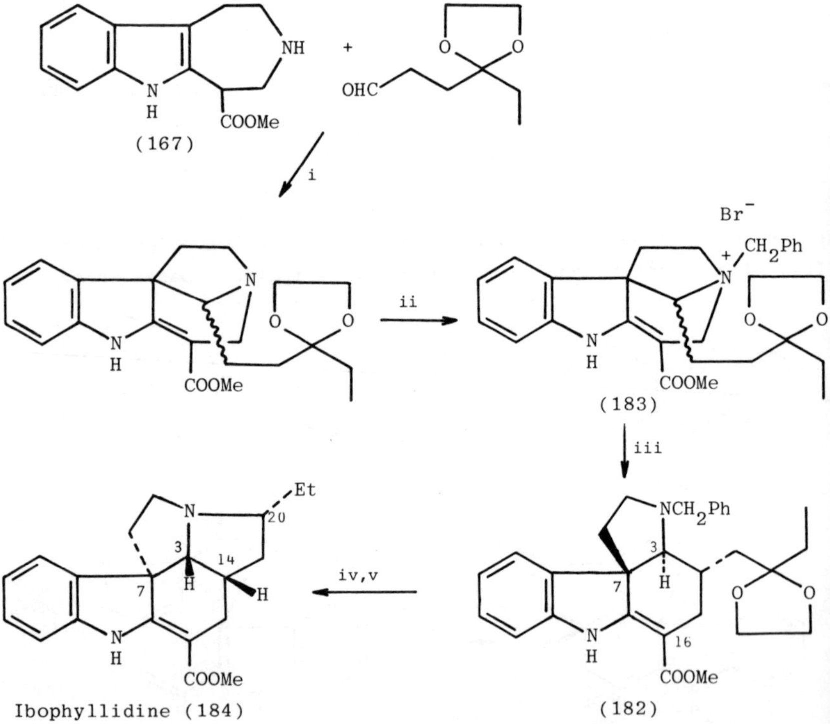

Reagents: i, MeOH, 20°C, 2h; ii, PhCH$_2$Br; iii, NEt$_3$, MeOH; iv, MeOH, 10% HCl, H$_2$O; v, H$_2$, Pd/C, AcOH.

Scheme 34

Indole Alkaloids

cyclization of the aminoketone, ensures the desired configuration at C-20 by delivery of hydrogen to the less hindered side of the molecule (Scheme 34).[139]

In an improved synthesis of pandoline and epipandoline, Kuehne et al.[128,cf.12j] have used the condensation of the chloro-lactol (185) with the indoloazepine (167) to generate the desired substitution in ring D. In other respects the synthesis follows the same route as the synthesis of 14-hydroxyvincadifformine (see Scheme 30).

Other recent synthetic work includes a synthesis[140] of pseudo-vincamone (not yet known as a natural product) and several 6-nor-20-desethylcatharanthine derivatives.[141]

5 Bisindole Alkaloids

The dimeric mould metabolite ditryptophenaline (186) has been very simply synthesised, albeit in only 3% yield, by the oxidative dimerization of cyclo-L-$\underline{N}$-methylphenylalanyl-L-tryptophan (187) by means of thallium trifluoroacetate (Scheme 35).[142] This synthesis incidentally establishes the absolute configuration of ditryptophenaline, which had previously only been assumed to have the absolute stereochemistry shown in (186).

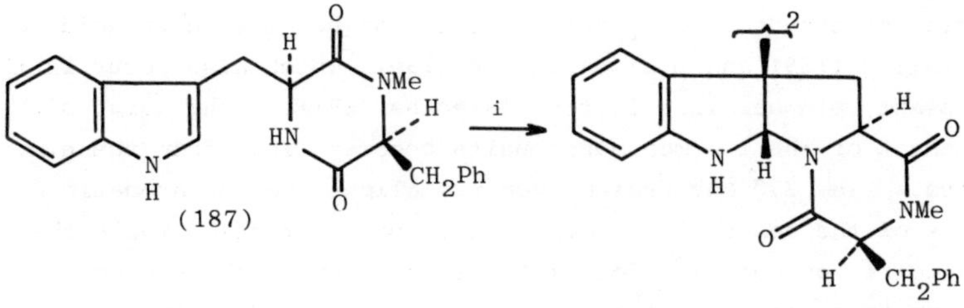

Ditryptophenaline (186)

Reagent: i, $Tl(CF_3CO_2)_3$, $BF_3 \cdot Et_2O$, MeCN, $0°$ C.

Scheme 35

Serpentinine has been found to occur in the root bark of Rauwolfia nitida.[60] The stem bark of Strychnos decussata and S. elaeocarpa contains bisnordihydrotoxiferine,[48a] while the stem bark of S. dolichothyrsa contains bisnor-C-alkaloid H, its $\underline{N}_b$-oxide and its $\underline{N}_b\underline{N}_b$,-bisoxide, and two new alkaloids, dolichocurine and dolichothine.[87] On the basis of spectroscopic data, dolichocurine is tentatively formulated as the bis-strychninoid base (188), in which case it originates, like bisnor-C-alkaloid H, from one molecule of Wieland-Gumlich aldehyde and one of 18-deoxy-Wieland-Gumlich aldehyde. Nothing is known at present concerning the structure of dolichothine.

Dolichocurine ? (188)

Plumocraline, $C_{42}H_{50}N_4O_5$, is a new alkaloid isolated from the root bark of New Caledonian <u>Alstonia</u> <u>plumosa</u> var. <u>communis</u> Boiteau forma <u>glabra</u> Boiteau.[143] Its u.v. spectrum is that of a substituted dihydroindole, and since it can be cleaved by acid to cabucraline (189) and pleiocarpamine (190), which also occur in the same plant, plumocraline is formulated as (191). The point of attachment of the two monomeric units becomes clear from the n.m.r. spectra. The 270 MHz proton spectrum allows the non-aromatic protons of the cabucraline component to be identified, while the aromatic region contains four multiplets owing to the pleiocarpamine protons, and two <u>singlets</u> owing to C-9 and C-12 in a 10-substituted cabucraline unit. The ^{13}C n.m.r. spectrum again allows the cabucraline signals to be identified; the signals owing to the pleiocarpamine unit resemble very closely those of the same unit in macrocarpamine, from which it is deduced that C-10 of cabucraline is attached to C-2 of pleiocarpamine. Confirmation of the structure thus deduced is provided by the partial synthesis of plumocraline (191) from pleiocarpamine and cabucraline in acid solution (Scheme 36).[143]

Flexicorine, a constituent of <u>Rauwolfia</u> <u>reflexa</u>, is unusual

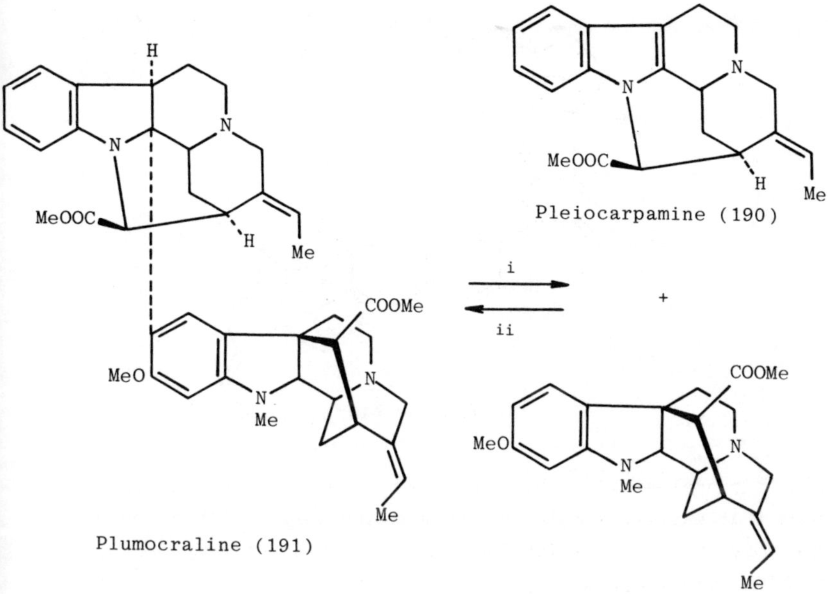

Indole Alkaloids

Pleiocarpamine (190)

Plumocraline (191)

Cabucraline (189)

Reagents: i, MeOH, anhydrous HCl; ii, MeOH, 2% HCl, 48 h.

Scheme 36

among the alkaloids in being a bright red solid.[144] The colour stems from an iminoquinone function, since the alkaloid can be reduced ($NaBH_4$) to a colourless 5-hydroxyindoline derivative, which readily re-oxidises in air. Inspection of the ^{13}C n.m.r. spectrum of flexicorine reveals that all the non-aromatic resonances of its congener, rauflexine, are present. The remaining non-aromatic signals are reminiscent of vincorine, the disparity in the C-6 resonances being attributed to a difference in configuration at C-16 in flexicorine and vincorine; a shift of the C-2' resonance is a consequence of the different state of oxidation of ring A'. Since the proton n.m.r. spectrum contains only four aromatic <u>singlets</u>, the points of attachment of the two components become self-evident, and flexicorine is thus formulated as (192).[144]

Details of the structure elucidation of gardmultine (193) by chemical and spectroscopic methods,[145a] and by X-ray crystal

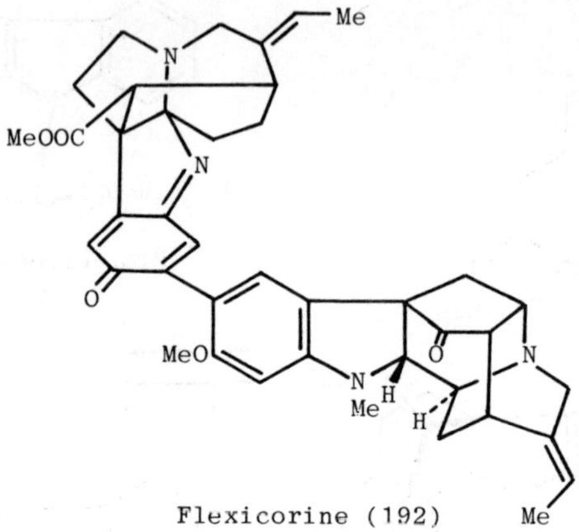

Flexicorine (192)

structure analysis,[145b] have been published. These studies have shown that earlier samples of gardmultine were contaminated with a demethoxygardmultine which lacks the aliphatic methoxy-group, and in which both ethylidene groups have the familiar E configuration.

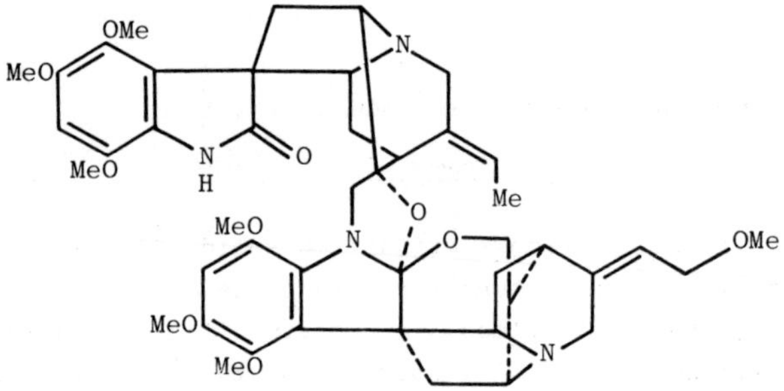

Gardmultine (193)

Four of the twenty-two alkaloids of the bark of Peschiera echinata are bisindole alkaloids; these are voacamine, N-demethylvoacamine, demethoxycarbonylvoacamine, and voacamidine.[50]

Ervatamia hainanensis is a shrub endemic to South China, which is used in popular medicine for the treatment of a variety of ailments (stomach disorders, dysentery, snake bites, rheumatism, hypertension, and liver infections). The roots of this plant contain a number of monomeric and dimeric alkaloids, of which the structures of three have been elucidated;[146] these are ervahanines A, B, and C. The presence of vobasine and coronaridine (194) components in these molecules was deduced from their mass and n.m.r. spectra, and confirmed by the partial synthesis of ervahanines A and B from vobasinol (195) and coronaridine (Scheme 37). The point of attachment of the vobasine component to coronaridine proved more difficult to establish, but was eventually achieved following a thorough analysis of the 400 MHz proton n.m.r. spectra, from which it was concluded that ervahanines A-C have the structures (196-198), respectively.[146]

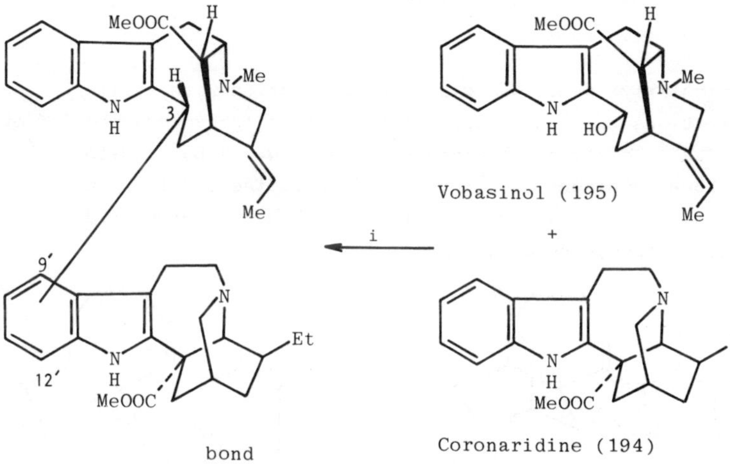

	bond
Ervahanine A (196)	3,11'
B (197)	3,10'
C (198)	3,12'

Reagent: i, MeOH, HCl.

Scheme 37

The stem bark of <u>Hunteria zeylanica</u> contains three bisindole alkaloids, namely norpleiomutine (199), pleiomutinine, and 19'-epipleiomutinine; of these, norpleiomutine is new.[59]

Two further bisindole alkaloids, which belong to the <u>Aspidosperma-iboga</u> group, have been extracted from the leaves of <u>Bonafousia tetrastachya</u>.[117] The structures of both alkaloids were deduced from a thorough analysis of their mass and

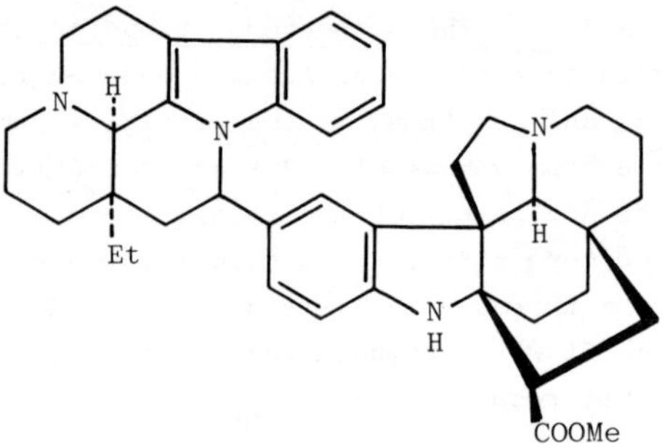

Norpleiomutine (199)

(particularly) n.m.r. spectra, in conjunction with earlier observations that this plant also contains 12-hydroxyvincadifformine (see above), and bisindole bases (e.g. isobonafousine)[12k] in which one component is 11-hydroxycoronaridine (170). On the basis of these considerations, tetrastachyne is formulated as (200) and tetrastachynine as (201).[117]

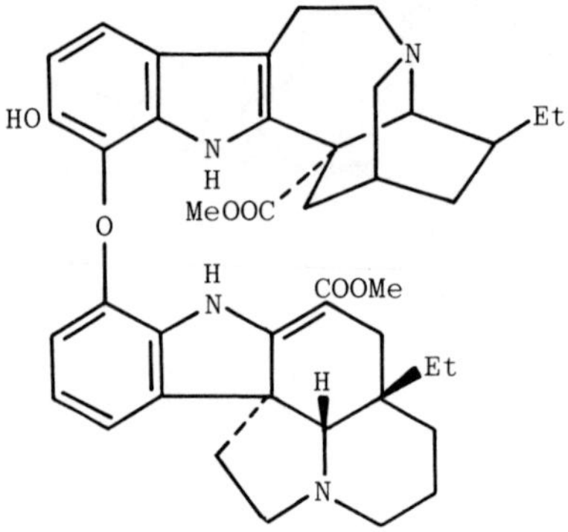

Tetrastachyne (200)

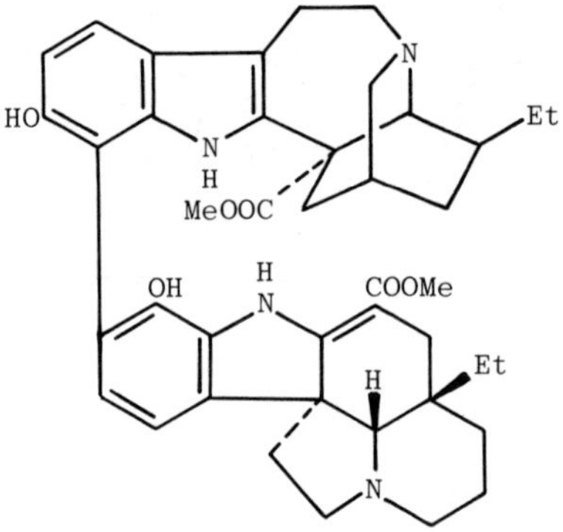

Tetrastachynine (201)

Details of the structure elucidation[12l,24f] of four alkaloids of the ervafoline group, and attempts to synthesise the complex ring systems of these alkaloids by a biomimetic approach,[12l] have now been published.[147] The remaining four bisindole alkaloids obtained from Stenosolen heterophyllus are ervafolidine and three relatives, and their structures have also been elucidated.[147] Ervafolidine contains one additional oxygen atom and two additional hydrogen atoms compared with ervafoline, to which it appears to be closely related according to spectral comparison. The additional oxygen in ervafolidine is contained in a secondary hydroxyl group, which, assuming a close biogenetic relationship between ervafoline and ervafolidine, suggests two possible structures for the latter. The structures were eventually established by X-ray crystal structure determination of 3-epiervafolidine, which revealed the structure (202); ervafolidine is thus (203). Similarly, the structure of 19'R-hydroxyervafolidine was shown to be (204). The remaining alkaloid (205) appears, on the basis of its proton and ^{13}C n.m.r. spectra, to be epimeric with 19'R-hydroxyervafolidine (204) at both positions 3 and 19'.

3-Epiervafolidine (202) R = H; 3S

Ervafolidine (203) R = H; 3R

19'R-Hydroxyervafolidine (204) R = OH; 3R

19'S-Hydroxyepiervafolidine (205) R = OH; 3S

In the vinblastine series, leurosidine N_b-oxide[148] and desacetoxyleurosine[149] have been isolated from the leaves of Catharanthus roseus. Several species of Streptomyces convert leurosine into a common metabolite which, following its production in quantity in resting cells of S. griseus, was identified as 10'-hydroxyleurosine (206).[150]

The biotransformation of 15',20'-anhydrovinblastine (207) in Catharanthus roseus cell suspension cultures (cell line 916) results in the formation of leurosine (208) (31%) and catharine (9%).[151]

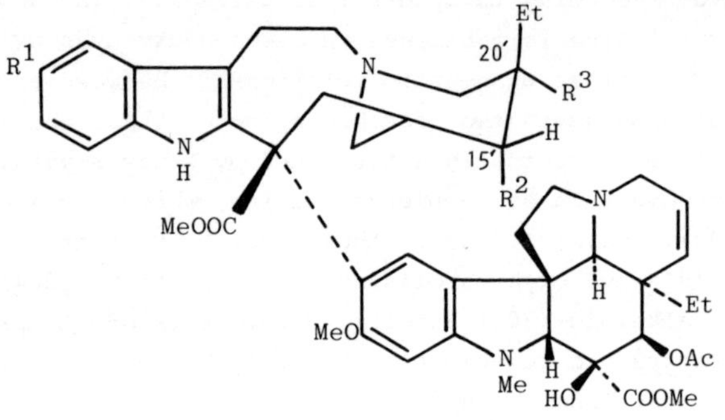

	R^1	R^2	R^3
10'-Hydroxyleurosine (206)	OH	—O—	
15',20'-Anhydrovinblastine (207)	H		$\Delta^{15',20'}$
Leurosine (208)	H	—O—	

Finally, a number of model vinblastine analogues, e.g. (209), have been synthesised by condensation of vindoline with an anilinoacrylate base related to vincadifformine, but lacking ring D.[152]

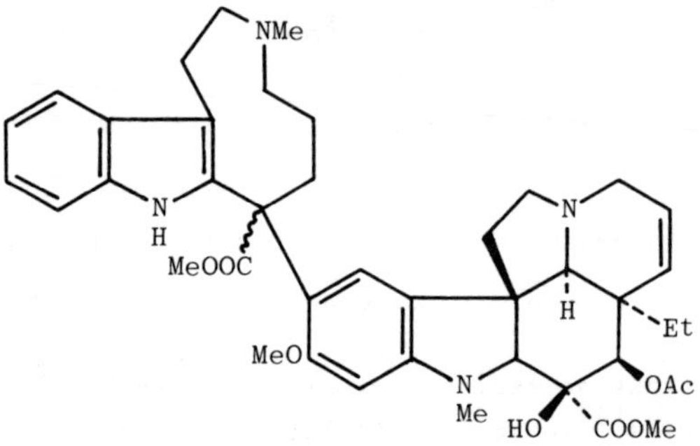

(209)

6 Biogenetically Related Quinoline Alkaloids

6.1 Cinchona Group. — In contrast to the bark, the leaves of

Cinchona species have not been extensively studied. One such study, on the leaves of C. succirubra from Thailand, has revealed the presence of quinine, quinidine, cinchonine, cinchonidine, and dihydroquinine;[153] this provides a contrast to the alkaloid content of C. ledgeriana leaves, which appear to contain bisindole alkaloids of the cinchophylline type.[12m]

The production of quinine and quinidine in C. ledgeriana and C. succirubra leaf, root, and suspension cultures has been investigated.[154] Some quinine and quinidine were obtained in the leaf organ cultures, but none in the root organ and suspension cultures.

X-ray crystal structure data for cinchonidine have been reported.[155]

A number of acrylonitrile - Cinchona alkaloid copolymers have been prepared in which the vinyl group of the alkaloid has been modified, and the ability of these copolymers to act as asymmetric catalysts in the Michael addition of methyl vinyl ketone to methyl indan-1-one-2-carboxylate has been studied.[156] In twelve cases, good enantiomeric yields, ranging from 41 to 56% ee, were obtained.

6.2 Camptothecin. — A new synthesis of (±)-10-hydroxy-camptothecin (210) and (±)-10-methoxycamptothecin (211) by the Shanghai group has been reported (Scheme 38).[157]

In an attempt to prepare less toxic camptothecin derivatives that might have clinical application, 7-hydroxymethylcamptothecin has been prepared by several radical processes, of which reaction of camptothecin with hydrogen peroxide in methanol-aqueous sulphuric acid at $0°$ C is typical.[158] 5-Hydroxycamptothecin was also prepared by the oxidation of camptothecin with ammonium persulphate in the presence of iron(II) sulphate and bromoacetic acid. Since the yield in the absence of this last ingredient was very low, it appears that the derived bromomethyl radical plays an important role in abstraction of hydrogen from position 5. Both 7-hydroxymethylcamptothecin and 5-hydroxycamptothecin were converted into ester or ether derivatives for pharmacological evaluation.[158] 7-Hydroxymethylcamptothecin was also converted into the related aldehyde by reaction with electrophilic reagents, e.g. aqueous sulphuric acid, or tosyl chloride in hot pyridine; no conventional oxidising agent appeared to be necessary.[158]

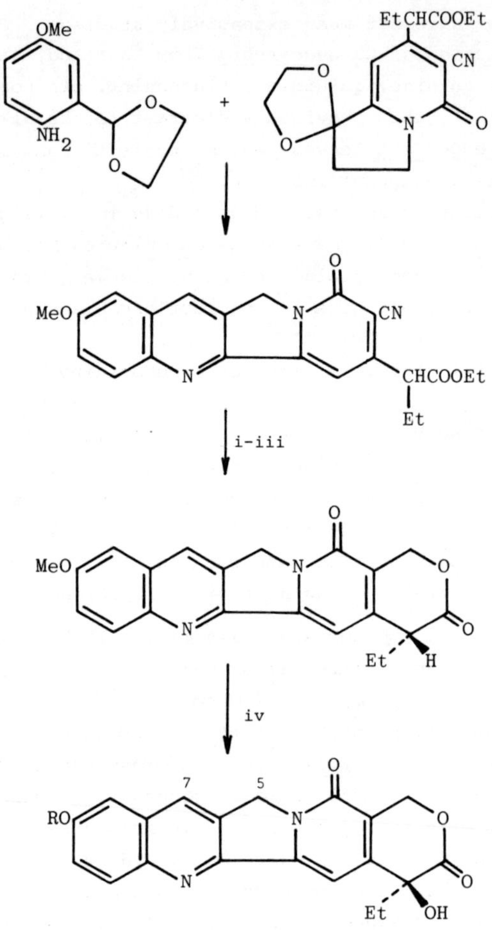

268 *The Alkaloids*

10-Methoxycamptothecin (211) R = Me

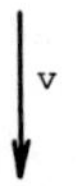

10-Hydroxycamptothecin (210) R = H

Reagents: i, Ni, H_2, Ac_2O, AcOH; ii, $NaNO_2$, Ac_2O, AcOH; iii, 10% H_2SO_4;
     iv, O_2, $CuCl_2$, Me_2NH; v, 48% HBr.

Scheme 38

References

1. 'The Alkaloids', Founding Editor R.H.F. Manske, ed. R. Rodrigo, Academic Press, New York, 1981, Vol. XX.

2. G.A. Cordell, 'Introduction to Alkaloids: A Biogenetic Approach', Wiley-Interscience, New York, 1981.

3. (a) D.G.I. Kingston and O. Ekundayo, J. Nat. Prod., 1981, 44, 509; (b) M. Lounasmaa and A. Nemes, Tetrahedron, 1982, 38, 223.

4. R.B. Sharma, R. Seth, F. Anwer, and R.S. Kapil, Indian J. Chem., Sect. B, 1981, 20, 701.

5. S. Kano, E. Sugino, S. Shibuya, and S. Hibino, J. Org. Chem., 1981, 46, 3856.

6. S. Takano, Y. Suzuki, and K. Ogasawara, Heterocycles, 1981, 16, 1479.

7. W.J. Keller and G.M. Hatfield, Phytochemistry, 1982, 21, 1415.

8. (a) Y. Koyama, K. Yokose, and L.J. Dolby, Agric. Biol. Chem., 1981, 45, 1285; (b) M. Noltemeyer, G. M. Sheldrick, H.U. Hoppe, and A. Zeeck, J. Antibiot., 1982, 35, 549; (c) T. Yoshioka, K. Mohri, Y. Oikawa, and O. Yonemitsu, J. Chem. Res. (S), 1981, 194; ibid. (M), 1981, 2252.

9. M.A. O'Leary and J.R. Hanson, Tetrahedron Lett., 1982, 23, 1855.

10. (a) N. Kyriadis, E.S. Waight, J.B. Day, and P.G. Mantle, Appl. Environ. Microbiol., 1981, 42, 61; (b) C.M. Maes, P.S. Steyn, and F.R. Van Heerden, J. Chromatogr., 1982, 234, 489.

11. D.O. Okorie, Phytochemistry, 1981, 20, 2575.

12. J.E. Saxton, in 'The Alkaloids', ed. M.F. Grundon (Specialist Periodical Reports), The Royal Society of Chemistry, London, 1982, Vol. 12, (a) p.165; (b) p.176; (c) p.184; (d) pp.194-198; (e) p.192; (f) p.187; (g) p.199; (h) p.224; (i) p.217; (j) p.234; (k) p.238; (l) p.240; (m) p.236.

13. R. Hocquemiller, G. Dubois, M. Leboeuf, A. Cavé, N. Kunesch, C. Riche, and A. Chiaroni, Tetrahedron Lett., 1981, 22, 5057.

14. D.H. Janzen, D.G. Lynn, L.E. Fellows, and W. Hallwachs, Phytochemistry, 1982, 21, 1035; (b) P. Wulff, J.S. Carlé and C. Christophersen, Comp. Biochem. Physiol., 1982, 71B, 523.

15. (a) T. Akizawa, K. Yamazaki, T. Yasuhara, T. Nakajima, M. Roseghini, G.F. Erspamer, and V. Erspamer, Biomed. Res., 1982, 3, 232; (b) M. Shimizu, M. Ishikawa, Y. Komoda, and T. Nakajima, Chem. Pharm. Bull., 1982, 30, 909; (c) R. Verpoorte, F.T. Joosse, H. Groenink, and A.B. Svendsen, Planta Med., 1981, 42, 32.

16. M. Leboeuf, A. Cavé, P. Forgacs, J. Provost, A. Chiaroni, and C. Riche, J. Chem. Soc., Perkin Trans. 1, 1982, 1205.

17. G. Baudoin, F. Tillequin, M. Koch, J. Pousset, and T. Sévenet, J. Nat. Prod., 1981, 44, 546.

18 (a) A. Harris, L.A. Anderson, and J.D. Phillipson, J. Pharm. Pharmacol., 1981, 33, Suppl. 17P; (b) S.A. Khan and K.M. Shamsuddin, Phytochemistry, 1981, 20, 2062; (c) E. Varga, K. Szendrei, J. Reisch, and G. Maróti, Fitoterapia, 1981, 52, 183.

19 (a) J. Kökösi, I. Hermecz, G. Szász, and Z. Mészáros, Tetrahedron Lett., 1981, 22, 4861; (b) K. Horváth-Dóra, G. Tóth, F. Hetényi, J. Tamás, and O. Clauder, Acta Chim. Acad. Sci. Hung., 1982, 109, 267.

20 G.W. Gribble and R.M. Soll, J. Org. Chem., 1981, 46, 2435.

21 T. Kosuge, K. Tsuji, and K. Hirai, Tetrahedron Lett., 1981, 22, 3417.

22 A. Probst and C. Tamm, Helv. Chim. Acta, 1981, 64, 2056.

23 T. Ohnuma, Y. Kimura, and Y. Ban, Tetrahedron Lett., 1981, 22, 4969.

24 See J.E. Saxton, in 'The Alkaloids', ed. M.F. Grundon, (Specialist Periodical Reports), The Royal Society of Chemistry, London, 1981, Vol. 11, (a) p.152; (b) p.153; (c) p.172; (d) p.162; (e) p.181; (f) p.192.

25 P. Wulff, J.S. Carlé, and C. Christophersen, J. Chem. Soc., Perkin Trans. 1, 1981, 2895.

26 J.S. Carlé and C. Christophersen, J. Org. Chem., 1981, 46, 3440.

27 (a) J.H. Cardellina, F.J. Marner, and R.E. Moore, Science, 1979, 204, 193; (b) N. Sakabe, H. Harada, Y. Hirata, Y. Tomiie, and I. Nitta, Tetrahedron Lett., 1964, 2523; (c) H. Fujiki, M. Mori, M. Nakayasu, M. Terada, T. Sugimura, and R.E. Moore, Proc. Nat. Acad. Sci. USA, 1981, 78, 3872.

28 K. Arai, S. Sato, S. Shimizu, K. Nitta, and Y. Yamamoto, Chem. Pharm. Bull., 1981, 29, 1510.

29 Y. Kimura, T. Hamasaki, H. Nakajima, and A. Isogai, Tetrahedron Lett., 1982, 23, 225.

30 G.A. Ellestad, P. Mirando, and M.P. Kunstmann, J. Org. Chem., 1973, 38, 4204.

31 M. Uramoto, M. Tanabe, K. Hirotsu, and J. Clardy, Heterocycles, 1982, 17, 349.

32 R. Ritchie and J.E. Saxton, Tetrahedron, 1981, 37, 4295.

33 M.L. Bianchi, N.C. Perellino, B. Gioia, and A. Minghetti, J. Nat. Prod., 1982, 45, 191.

34 S.M. Atwell and P.G. Mantle, Experientia, 1981, 37, 1257.

35 A.P. Kozikowski and M.N. Greco, Tetrahedron Lett., 1982, 23, 2005.

36 (a) M. Somei and M. Tsuchiya, Chem. Pharm. Bull., 1981, 29, 3145; (b) For conversion of (39) into secoagroclavine see M. Somei, F. Yamada, Y. Karasawa, and C. Kaneko, Chem. Lett., 1981, 615.

37 A.P. Kozikowski and Y.Y. Chen, J. Org. Chem., 1981, 46, 5248.

38 (a) J. Rebek and Y.K. Shue, Tetrahedron Lett., 1982, 23, 279; (b) J. Am. Chem. Soc., 1980, 102, 5426.

39 M. Natsume and H. Muratake, Heterocycles, 1981, 16, 1481.

40 (a) A. M. Crider, R. Grubb, K. A. Bachmann, and A. K. Rawat, J. Pharm. Sci., 1981, 70, 1319; (b) J. Šmidrkal and M. Semonský, Collect. Czech. Chem. Commun., 1982, 47, 622, 625; (c) J. Beneš and M. Semonský, ibid., p. 1235.

41 I. R. C. Bick and M. A. Hai, Heterocycles, 1981, 16, 1301.

42 R. Kyburz, E. Schöpp, I. R. C. Bick, and M. Hesse, Helv. Chim. Acta, 1981, 64, 2555.

43 I. R. C. Bick and M. A. Hai, Tetrahedron Lett., 1981, 22, 3275.

44 A. Jössang, H. Jacquemin, J. L. Pousset, and A. Cavé, Planta Med., 1981, 43, 301.

45 J. D. Phillipson, S. R. Hemingway, and C. E. Ridsdale, J. Nat. Prod., 1982, 45, 145.

46 J. Levesque, R. Jacquesy, and J. P. Foucher, Tetrahedron, 1982, 38, 1417.

47 F. Hotellier, P. Delaveau, and J. L. Pousset, C. R. Hebd. Seances Acad. Sci., Ser. 2, 1981, 293, 577.

48 (a) W. N. A. Rolfsen, A. A. Olaniyi, R. Verpoorte, and L. Bohlin, J. Nat. Prod. 1981, 44, 415; (b) K. H. C. Baser and N. G. Bisset, Phytochemistry, 1982, 21, 1423.

49 F. Ladhar, M. Damak, A. Ahond, C. Poupat, P. Potier, and C. Moretti, J. Nat. Prod., 1981, 44, 459.

50 N. Ghorbel, M. Damak, A. Ahond, E. Philogene, C. Poupat, P. Potier, and H. Jacquemin, J. Nat. Prod., 1981, 44, 717.

51 S. Mukhopadhyay, G. A. Handy, S. Funayama, and G. A. Cordell, J. Nat. Prod., 1981, 44, 696.

52 N. Petitfrère-Auvray, J. Vercauteren, G. Massiot, G. Lukacs, T. Sévenet, L. Le Men-Olivier, B. Richard, and M. J. Jacquier, Phytochemistry, 1981, 20, 1987.

53 C. Galeffi and G. B. Marini-Bettolo, Tetrahedron, 1981, 37, 3167.

54 P. K. Lala, J. Inst. Chem. (India), 1981, 53, 30; Chem. Abstr., 1981, 95, 81293.

55 M. Tits, M. Franz, D. Tavernier, and L. Angenot, Planta Med., 1981, 42, 371; Chem. Abstr., 1981, 95, 147177.

56 B. A. Akinloye and W. E. Court, J. Ethnopharmacol., 1981, 4, 99.

57 J. D. Phillipson and N. Supavita, J. Pharm. Pharmacol., 1981, 33, Suppl. 13P.

58 A. Malik, N. Afza, and S. Siddiqui, Heterocycles, 1981, 16, 1727.

59 C. Lavaud, G. Massiot, J. Vercauteren, and L. Le Men-Olivier, Phytochemistry, 1982, 21, 445.

60 M. A. Amer and W. E. Court, Phytochemistry, 1981, 20, 2569.

61 A. Chatterjee, D. J. Roy, and S. Mukhopadhyay, Phytochemistry, 1981, 20, 1981.

62 W. Kohl, B. Witte, and G. Höfle, Z. Naturforsch., Teil. B, 1981, 36, 1153.

63 J. P. Kutney, L. S. L. Choi, P. Kolodziejczyk, S. K. Sleigh, K. L. Stuart, B. R. Worth, W. G. W. Kurz, K. B. Chatson, and F. Constabel, J. Nat. Prod., 1981, 44, 536; Helv. Chim. Acta, 1981, 64, 1837.

64 (a) R. Uusvuori and M. Lounasmaa, Planta Med., 1981, 41, 406; (b) J. B. Del Castillo, M. T. M. Ferrero, J. L. M. Ramon, F. R. Luis, P. V. Bueno and P. Joseph-Nathan, An. Quim., 1982, 78, 126.

65 M. J. Calverley, J. Chem. Soc., Chem. Commun., 1981, 1209.

66 C. Kan, S. K. Kan, M. Lounasmaa, and H. P. Husson, Acta Chem. Scand., Sect. B, 1981, 35, 269.

67 M. Hämeilä and M. Lounasmaa, Acta Chem. Scand., Sect. B, 1981, 35, 217.

68 L. Calabi, B. Danieli, G. Lesma, and G. Palmisano, Tetrahedron Lett., 1982, 23, 2139.

69 T. Kametani, T. Suzuki, S. Kamada, and K. Unno, Heterocycles, 1982, 19, 815.

70 E. Frostell, R. Jokela, and M. Lounasmaa, Acta Chem. Scand., Sect. B, 1981, 35, 671.

71 Atta-ur-Rahman and M. Ghazala, J. Chem. Soc., Perkin Trans. 1, 1982, 59.

72 R. Jokela and M. Lounasmaa, Tetrahedron, 1982, 38, 1015.

73 J. Bosch, M. Feliz, and M. L. Bennasar, Heterocycles, 1982, 19, 853.

74 S. Takano, N. Tamura, and K. Ogasawara, J. Chem. Soc., Chem. Commun., 1981, 1155.

75 S. Takano, M. Yonaga, and K. Ogasawara, Synthesis, 1981, 265.

76 G. N. Saunders, R. G. Hamilton, and S. McLean, Tetrahedron Lett., 1982, 23, 2359.

77 T. Kametani, N. Kanaya, and M. Ihara, Heterocycles, 1981, 16, 925; T. Kametani, N. Kanaya, H. Hino, S. P. Huang, and M. Ihara, J. Chem. Soc., Perkin Trans. 1, 1981, 3168.

78 T. Kametani, N. Kanaya, T. Honda, and M. Ihara, Heterocycles, 1981, 16, 1937.

79 S. Takano, K. Shibuya, M. Takahashi, S. Hatakeyama, and K. Ogasawara, Heterocycles, 1981, 16, 1125.

80 D. Herlem, A. Florès-Parra, F. Khuong-Huu, A. Chiaroni, and C. Riche, Tetrahedron, 1982, 38, 271.

81 (a) J. Gutzwiller, G. Pizzolato, and M. R. Uskoković, J. Am. Chem. Soc., 1971, 93, 5907; (b) Helv. Chim. Acta, 1981, 64, 1663.

82 S. Sakai, N. Saito, N. Hirose, and E. Yamanaka, Heterocycles, 1982, 17, 99.

83 T. Naito, Y. Tada, Y. Nishiguchi, and I. Ninomiya, Heterocycles, 1982, 18, 213.

84 T. Kametani, T. Suzuki, and K. Unno, Tetrahedron, 1981, 37, 3819.

85 E. Wenkert, J. St. Pyrek, S. Uesato, and Y. D. Vankar, J. Am. Chem. Soc., 1982, 104, 2244.

86 A. I. Meyers and S. Hellring, J. Org. Chem., 1982, 47, 2229.

87 R. Verpoorte, M. J. Verzijl, and A. B. Svendsen, Planta Med., 1982, 44, 21.

88 H. Achenbach, B. Raffelsberger, and I. Addae-Mensah, Liebigs Ann. Chem., 1982, 830.

89 G. Lewin, O. Tamini, P. Cabalion, and J. Poisson, Ann. Pharm. Fr., 1981, 39, 273.

90 M. Quaisuddin, Bangladesh J. Sci. Ind. Res., 1980, 15, 35; Chem. Abstr., 1982, 96, 3651.

91 M. A. Khan, H. Horn, and W. Voelter, Z. Naturforsch., Teil. B, 1982, 37, 494.

92 L. L. Shimolina and S. A. Minina, Khim. Prir. Soedin., 1981, 807.

93 M. R. Yagudaev, Khim. Prir. Soedin., 1981, 608; Chem. Abstr., 1982, 96, 104570.

94 B. Danieli, G. Palmisano, and G. S. Ricca, Tetrahedron Lett., 1981, 22, 4007.

95 A. Koskinen and M. Lounasmaa, Heterocycles, 1982, 19, 851.

96 C. Liu, Q. Wang, and C. Wang, J. Am. Chem. Soc., 1981, 103, 4634.

97 A. Malik, N. Afza, N. Sultana, and S. Siddiqui, Heterocycles, 1981, 16, 1101.

98 A. Husson, Y. Langlois, C. Riche, and H. P. Husson, Tetrahedron, 1973, 29, 3095.

99 C. Thal, M. Dufour, P. Potier, M. Jaouen, and D. Mansuy, J. Am. Chem. Soc., 1981, 103, 4956.

100 I. S. Cloudsdale, A. F. Kluge, and N. L. McClure, J. Org. Chem., 1982, 47, 919.

101 M. Sainsbury, D. Watkins, and D. K. Weerasinghe, Org. Magn. Res., 1982, 18, 117.

102 C. Mirand, G. Massiot, L. Le Men-Olivier, and J. Lévy, Tetrahedron Lett., 1982, 23, 1257.

103 J. T. Edward, P. G. Farrell, and S. A. Samad, Bangladesh J. Sci. Ind. Res., 1980, 15, 169.

104   F. Toda, K. Tanaka, and H. Ueda, Tetrahedron Lett., 1981, 22, 4669.

105   S. Takano, M. Hirama, and K. Ogasawara, Tetrahedron Lett., 1982, 23, 881.

106   D. Schumann and H. Schmid, Helv. Chim. Acta, 1963, 46, 1996.

107   Y. Ban, K. Yoshida, J. Goto, and T. Oishi, J. Am. Chem. Soc., 1981, 103, 6990.

108   B. A. Dadson, J. Harley-Mason, and G. H. Foster, J. Chem. Soc., Chem. Commun., 1968, 1233.

109   M. Sainsbury, D. Weerasinghe, and D. Dolman, J. Chem. Soc., Perkin Trans. 1, 1982, 587.

110   S. Kano, E. Sugino, S. Shibuya, and S. Hibino, J. Org. Chem., 1981, 46, 2979.

111   M. J. E. Hewlins, A. H. Jackson, A. M. Oliveira-Campos, and P. V. R. Shannon, Chem. Ind. (London), 1981, 338; J. Chem. Soc., Perkin Trans. 1, 1981, 2906.

112   W. R. Ashcroft, M. G. Beal, and J. A. Joule, J. Chem. Soc., Chem. Commun., 1981, 994.

113   (a) G. N. Taylor, J. Chem. Res. (S), 1981, 332; (b) Y. Oikawa, M. Tanaka, H. Hirasawa, and O. Yonemitsu, Chem. Pharm. Bull., 1981, 29, 1606; (c) M. J. Wanner, G. J. Koomen, and U. K. Pandit, Heterocycles, 1982, 17, 59.

114   J. P. Kutney, M. Noda, N. G. Lewis, B. Monteiro, D. Mostowicz, and B. R. Worth, Heterocycles, 1981, 16, 1469.

115   M. Feliz, J. Bosch, D. Mauleón, M. Amat, and A. Domingo, J. Org. Chem., 1982, 47, 2435.

116   A. Rabaron, L. Le Men-Olivier, J. Lévy, T. Sévenet, and M. Plat, Ann. Pharm. Fr., 1981, 39, 369.

117   M. Damak, A. Ahond, and P. Potier, Bull. Soc. Chim. Fr., Part 2, 1981, 213.

118   C. Wei-shin, L. Sao-han, A. Kirfel, G. Will, and E. Breitmaier, Liebigs Ann. Chem., 1981, 1886.

119   N. Rodier, S. Baassou, H. Mehri, and M. Plat, Acta Crystallogr., Sect. B, 1982, 38, 863.

120   J. Hájiček and J. Trojánek, Tetrahedron Lett., 1982, 23, 365.

121   (a) B. Danieli, G. Lesma, G. Palmisano, and B. Gabetta, J. Chem. Soc., Chem. Commun., 1981, 908; (b) L. Calabi, B. Danieli, G. Lesma, and G. Palmisano, J. Chem. Soc., Perkin Trans. 1, 1982, 1371.

122   E. Wenkert, T. D. J. Halls, L. D. Kwart, G. Magnusson, and H. D. H. Showalter, Tetrahedron, 1981, 37, 4017.

123   S. Takano, M. Yonaga, and K. Ogasawara, J. Chem. Soc., Chem. Commun., 1981, 1153.

124 T. Gallagher, P. Magnus, and J. C. Huffman, J. Am. Chem. Soc., 1982, 104, 1140; see also T. Gallagher and P. Magnus, Tetrahedron, 1981, 37, 3889.

125 A. J. Pearson, Tetrahedron Lett., 1981, 22, 4033; J. Chem. Soc., Perkin Trans. 1, 1979, 1255; A. J. Pearson and D. C. Rees, J. Am. Chem. Soc., 1982, 104, 1118.

126 S. J. Veenstra and W. N. Speckamp, J. Am. Chem. Soc., 1981, 103, 4645.

127 D. Cartier, D. Patigny, and J. Lévy, Tetrahedron Lett., 1982, 23, 1897.

128 M. E. Kuehne, F. J. Okuniewicz, C. L. Kirkemo, and J. C. Bohnert, J. Org. Chem., 1982, 47, 1335.

129 M. Natsume and I. Utsunomiya, Heterocycles, 1982, 17, 111.

130 T. Imanishi, K. Miyashita, A. Nakai, M. Inoue, and M. Hanaoka, Chem. Pharm. Bull., 1982, 30, 1521.

131 G. Massiot, F. S. Oliveira, and J. Lévy, Tetrahedron Lett., 1982, 23, 177.

132 B. Danieli, G. Lesma, and G. Palmisano, Gazz. Chim. Ital., 1981, 111, 257.

133 J. Hájiček and J. Trojánek, Cesk. Farm., 1981, 30, 340.

134 K. Irie and Y. Ban, Heterocycles, 1982, 18, 255.

135 L. E. Overman, M. Sworin, L. S. Bass, and J. Clardy, Tetrahedron, 1981, 37, 4041.

136 R. M. Wilson, R. A. Farr, and D. J. Burlett, J. Org. Chem., 1981, 46, 3293.

137 T. Imanishi, N. Yagi, H. Shin, and M. Hanaoka, Tetrahedron Lett., 1981, 22, 4001.

138 H. Tomisawa, H. Hongo, H. Kato, K. Sato, and R. Fujita, Heterocycles, 1981, 16, 1947; 1982, 19, 174.

139 M. E. Kuehne and J. C. Bohnert, J. Org. Chem., 1981, 46, 3443.

140 G. Massiot, F. S. Oliveira, and J. Lévy, Bull. Soc. Chim. Fr., Part 2, 1982, 185.

141 R. J. Sundberg and J. D. Bloom, J. Org. Chem., 1981, 46, 4836.

142 M. Nakagawa, H. Sugumi, S. Kodato, and T. Hino, Tetrahedron Lett., 1981, 22, 5323.

143 G. Massiot, J. Vercauteren, M. J. Jacquier, J. Lévy, and L. Le Men-Olivier, C. R. Hebd. Seances Acad. Sci., Ser. 2, 1981, 292, 191.

144 A. Chatterjee, A. K. Ghosh, and E. W. Hagaman, J. Org. Chem., 1982, 47, 1732.

145 (a) S. Sakai, N. Aimi, K. Yamaguchi, E. Yamanaka, and J. Haginiwa, J. Chem. Soc., Perkin Trans. 1, 1982, 1257; (b) J. Silverton and T. Akiyama, ibid., p. 1263.

146 X. Z. Feng, C. Kan, H. P. Husson, P. Potier, S. K. Kan, and M. Lounasmaa, J. Nat. Prod., 1981, 44, 670.

147 A. Henriques, C. Kan, A. Chiaroni, C. Riche, H. P. Husson, S. K. Kan, and M. Lounasmaa, J. Org. Chem., 1982, 47, 803.

148 S. Mukhopadhyay and G. A. Cordell, J. Nat. Prod., 1981, 44, 611.

149 A. De Bruyn, L. De Taeye, R. Simonds, M. Versele, and C. De Pauw, Bull. Soc. Chim. Belg., 1982, 91, 75.

150 J. P. Schaumberg and J. P. Rosazza, J. Nat. Prod., 1981, 44, 478.

151 J. P. Kutney, B. Aweryn, L. S. L. Choi, P. Kolodziejczyk, W. G. W. Kurz, K. B. Chatson, and F. Constabel, Heterocycles, 1981, 16, 1169; Helv. Chim. Acta, 1982, 65, 1271.

152 M. Döé de Maindreville and J. Lévy, Bull. Soc. Chim. Fr., Part 2, 1981, 179.

153 A. T. Keene, L. A. Anderson, and J. D. Phillipson, J. Pharm. Pharmacol., 1981, 33 Suppl., 15P.

154 E. J. Staba and A. C. Chung, Phytochemistry, 1981, 20, 2495.

155 B. J. Oleksyn, Acta Crystallogr., Sect. B, 1982, 38, 1832.

156 N. Kobayashi and K. Iwai, J. Polymer Sci., Polymer Lett., 1982, 20, 85.

157 J. C. Cai, M. G. Yin, A. Z. Min, D. W. Feng, and X. X. Zhang, Hua Hsueh Hsueh Pao, 1981, 39, 171; Chem. Abstr., 1981, 95, 133209.

158 T. Miyasaka, S. Sawada, and K. Nokata, Heterocycles, 1981, 16, 1713, 1719.

13
Lycopodium Alkaloids

BY W. A. AYER

Relatively little work has been reported in the period since the last review (cf. Vol. 11, p. 199). An interesting review of the total syntheses of the alkaloids, including annotinine, lycopodine and related alkaloids, luciduline, and serratinine, has appeared.[1] A full account of Heathcock's synthesis of lycopodine (cf. Vol. 10, p. 207, Vol. 11, p. 200), lycodine (cf. Vol. 11, p. 200) and lycodoline (1) has been published.[2] The synthesis of lycodoline

(1)   (2)   (3) R = OH
                   (4) R = H

(1), first announced in 1981,[3] utilizes an aerial oxidation similar to that observed with lucidine B (cf. Vol. 10, p. 206) to introduce the C-12 hydroxyl group. Acid hydrolysis of the ketal (2) (cf. Vol. 11, p. 206), followed by basification and exposure of the resulting imine to oxygen, provided the hydroperoxide (3), which was subsequently reduced to the corresponding alcohol (4). Compound (4) was cyclized to the tricyclic compound (5) in a rather unique manner, by heating (4) in a 5:1 mixture of toluene and 3-bromo-1-propanol. Apparently the bromopropanol undergoes a slow base-catalyzed elimination of HBr and it is necessary to have only a catalytic amount of HBr present at any particular moment to catalyze enolization of unprotonated (4). Treatment of (5) with 3-iodo-1-propanol gives the alcohol (6). Modified Oppenauer oxidation of (6) gave the corresponding aldehyde which underwent immediate aldol condensation to provide dehydrolycodoline

(5) R = H
(6) R = CH$_2$CH$_2$CH$_2$OH
(6a) R = COCH$_2$CH$_2$OBz

(7)

(8) R = H
(9) R = Ac

(7) which on catalytic hydrogenation furnished lycodoline (1).[2] A formal total synthesis of lycodoline has been reported by Kim et al.[4] The previously reported (cf. Vol. 10, p. 209) tricyclic ketone (5) was condensed with 3-benzyloxypropionyl chloride to give the amide (6a). Reduction of (6a) with LAH followed by Jones oxidation and hydrogenolysis of the benzyl ether gave the alcohol (6), an intermediate in Heathcock's total synthesis.

A total synthesis of N$_\alpha$-methylphlegmarine (8) (cf. Vol. 10, p. 205) and its N$_\beta$-acetyl derivative (9), which defines for the first time the relative stereochemistry at four of the five chiral centers in these alkaloids, has been reported,[5] and a ^{13}C nmr study of the synthetic compounds allows the assignment of the stereochemistry of the fifth chiral center.[6] Biogenetic considerations (cf. Vol. 10, p. 206 and reference 4 herein) suggested that the phlegmarines should contain a cis-decahydroquinoline ring system and the initial synthetic studies were so directed. During these studies it became clear, from comparison of ^{1}H nmr spectra of synthetic cis-decahydroquinolines of general structure (9) and the natural products, that the phlegmarines did not contain a cis-decahydroquinoline system.[5] The appropriate trans-decahydroquinoline derivative (11) was obtained by lithium-ammonia reduction of the ketoamide (10), which in turn is available in high yield by acid-catalyzed cyclization of 2-(2-cyanoethyl)-5-methyl-1,3-cyclohexanedione (Scheme 1). Condensation of (11) with 2-trimethylsilylmethylpyridine gave (12) and this,

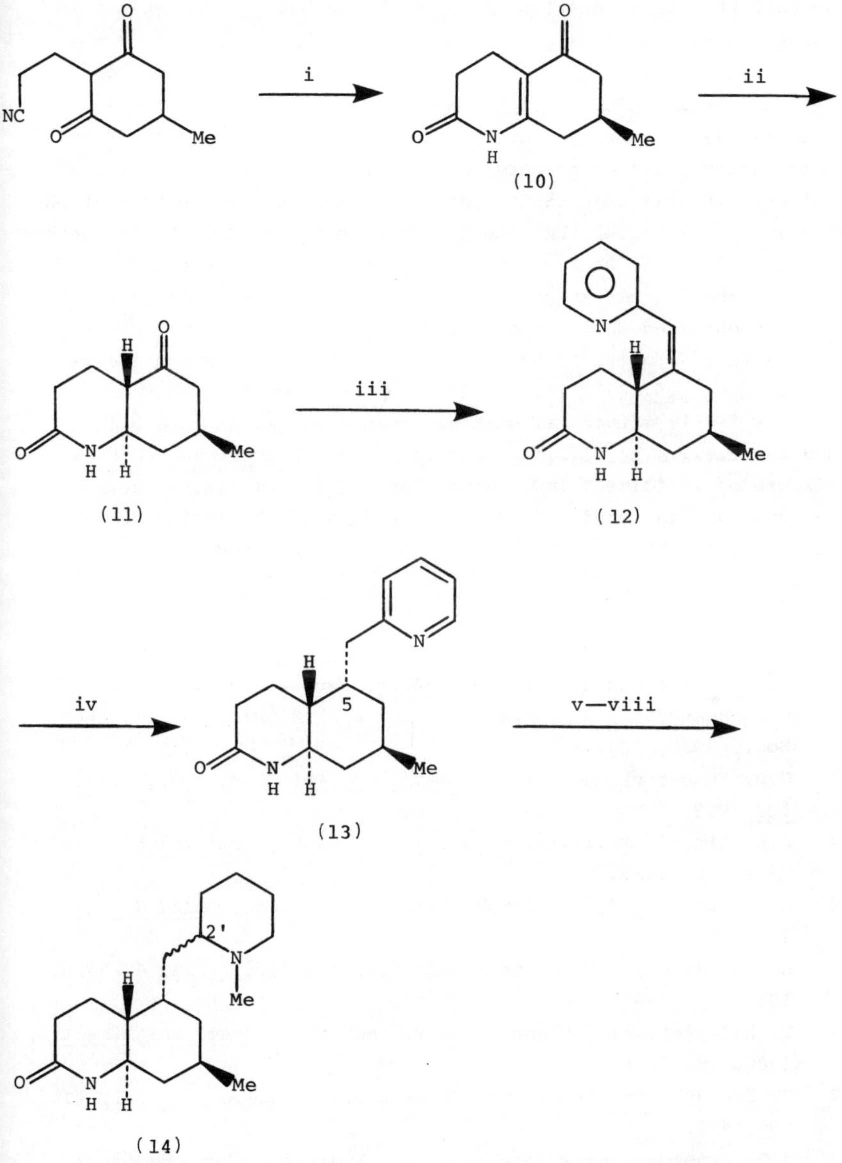

Reagents: i, HOH-HOAc (1:1); ii, Li/NH$_3$; iii, 2-trimethylsilylmethylpyridine, Bu^{n}Li; iv, H$_2$/Pt; v, m-CPBA; vi, PCl$_3$; vii, CH$_3$I; viii, H$_2$/Pt

Scheme 1

on catalytic hydrogenation, furnished (13) and its C-5 epimer in approximately equal amounts. The epimers were separable in the form of their N-oxides. Compound (13) was transformed into its methiodide and reduced to give an inseparable mixture of the C-2' epimers (14). Reduction of the mixture (14) with LAH followed by acetylation provided N-acetylphlegmarine (9) along with its C-2' epimer. In this case the epimers were separable by chromatography. The synthesis established the relative configuration at all centers except C-2'. The stereochemistry at C-2' was assigned by comparison of the ^{13}C nmr spectra of the various C-5 and C-2' stereoisomers obtained during the synthetic studies.[6]

The full details of the synthesis of fawcettimine and deoxyserratinine (cf. Vol. 11, p. 201) have appeared.[7]

Very little structural work has been reported in the period under review. The alkaloid serratanine has been shown[8] to be a mixture of lucidine B and oxolucidine B (cf. Vol. 10, p. 206). A chemotaxonomical study of the distribution of the lysine-derived alkaloids in the genus Lycopodium has been reported.[9]

References

1. Y. Inubushi and T. Harayama, Heterocycles, 1981, 15, 611.
2. C.H. Heathcock, E.F. Kleinman, and E.S. Binkley, J. Am. Chem. Soc., 1982, 104, 1054.
3. C.H. Heathcock and E.F. Kleinman, J. Am. Chem. Soc., 1981, 103, 222.
4. S.W. Kim, I. Fujii, K. Nagao, and Y. Ozaki, Heterocycles, 1981, 16, 1515.
5. A. Leniewski, J. Szychowski, and D.B. MacLean, Can. J. Chem., 1981, 59, 2479.
6. A. Leniewski, D.B. MacLean, and J.K. Saunders, Can. J. Chem., 1981, 59, 2695.
7. T. Harayama, M. Takatani, and Y. Inubushi, Chem. Pharm. Bull., 1980, 28, 2394.
8. Y. Inubushi, H. Ishii, and T. Harayama, Yakagaku Zasshi, 1980, 100, 672.
9. J.C. Braekman, L. Nyembo, and J.J. Symoens, Phytochemistry, 1980, 19, 803.

14
Diterpenoid Alkaloids

BY S. W. PELLETIER AND S. W. PAGE

1 Introduction

Continued interest in the traditional medicines utilizing plants containing diterpenoid alkaloids has resulted in numerous publications during the past year. As new plant species are recognized and studied, this increased activity should continue.

Not unexpected in an area of very active research, there have been some duplications of work and some confusion in nomenclature. With regard to trivial names, we have endorsed the use of the first reported name for a given new alkaloid.

Several reviews of recent work in the Peoples Republic of China have appeared.[1,2,3] Since some of the work reviewed was either previously unpublished or in relatively inaccessible literature, some of these results are discussed in this review under the appropriate species.

Wang[4] has reviewed the ^{13}C n.m.r. spectral data for 186 C_{19}- and C_{20}- diterpenoid alkaloids and derivatives. Chemical shift ranges for the common substituent groups and additivity relationships for substituents at various positions are presented.

A number of studies of the pharmacology and toxicology of the diterpenoid alkaloids have appeared.[5,6,7,8,9,10] These will not be reviewed, but are cited for reference. Of particular interest is a review of the toxicity data and structure-toxicity relationships for C_{19}- and C_{20}- alkaloids of *Aconitum* species.[11]

Sawada and coworkers[12] have reported preliminary studies on the induction of callus tissues from *Aconitum ibukiense* var. *eizaenense* Nakai. The production of alkaloids by these callus tissues was indicated by a positive Dragendorff test of the weak base fraction of an extract. These studies are significant in that such methods might be useful for the production of alkaloids from rare species of plants.

No reports of new work on the *Daphniphyllum* alkaloids were available to our laboratories.

The numbering systems used in this review follow the previous conventions for the aconitine, lycoctonine, atisine, and veatchine skeletons, as indicated in structures A, B, C, and D, respectively.

2 Structural Elucidations and General Studies

Configuration of the C(1)-Oxygen Function of the Lycoctonine Alkaloids.- The recent revision[13] of the structures of thirty-seven lycoctonine-related alkaloids based on several chemical correlations and on the X-ray crystallographic analysis of browniine perchlorate and dictyocarpine (cf. this Review, vol. 12, 1982, p. 250.) is augmented by additional X-ray crystallographic data. The full reports of the structure determinations of delphinifoline (1)[14] and the ketolactam (2)[15] derived from lycoctonine (3) have appeared. The study of the formation of (2) is reviewed under *Chemical Studies*. All these studies make certain the α-configuration of the C(1) oxygen functions in lycoctonine and its relatives.

Delphinifoline had been isolated as a minor alkaloid from *Aconitum delphinifolium* DC.[16] The structure (1) was refined to a final R value of 0.038.[14] The gross features of this structure are comparable to the previously reported structures for other C_{19}-lycoctonine alkaloids: an extremely rigid skeleton with rings A and D in boat forms, rings B and E in distorted chairs, ring C in an envelope with C(14) at the flap, and ring F in a half-chair. The hydroxyl group at C(1) is clearly on the same side of the ring as the N-bridge.

In a paper submitted prior to the publication of these structure revisions, Salimov and coworkers[17] incorrectly revised the structures of delcorine (4) and delcoridine (5) to reflect the β-configuration for the C(1)-methoxyl group. (The correct α-configuration is used in the structures in this review.) This revision was based on the chemical correlation of these alkaloids with delphatine (6), in which the configuration of the C(1)-methoxyl group had been incorrectly assigned.[18] Methylation of delcorine (4) with methyl iodide/sodium hydride gave (7), which, on heating with 10% H_2SO_4 for 10 hrs., led to delphatine (6). Methylation of delcoridine (5) gave delcorine (4). The α-configuration for the C(1)-methoxyl group in delphatine (6) has been established by chemical correlations with delsoline (8), lycoctonine (3), and browniine (9), whose structure has been confirmed by an X-ray crystallographic structure determination.[13] Therefore, the orginally assigned structures for delcorine

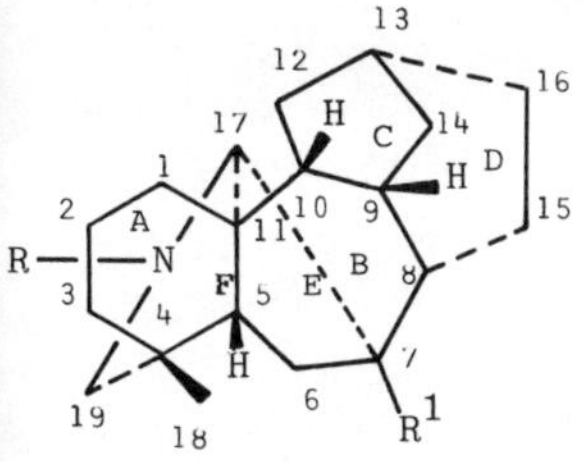

Diterpenoid Alkaloids

A) R^1 = H; Aconitine skeleton
B) R^1 = OR^2; Lycoctonine skeleton

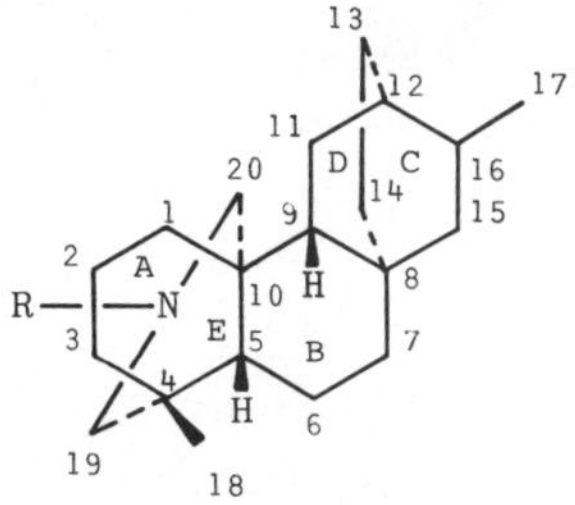

(C) Atisine skeleton

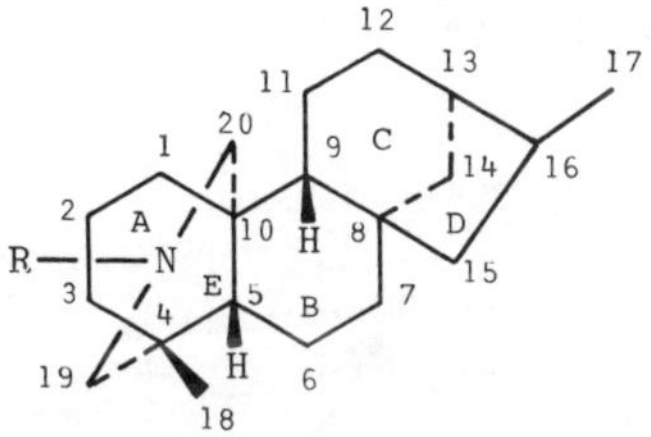

(D) Veatchine skeleton

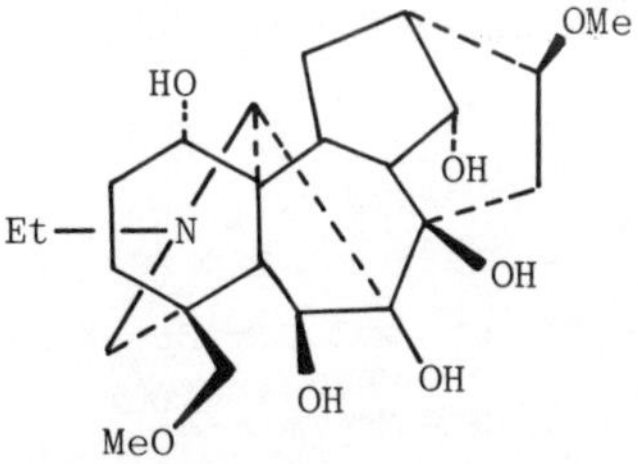

Delphinifoline (1)

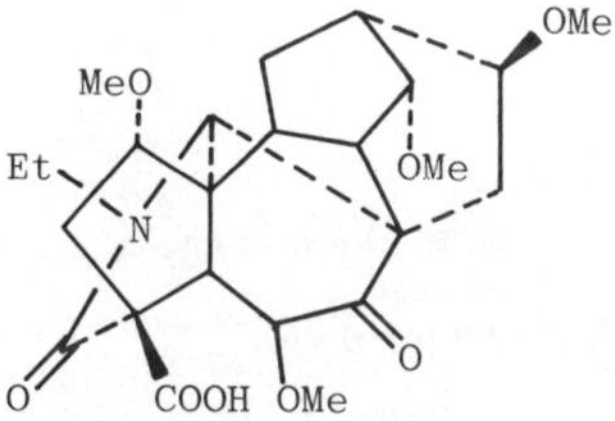

(2)

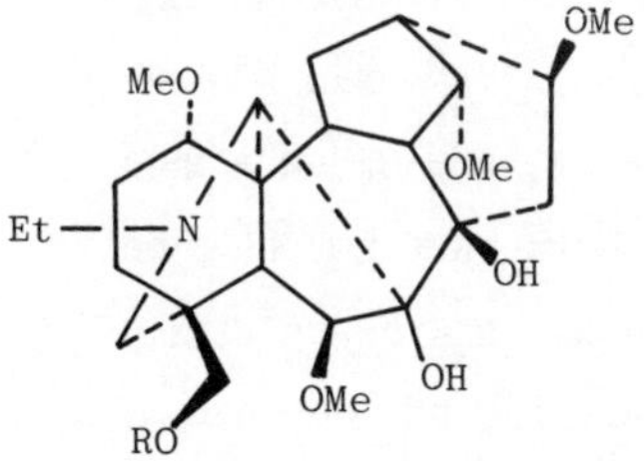

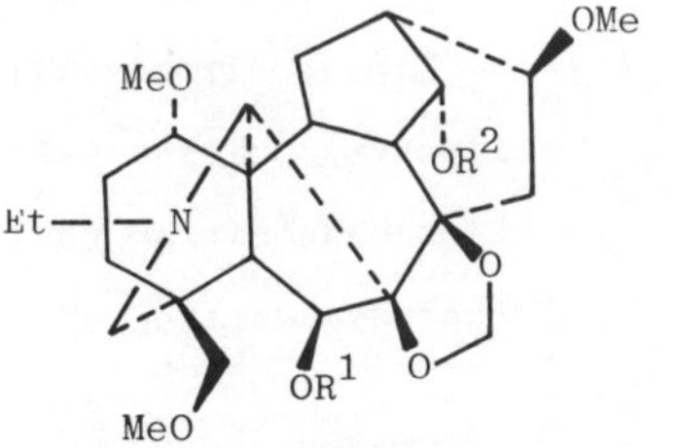

Lycoctonine (3) R = H
Avadharidine (49) R = -C(=O)– [with HNCO(CH$_2$)$_2$CONH$_2$ substituent]

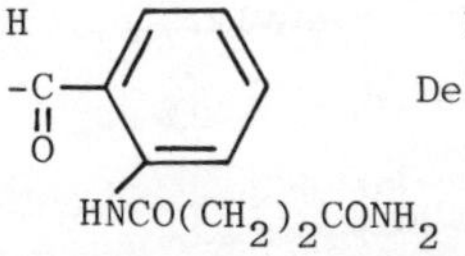

Delcorine (4) R^1 = H; R^2 = Me
Delcoridine (5) R^1 = R^2 = H
(7) R^1 = R^2 = Me

Delphatine (6) $R^1 = R^2 =$ Me
Delsoline (8) $R^1 =$ H, $R^2 =$ Me
Browniine (9) $R^1 =$ Me, $R^2 =$ H

Mesaconitine (10) $R^1 =$ OH, $R^2 =$ Ac, $R^3 =$ Bz, $R^4 =$ Me
Deoxyaconitine (11) $R^1 =$ H, $R^2 =$ Ac, $R^3 =$ Bz, $R^4 =$ Et
Hypaconitine (12) $R^1 =$ H, $R^2 =$ Ac, $R^3 =$ Bz, $R^4 =$ Me
Jesaconitine (13) $R^1 =$ OH, $R^2 =$ Ac, $R^3 =$ As, $R^4 =$ Et
Aconitine (14) $R^1 =$ OH, $R^2 =$ Ac, $R^3 =$ Bz, $R^4 =$ Et
Lipoaconitine (20) $R^1 =$ OH, $R^2 =$ FA, $R^3 =$ Bz, $R^4 =$ Et
Lipohypaconitine (21) $R^1 =$ H, $R^2 =$ FA, $R^3 =$ Bz, $R^4 =$ Me
Lipomesaconitine (22) $R^1 =$ OH, $R^2 =$ FA, $R^3 =$ Bz, $R^4 =$ Me
Lipodeoxyaconitine (23) $R^1 =$ H, $R^2 =$ FA, $R^3 =$ Bz, $R^4 =$ Et
(47) $R^1 =$ OAc, $R^2 =$ Ac, $R^3 =$ Bz, $R^4 =$ Et

(FA = Linoleoyl, palmitoyl, oleoyl, stearoyl, or linolenoyl)
(As = anisoyl)

(4)[19] and delcoridine (5)[20] are correct.

High-Performance Liquid Chromatographic (HPLC) Methods for the Determination of Aconitine Alkaloids.- The crude drug "bushi" is a traditional medicine used in large quantities in East Asia. This drug is prepared by processing roots of several *Aconitum* species. Since the methods of processing have not been standardized, the alkaloid composition, and consequently the toxicity or therapeutic efficacy, may vary. Hikino and coworkers[21] have reported two rapid HPLC methods for determination of these alkaloids in commercial preparations. Separations of mesaconitine (10), deoxyaconitine (11), hypaconitine (12), jesaconitine (13), aconitine (14), and their benzoyl analogs were achieved with reverse phase and ion-pair methods. The reverse phase methods employed an ODS chemically bonded silica gel with a phosphate buffer (pH 2.7) and tetrahydrofuran (89:11) as the mobile phase with a variable-wavelength ultraviolet detector at 254 nm. The ion-pair method used a phosphate buffer (pH 2.7)- tetrahydrofuran (85:15) mixture with 0.01 M sodium hexanesulphonate. The detection limit for mesaconitine was 12 ng. A comparison of these two methods showed good agreement. The application of these methods to several commercial aconitine preparations indicated considerable variations in alkaloid composition. See also: Alkaloids of *Aconitum subcuneatum* below.

Alkaloids of *Aconitum carmichaeli*. - In chemical characterization studies of crude and processed drugs, Kitagawa et al.[22] have reported their work on "chuan-wu", from Si Chuan, China, which was reported to be tubers of *A. carmichaeli*. The known alkaloids aconitine (14), hypaconitine (12), mesaconitine (10), talatizamine (15), 14-acetyltalatizamine (16), isotalatizidine (17), karakoline (18), and neoline (19) were isolated from these roots and identified by direct comparisons or comparisons of the physical data with authentic samples. In addition, fatty acid esters of aconitine [lipoaconitine (20)], hypaconitine [lipohypaconitine (21)], mesaconitine [lipomesaconitine (22)], and deoxyaconitine [lipodeoxyaconitine (23)] were identified for the first time. These structures were assigned from the spectral data (including ^{1}H and ^{13}C n.m.r. and mass spectra) of the parent esters and studies of their hydrolysis products.

In chemical and pharmacological investigations of the Japanese drug "sen-bushi", the roots of *A. carmichaeli*, Japanese workers[23] have identified the known alkaloid constituents aconitine (14), hypaconitine (12), mesaconitine (10), talatizamine (15), isotalatizidine (17), karakoline (18), 14-acetyltalatizamine (16), and three new bases: senbusine A (24), $C_{23}H_{37}NO_6$, amorphous powder; senbusine B (25), $C_{23}H_{37}NO_6$, amorphous powder; and senbusine C (26), $C_{24}H_{39}NO_7$, which they reported as an amorphous powder. These assignments were based on ^1H and ^{13}C n.m.r. spectral comparisons of (17), (18), (16), (24), (25), and (26) and the triacetate derivatives of (24) and (26).

A report[24] of the alkaloids of the roots of *A. carmichaeli* Debx. describes the isolation and structure elucidation of "fuziline", $C_{24}H_{39}NO_7$, m.p. 214-216°C., and the known bases aconitine (14), hypaconitine (12), mesaconitine (10), neoline (19), and songorine (27). The alkaloid contents of these plants were found to vary considerably with growing location.[25] From ^1H and ^{13}C n.m.r. spectral comparisons with neoline (19) and isodelphinine (28), structure (26), identical with that for senbusine C, was assigned for "fuziline".[24] This structure was confirmed by a single-crystal X-ray analysis.[24] Curiously, reference 25 utilizes a reproduction of the figure for the structure fuziline appearing in reference 24, but does not even cite reference 24!

Zhu and Zhu[1] discussed the isolation of six alkaloids from the roots of *A. carmichaeli* (which they referenced as the traditional Chinese drugs Chuan-wu and Fu-tzu). Aconitine (14), hypaconitine (12), mesaconitine (10), talatizamine (15), isotalatizamine, and isotalatizidine (17) were identified. The base previously named chuan-wu base A is identical with isotalatizidine.

Alkaloids of *Aconitum crassicaule*.- The roots of *A. crassicaule* are a traditional medicine in the Yunnan Province of China, often prescribed for bruises and injuries. Wang and Fang[26,27] have reported the isolation of four diterpenoid alkaloids from these materials: chasmanine (29), yunaconitine (30), and two new bases, crassicauline A, $C_{35}H_{49}NO_{10}$, m.p. 162.5-164.5°C., and crassicauline B, $C_{27}H_{31}NO_4$, m.p. 311-315°C. Structure (31) was assigned for crassicauline A from the ^1H n.m.r. spectral comparisons and the conversion of (31) to bikhaconine (32) and its triacetate derivative (33). The u.v., i.r., and ^1H n.m.r. spectral data for crassicauline B indicated the presence of two hydroxyl groups, a benzoyloxyl group, and an exocyclic double bond. These data indicated that this alkaloid is similar in structure, but not identical, with isohypognavine (34).

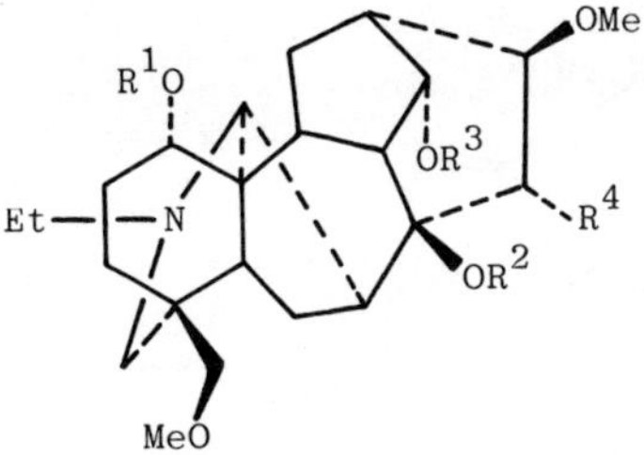

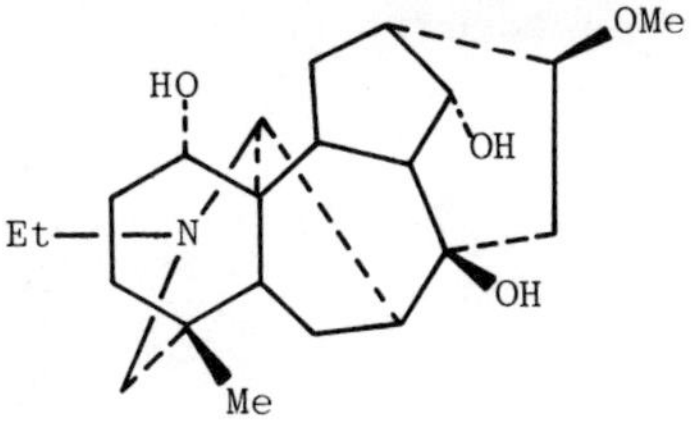

Diterpenoid Alkaloids

Talatizamine (15) R^1 = Me, R^2 = R^3 = R^4 = H

14-Acetyl-
  talatizamine (16) R^1 = Me, R^2 = R^4 = H, R^3 = Ac

Isotalatizidine (17) R^1 = R^2 = R^3 = R^4 = H

Senbusine B (25) R^1 = R^2 = R^3 = H, R^4 = OH

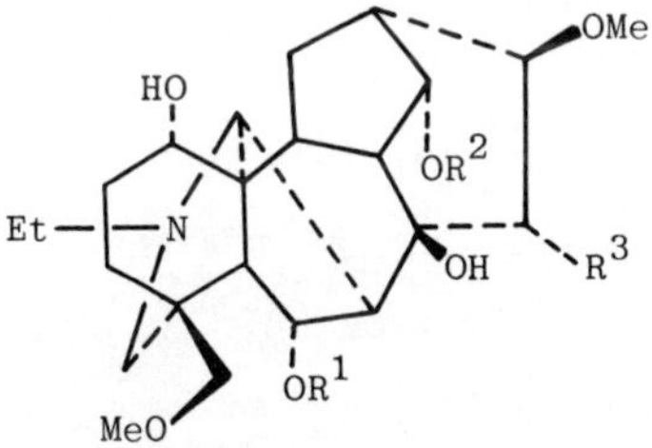

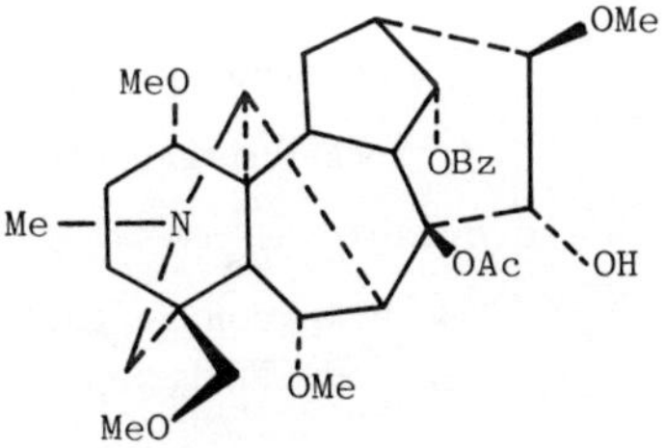

Karakoline (18)

Neoline (19) R^1 = Me, R^2 = R^3 = H

Senbusine A (24) R^1 = R^2 = R^3 = H

Senbusine C (26) R^1 = Me, R^2 = H, R^3 = OH
  (Fuziline)

14-Acetylneoline (61) R^1 = Me, R^2 = Ac, R^3 = H

Isodelphinine (28)

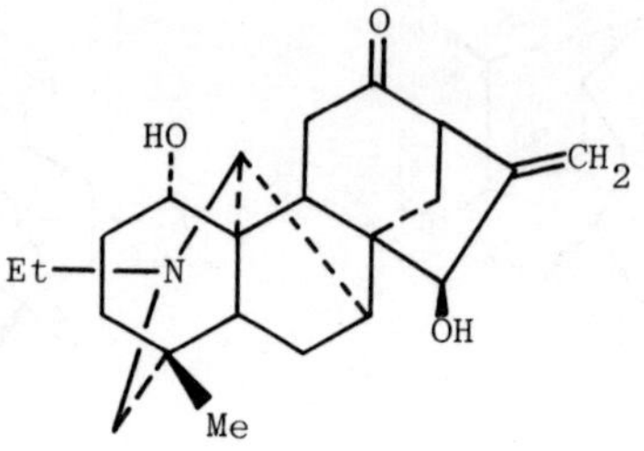

Songorine (27)

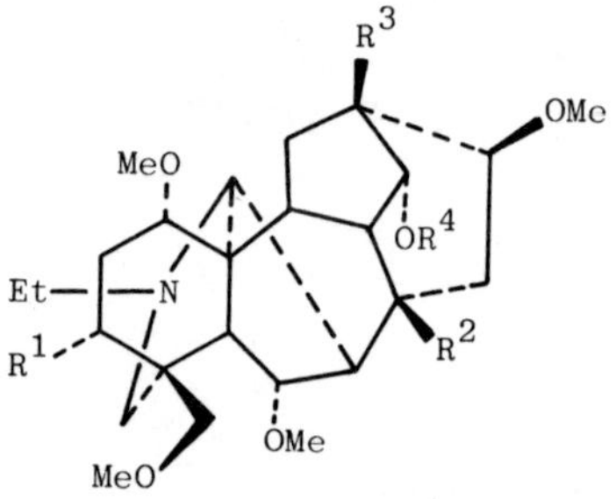

Chasmanine (29) $R^1 = R^3 = R^4 = H$, $R^2 = OH$
Yunaconitine (30) $R^1 = R^3 = OH$, $R^2 = OAc$, $R^4 = As$
Crassicauline A (31) $R^1 = H$, $R^2 = OAc$, $R^3 = OH$, $R^4 = As$
Bikhaconine (32) $R^1 = R^4 = H$, $R^2 = R^3 = OH$
(33) $R^1 = H$, $R^2 = R^3 = OAc$, $R^4 = Ac$
Vilmorrianine A (42) $R^1 = OH$, $R^2 = OAc$, $R^3 = H$, $R^4 = As$
Vilmorrianine C (43) $R^1 = R^3 = H$, $R^2 = OAc$, $R^4 = As$
(45) $R^1 = R^2 = OAc$, $R^3 = OH$, $R^4 = As$
(46) $R^1 = R^2 = OAc$, $R^3 = H$, $R^4 = As$

(As = anisoyl)

[structure (47) is with (10)]

Alkaloids of *Aconitum delavayi*.- The structures of delavaconitine (35), $C_{29}H_{39}NO_6$, m.p. 154°C, and isoaconitine (36), $C_{25}H_{49}NO_{11}$, m.p. 144-146°C, have been reported.[1] Pyrolysis of diacetyldelavaconitine (37) afforded the pyroderivative (38). When treated with acid, (38) was isomerized to acetyl isopyrodelavaconitine (39). From these transformations and the spectral data, structure (35) was assigned for delavaconitine. Alkaline hydrolysis of isoaconitine gave the amine alcohol (40). Comparisons of the physical and chemical properties, particularly the n.m.r. spectral data, of (40) and its acetate (41) with those of pseudaconine and its acetate derivative, respectively, indicated their structural identities and permitted assignment of structure (36) for isoaconitine.

Alkaloids of *Aconitum episcopale* Levl.- Yang and coworkers[28] have reported a ^{13}C n.m.r. study of several diterpenoid alkaloids isolated from *Aconitum* species used in traditional medicine in the Yunnan Province in China. The ^{13}C n.m.r. chemical shifts for yunaconitine (30), vilmorrianine A (42), karakoline (vilmorrianine B) (18), vilmorrianine C (43), vilmorrianine D (44), 3-acetylyunaconitine (45), and 3-acetylvilmorrianine A (46) and 3-acetylaconitine (47) were assigned on the basis of single-frequency off-resonance decoupling, additivity relationships, and comparisons of the spectra of similar alkaloids. From these data, the previous[29] assignments for the chemical shifts of C(2) and C(12) of aconitine (14) were reversed. In these studies, a new alkaloid, scopaline, $C_{21}H_{33}NO_4$, m.p. 167-169°C, was isolated from the roots of *A. episcopale* and was assigned structure (48) from the ^1H and ^{13}C n.m.r, m.s., and i.r. spectral data.

Alkaloids of *Aconitum finetianum* Hand-Mazz.- Two research groups in the Peoples Republic of China have reported studies on the alkaloids of the roots of *A. finetianum*. Chen and coworkers[30] isolated delsoline (8), avadharidine (49), lycoctonine (3), and two unknown alkaloids, $C_{27}H_{21}N_3O_5$, m.p. 141-142°C, and $C_{22}H_{31}NO_3$, m.p. 221-226°C, by a combination of pH extraction and chromatography. Delsoline (8) was reported to be useful in the treatment of acute dysentery and enteritis and as a smooth muscle relaxant. Jiang and coworkers[31,32] reported the isolation of lappaconitine (50), ranaconitine (51), and a new base, finaconitine, $C_{32}H_{44}N_2O_{10}$, m.p. 220-221°C, from this species. Structure (52), corresponding to 10β-hydroxyranaconitine, was assigned for finaconitine from the i.r.,

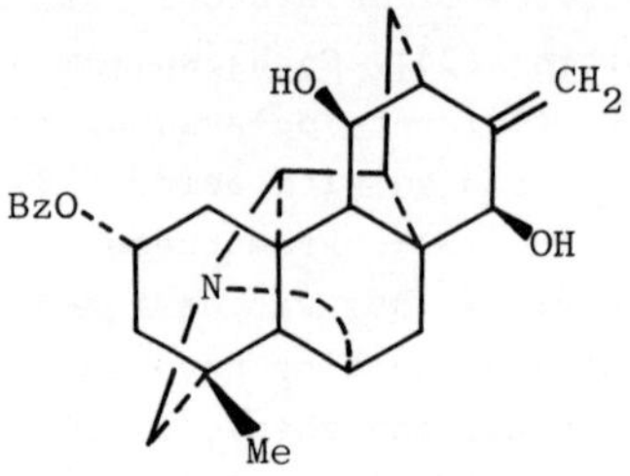

Isohypognavine (34)

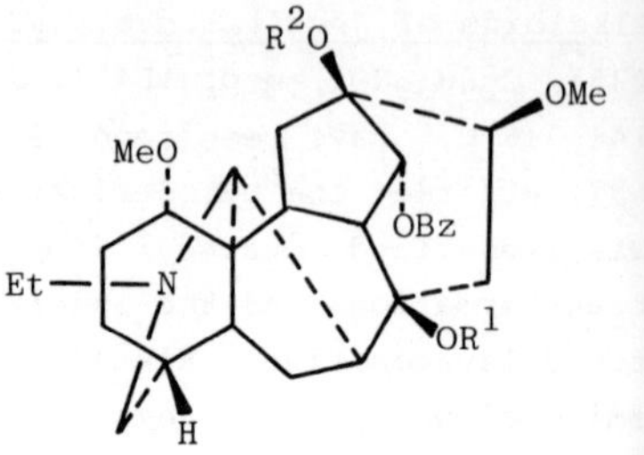

Delavaconitine (35) $R^1 = R^2 = H$
(37) $R^1 = R^2 = Ac$

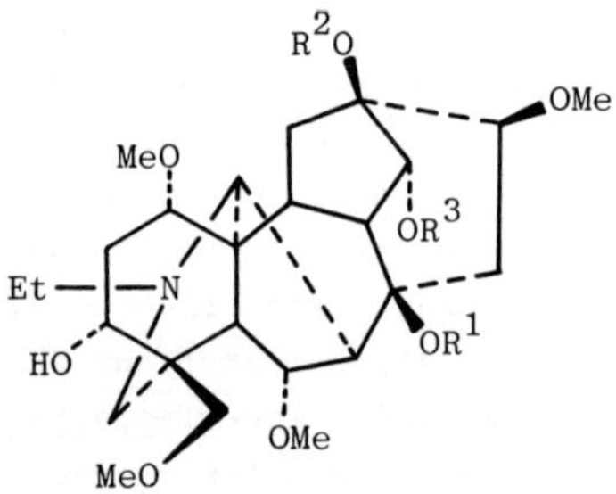

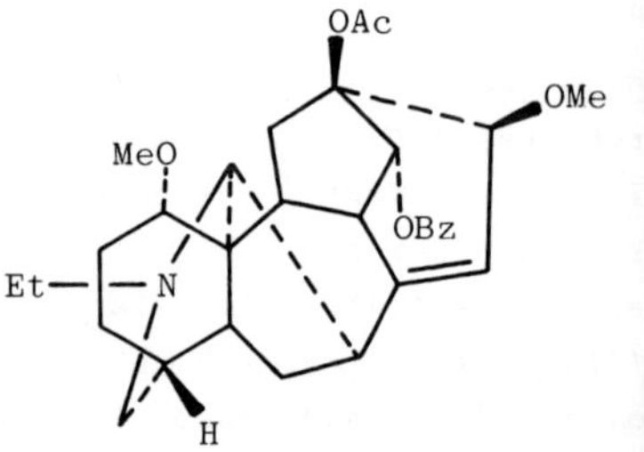

Isoaconitine (36) $R^1 = Ac$, $R^2 = H$, $R^3 = As$
Pseudaconine (40) $R^1 = R^2 = R^3 = H$
(41) $R^1 = R^2 = H$, $R^3 = Ac$

(As = anisoyl)

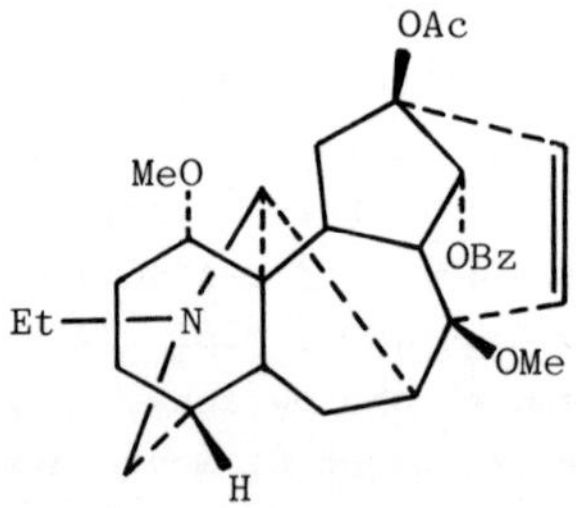

(38)

(39)

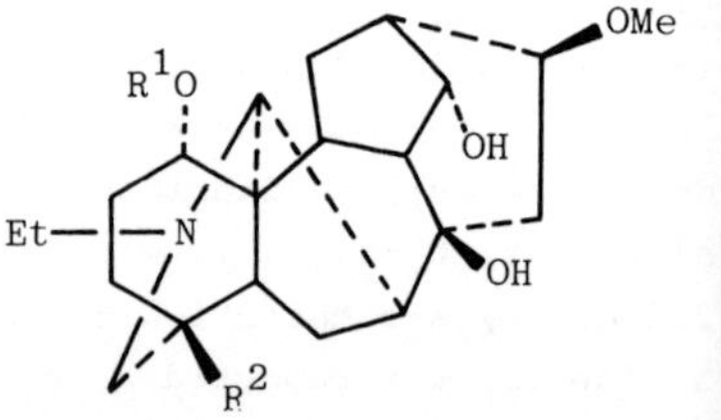

Vilmorrianine D (44) $R^1 = R^2 = Me$
Scopaline (48) $R^1 = R^2 = H$

[structure (49) is with (3)]

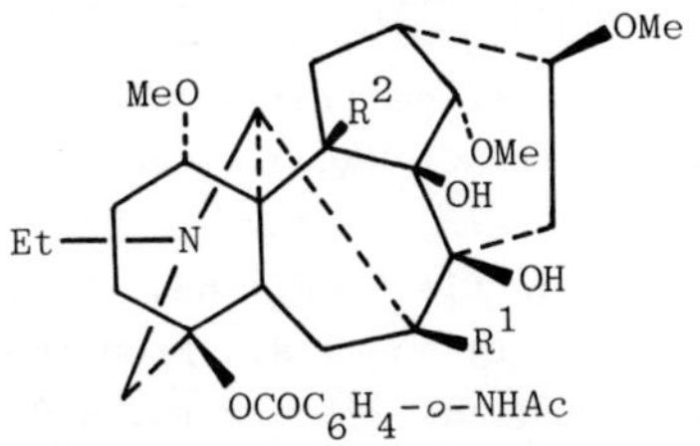

Diterpenoid Alkaloids

Lappaconitine (50) R^1 = R^2 = H
Ranaconitine (51) R^1 = OH, R^2 = H
Finaconitine (52) R^1 = R^2 = OH

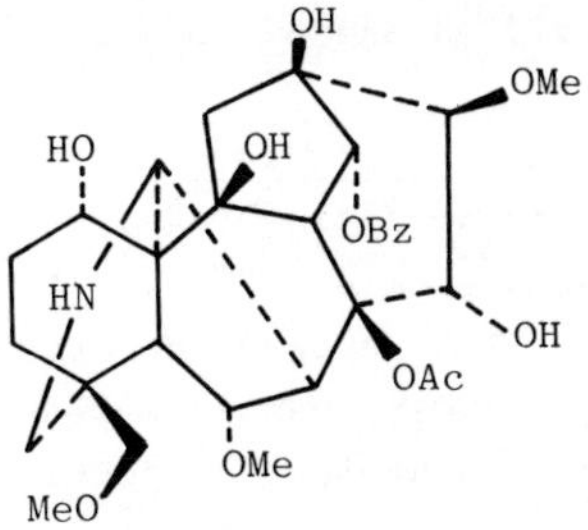

Flavaconitine (53)

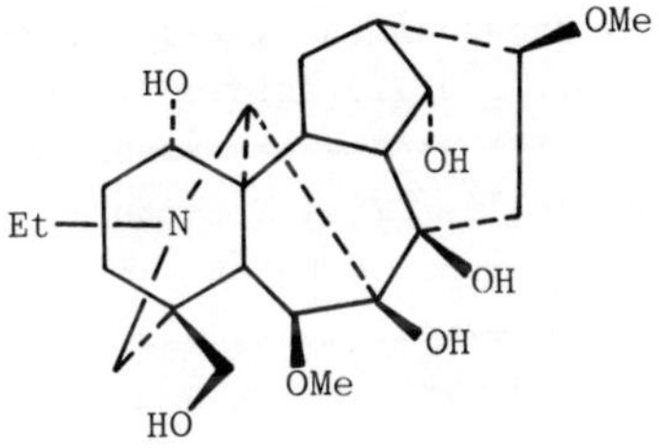

Takaosamine (54)

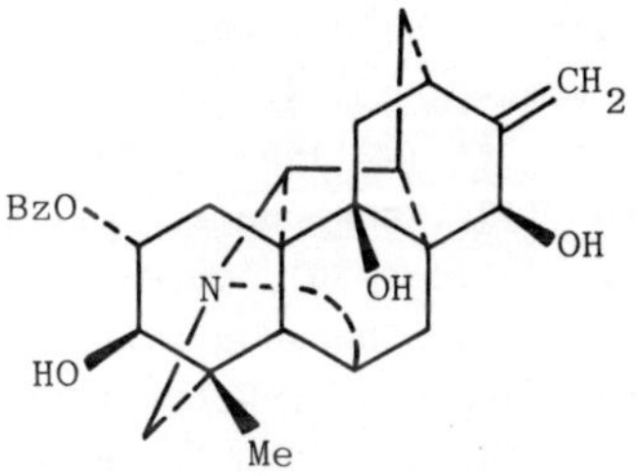

Ignavine (55)

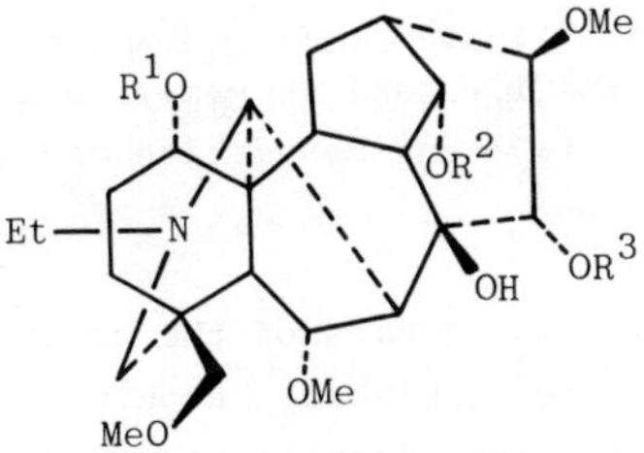

(56) R^1 = R^2 = R^3 = H
(57) R^1 = R^2 = R^3 = Ac

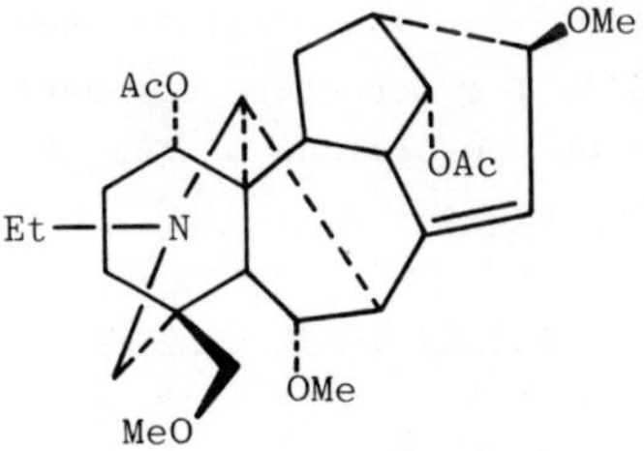

(58)

u.v., ^{1}H and ^{13}C n.m.r., and mass spectral data.

Alkaloids of *Aconitum flavum* Hand-Mazz.- Aconitine (14) and 3-acetylaconitine (47), $C_{36}H_{47}NO_{12}$, m.p. 196-197°C, have been isolated from extracts of this species.[33,34]

In a subsequent study[35], a new alkaloid, flavaconitine, $C_{31}H_{41}NO_{11}$, was isolated from the roots of this species. From the physicochemical properties and i.r., u.v., ^{1}H and ^{13}C n.m.r., and mass spectral data, structure (53) was assigned for flavaconitine.

Alkaloids of *Aconitum japonicum* Thumb.- In investigations[36,37] of the alkaloids of the roots of *A. japonicum* collected in the southern part of the Boso Peninsula, Japan, ten alkaloids were identified: aconitine (14), mesaconitine (10), hypaconitine (12), deoxyaconitine (11), 14-acetyltalatizamine (16), isotalatizidine (17), neoline (19), takaosamine (54), ignavine (55), and a new base, 15α-hydroxyneoline (56), $C_{24}H_{39}NO_7$, m.p. 206-207°C. This structural assignment was based primarily on the ^{13}C n.m.r. data of (56) and its triacetate derivative (57) and on the chemical conversion of neoline (19) to (56) via structures (58) and (59).

Alkaloids of *Aconitum jinyangense* W.T. Wang.- Chen and Song[38] have reported the isolation of four alkaloids from this new *Aconitum* species. They isolated denudatine (60), 14-acetylneoline (61), an unknown base, $C_{22}H_{33}NO_2$, m.p. 198-200°C, and a new naturally occurring alkaloid, jynosine (62), $C_{24}H_{35}NO_3$, m.p.-HClO$_4$ 254-256°C. The structure was assigned from the i.r. and ^{1}H n.m.r. data and by conversions of (62) to denudatine (60) and denudatine diacetate (63).

Alkaloids of *Aconitum karakolicum* Rapcs.- In studies of the aerial parts of *A. karakolicum*, five alkaloids were isolated: aconitine (14), songorine (27), the aporphine isoboldine, napelline $\underline{N}$-oxide (64), and 12-acetylnapelline $\underline{N}$-oxide (65), $C_{24}H_{35}NO_5$, m.p. 235°C.[39] Structure (65) was assigned from the ^{1}H n.m.r., i.r., and mass spectral data.

Alkaloids of *Aconitum koreanum (coreanum)*.- Further studies of the alkaloids of this species, which is used as a traditional medicine for its expectorant and analgesic activities, have been reported.[1,40] Two new alkaloids, guan-fu base F, $C_{26}H_{35}NO_6$, m.p.

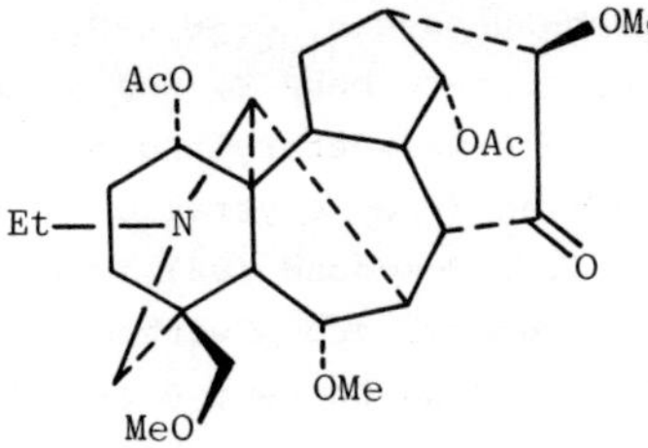

Diterpenoid Alkaloids

(59)

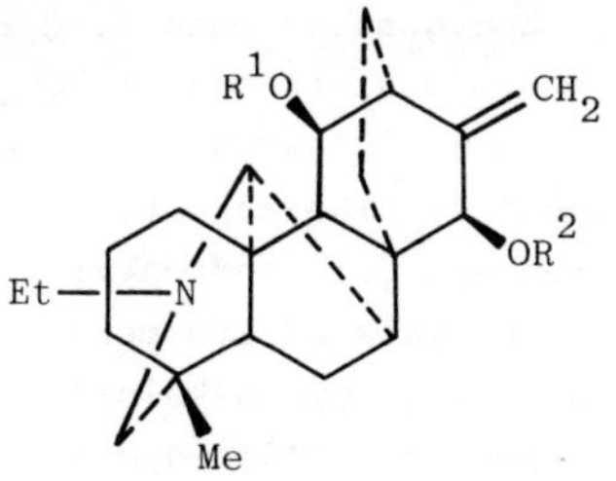

Denudatine (60) $R^1 = R^2 = H$

Jynosine (62) $R^1 = H$, $R^2 = Ac$

(63) $R^1 = R^2 = Ac$

[structure (61) is with (19)]

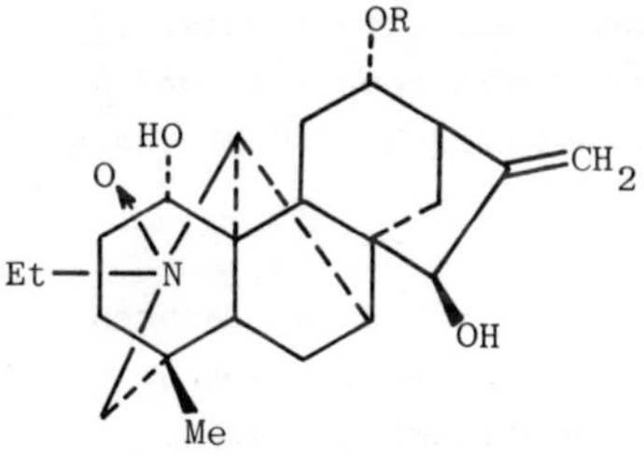

Napelline
N-oxide (64) R = H

(65) R = Ac

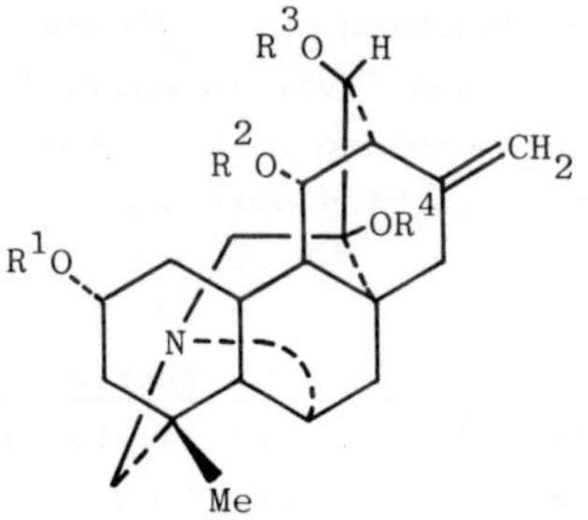

Guan-fu base G (66) $R^1 = R^2 = R^3 = Ac$, $R^4 = H$

(67) $R^1 = R^2 = R^3 = R^4 = H$

(68) $R^1 = R^2 = R^3 = R^4 = Ac$

Guan-fu base A (69) $R^1 = R^2 = Ac$, $R^3 = R^4 = H$

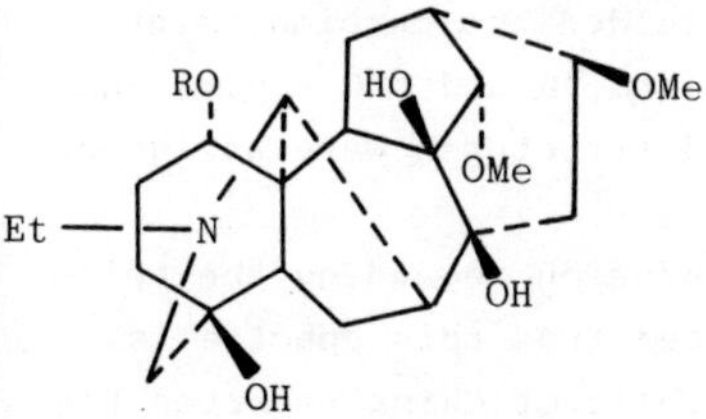

Lappaconidine (70) R = H

Lappaconine (71) R = Me

(72) R = Me; unspecified -OH

184°C, and guan-fu base G, $C_{26}H_{33}NO_7$, m.p. 178°C, were isolated in addition to guan-fu base A, $C_{24}H_{31}NO_6$, guan-fu base B, $C_{22}H_{29}NO_5$, guan-fu base C, $C_{22}H_{33}NO_2$, guan-fu base D, guan-fu base E, and hypaconitine (12). The structure of guan-fu base G was determined to be (66) from spectral and chemical studies and an X-ray crystallographic analysis of its methiodide derivative. Compound (66) is the first hetisine-type alkaloid with a C(13) hydroxyl group without an oxygen group at C(15). Hydrolysis of (66) and guan-fu base A gave the same amino-alcohol (67). Acetylation of either guan-fu bases A or G afforded (68). N.m.r. studies indicated that the hydroxyl group and an acetyl group of guan-fu base A must be vicinal to C(12). Therefore, guan-fu base A was assigned structure (69).

Alkaloids of *Aconitum leucostomum*.- Plugar and coworkers[41] have published the results of gas chromatographic-mass spectral studies of the alkaloids of *A. leucostomum*. The diterpenoid alkaloids lappaconidine (70), lappaconine (71), and hydroxylappaconine (72) and the aporphine alkaloids corydine, O-methylarmepavine, and N-demethylcolletine were quantitatively determined by GLC-MS analyses.

Alkaloids of *Aconitum monticola* Steinb.- Researchers at Tashkent, USSR,[42] have reported the isolation of two new alkaloids from the roots and epigeal parts of *A. monticola*. Monticamine (73), $C_{22}H_{33}NO_5$, m.p. 163-164°C, afforded the monoacetate (74) when acetylated with acetic anhydride in pyridine and (75) when reduced with Raney nickel. Monticoline (76), $C_{22}H_{33}NO_6$, m.p. 166-167°C, when reduced with Raney nickel, gave dihydromonticoline (77). Acetylation of (77) with acetic anhydride in pyridine gave (78). Oxidation of (77) with $KMnO_4$ in aqueous acetone yielded the carbinolamine ether (79). From these transformations, and ^1H and ^{13}C n.m.r. and mass spectral comparisons, the indicated structures were assigned.

Alkaloids of *Aconitum nagarum*.- Some confusion regarding the trivial names of several new alkaloids isolated from this species is evident. Researchers in the Peoples Republic of China reported the isolation of a "new" alkaloid from the roots of *A. nagarum* Stapf. var. *lasiandrum*, collected in the Yunnan Province,[43,44] which they named "nagarine" (80), $C_{34}H_{47}NO_7$, m.p. 198-200°C. These results were also discussed in a recent review.[1] However, this "nagarine" is identical in melting point, molecular formula, and struc-

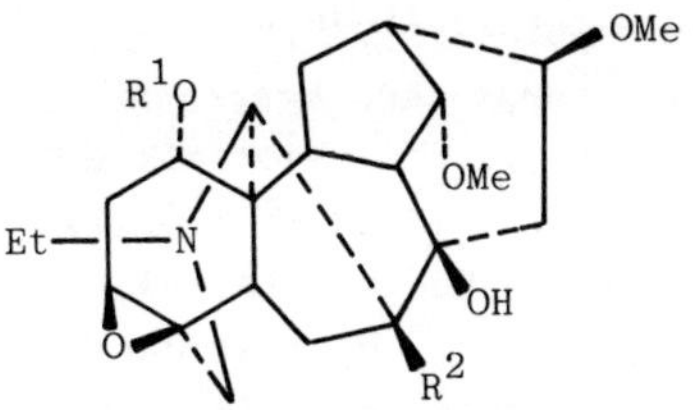

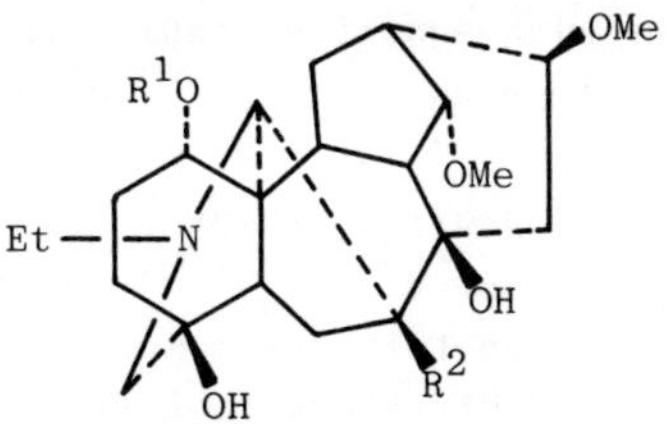

Monticamine (73) $R^1 = R^2 = H$
(74) $R^1 = Ac, R^2 = H$
Monticoline (76) $R^1 = H, R^2 = OH$

(75) $R^1 = R^2 = H$
(77) $R^1 = H, R^2 = OH$
(78) $R^1 = Ac, R^2 = OH$

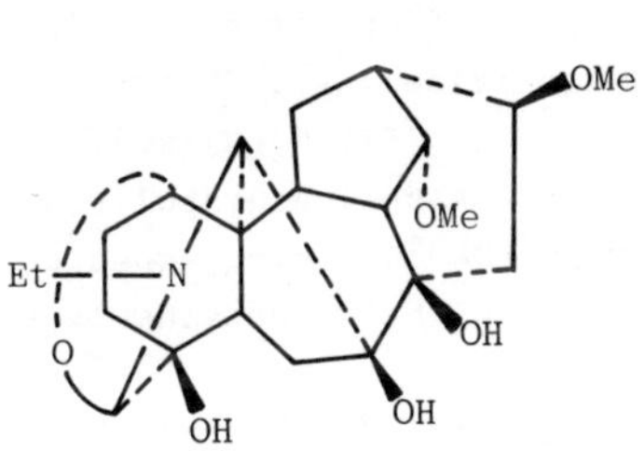

(79)

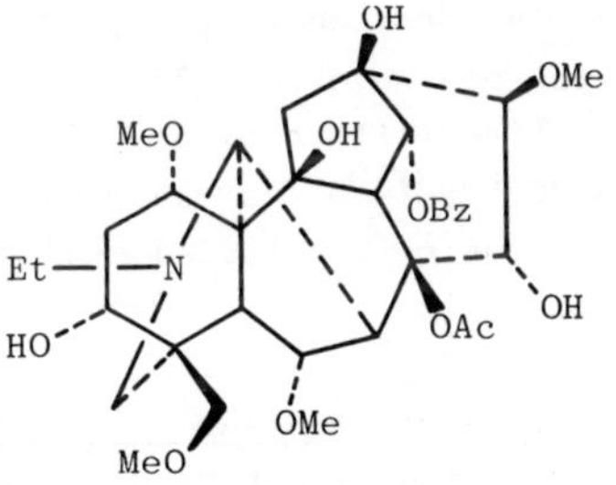

Aconifine (80)

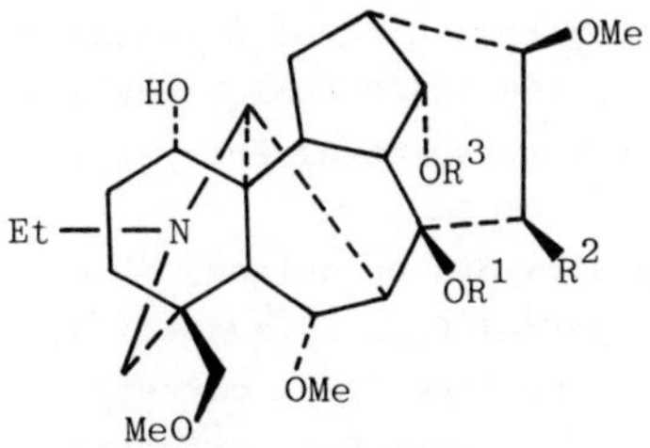

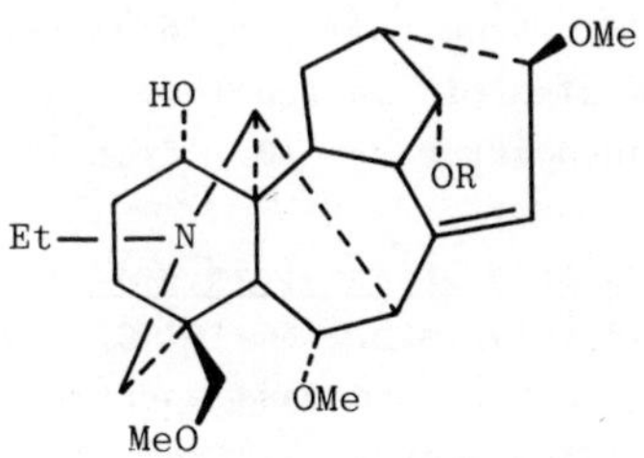

Nagarine (81) $R^1 = H, R^2 = OH, R^3 = H$
Delphisine (82) $R^1 = Ac, R^2 = H, R^3 = Ac$

Pyrodelphisine (83) R = Ac
Pyroneoline (84) R = H

ture with aconifine (80) (10β-hydroxyaconitine).[45,46,47]

In studies of the alkaloids of *A. nagarum* var. *heterotrichum* f. *dielsianum* W. T. Wang, aconitine (14), the major alkaloid, 3-deoxyaconitine (11), and a new base, $C_{24}H_{39}NO_7$, m.p. 190-191°C, named "nagarine", were isolated.[48] Structure (81) was assigned for this new compound from the ^{13}C n.m.r. spectral data. This structure was confirmed by a partial synthesis of (81) from delphisine (82). Pyrolysis of (82) gave pyrodelphisine (83), which when hydrolysed afforded pyroneoline (84). Treatment of (84) with osmium tetroxide in pyridine/dioxane followed by aqueous sodium bisulfite gave 15β-hydroxyneoline (81), which was identical in all respects with nagarine.[48]

In view of the prior use of the trivial name aconifine for the alkaloid isolated by Wang and coworkers, the name "nagarine" should be used only to identify alkaloid (81).[47]

In addition to (80), Wang and coworkers[1,49] reported the isolation of songorine (27), bullatine A, aconitine (14), deoxyaconitine (11), neoline (19), and bullatine C (61), $C_{26}H_{41}NO_7$, m.p. 198-202°C, from *A. nagarum* var. *lasiandrum*. Alkaline hydrolysis of bullatine C afforded neoline (19). From the ^1H n.m.r. spectral data, bullatine C was shown to be identical with 14-acetylneoline.

Alkaloids of *Aconitum paniculatum* Lam.- Katz and Staehelin[50] have reported the structure elucidation of paniculatine (85), $C_{31}H_{35}NO_7$, m.p. 265-268°C, an alkaloid first isolated from this plant some sixty years earlier. This structure was assigned from ^1H and ^{13}C n.m.r. and mass spectral studies of paniculatine and its dehydro derivative (86), prepared by oxidation with CrO_3 in glacial acetic acid. From this species[51] these investigators have also isolated a new alkaloid panicutine, $C_{23}H_{29}NO_4$, m.p. 160-165°C, which was assigned structure (87) from ^1H and ^{13}C n.m.r. spectral analysis.

Alkaloids of *Aconitum pendulum*.- A new alkaloid penduline, $C_{34}H_{47}NO_9$, m.p. 166-167°C, has been reported from this species.[1] From n.m.r. and mass spectral studies, structure (88), corresponding to 3,3-dideoxyaconitine, was tentatively assigned for penduline. 3-Acetylaconitine (47) and deoxyaconitine (11) were also isolated from these plants. Acetylation of aconitine (14) with acetic anhydride and pyridine at room temperature afforded (47).

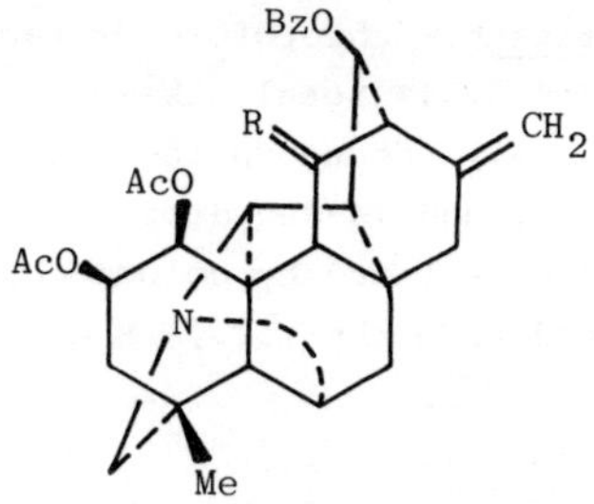

Diterpenoid Alkaloids

Paniculatine (85) R = α-OH,H
(86) R = O

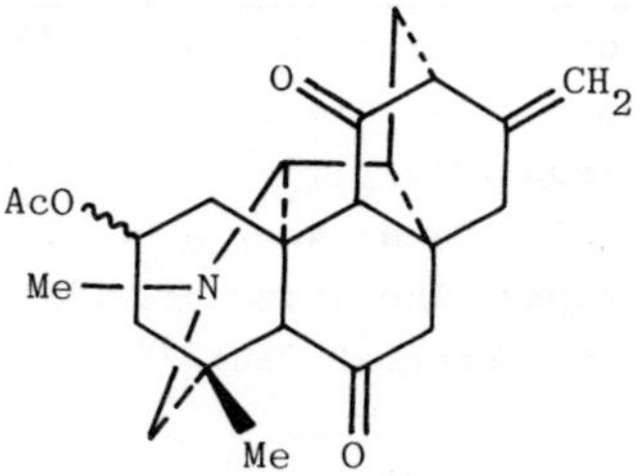

Panicutine (87)

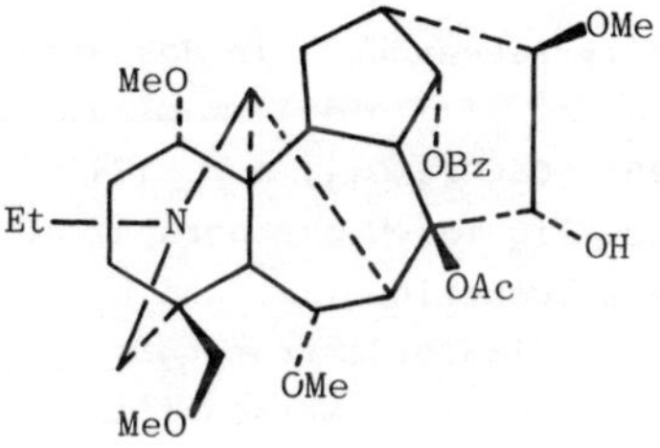

Penduline (88)

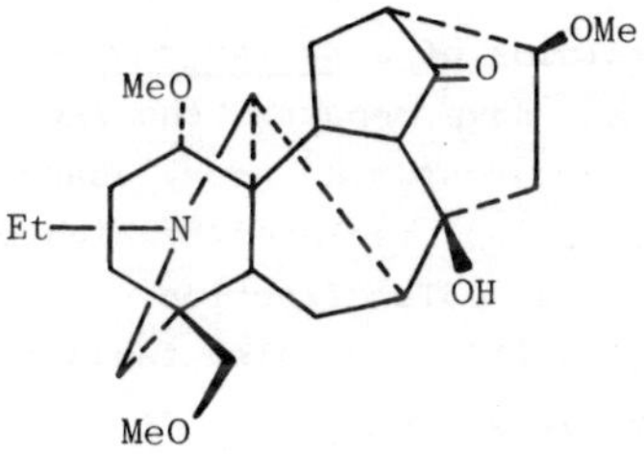

14-Dehydrotalatizamine (89)

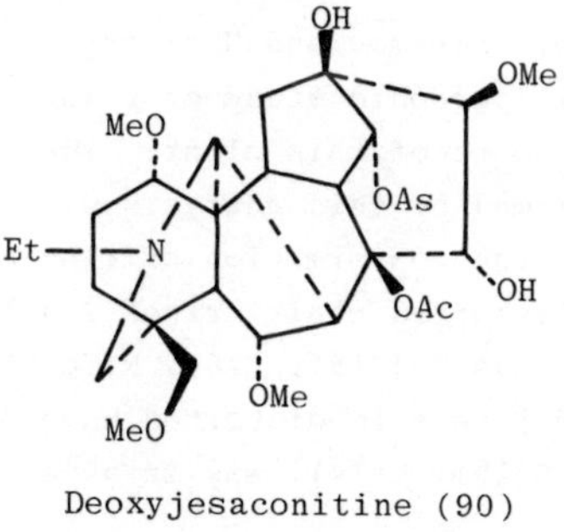

Deoxyjesaconitine (90)

(As = anisoyl)

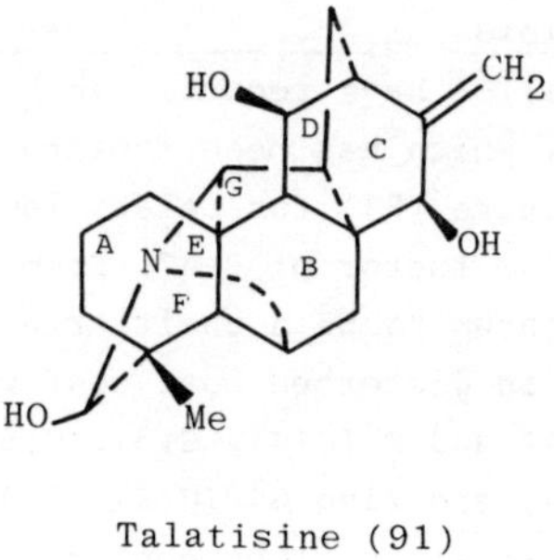

Talatisine (91)

Alkaloids of *Aconitum saposhnikovii* B. Fedtsch.- Chloroform extraction of plants of *A. saposhnikovii* afforded 0.54% total alkaloids.[39] From this extract, talatizamine (15), the aporphine alkaloid isoboldine, 14-acetyltalatizamine (16), and 14-dehydrotalatizamine (89), $C_{24}H_{37}NO_5$, m.p. 128-130°C, were isolated by column chromatography. The structure (89) was assigned from the i.r., ^{1}H n.m.r., and mass spectral data.[39]

Alkaloids of *Aconitum sinomontanum* Nakai.- A study of the alkaloids of the roots of this plant, which has antitumor activity, has been reported.[52] Lappaconitine (50), ranaconitine (51), and an unknown alkaloid, m.w. 343, m.p. 244°C, were isolated by chloroform extraction and chromatography on aluminum oxide.

Alkaloids of *Aconitum subcuneatum* Nakai.- Researchers in Hokkaido, Japan, have reported the isolation of a new alkaloid from plants of *A. subcuneatum*.[53] Deoxyjesaconitine (90), $C_{35}H_{49}NO_{11}$, m.p. 174-176°C, was isolated by high-performance liquid chromatography of the less polar fraction, from which deoxyaconitine (11) and hypaconitine (12) were also isolated. The major alkaloids from this species were the more polar alkaloids aconitine (14), mesaconitine (10), and jesaconitine (13). The ^{1}H and ^{13}C n.m.r. and i.r. data were used for the structural assignments. The ^{13}C n.m.r. chemical shifts are reported.

Alkaloids of *Aconitum talassicum* M. Pop.- Karimov and Zhamierashvili[54] have reported an X-ray crystallographic study of talatisine, which has been isolated from the roots of this plant. The structure (91) for talatisine was confirmed by this analysis with a final R factor of 0.07. From this study the conformation of ring A was shown to be a chair, ring B was a distorted chair, rings C and D were in distorted boat conformations, rings E [C(5), C(6), N, C(20), C(10)] and F [C(4), C(5), C(6), N, C(19)] were in distorted envelope forms, and ring G [C(8), C(9), C(10), C(20), C(14)] was in an almost ideal envelope conformation.

Alkaloids of *Aconitum yesoense* Nakai.- Takayama and coworkers[55] have reported an extensive reinvestigation of the alkaloids of this species. Kobusine (92), pseudokobusine (93), lucidusculine (94), neoline (19), chasmanine (29), mesaconitine (10), and jesaconitine (13) had previously been reported from *A. yesoense*. These com-

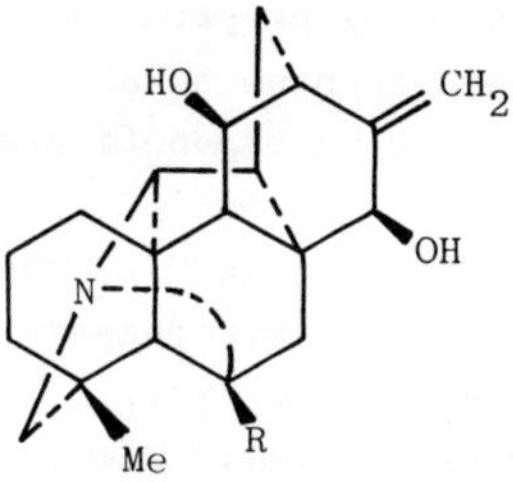

Kobusine (92) R = H
Pseudokobusine (93) R = OH

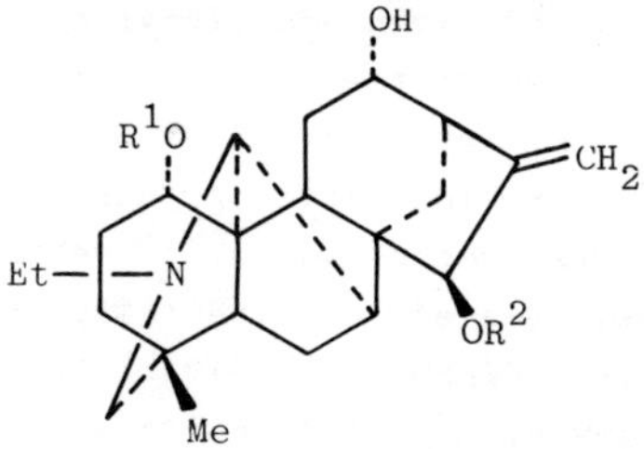

Lucidusculine (94) R^1 = H, R^2 = Ac
Luciculine (95) R^1 = R^2 = H
1-Acetyl-luciculine (96) R^1 = Ac, R^2 = H

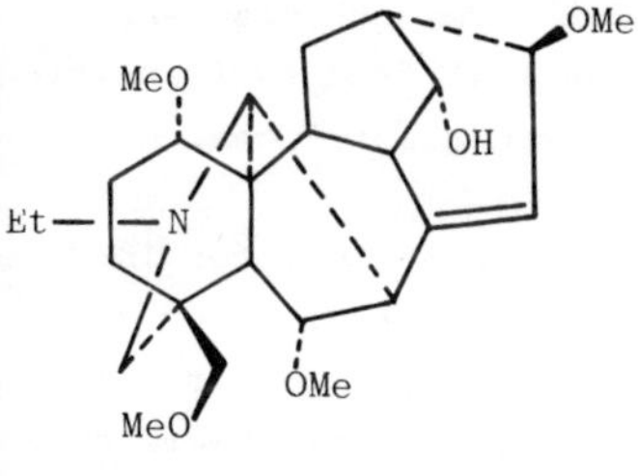

Pyrochasmanine (97)

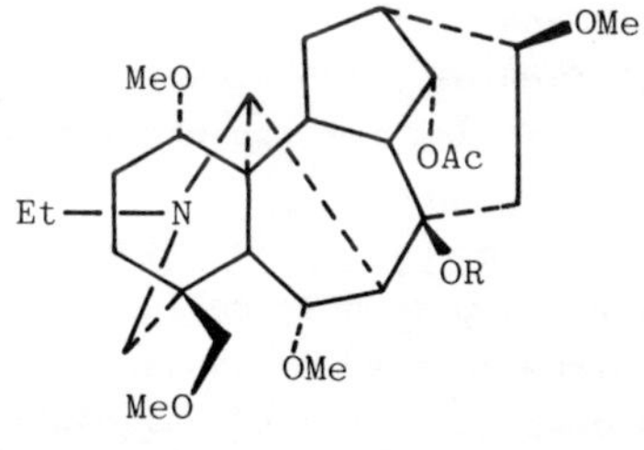

Ezochasmaconitine (98) R = Bz
Anisoezochasmaconitine (99) R = As
(As = anisoyl)

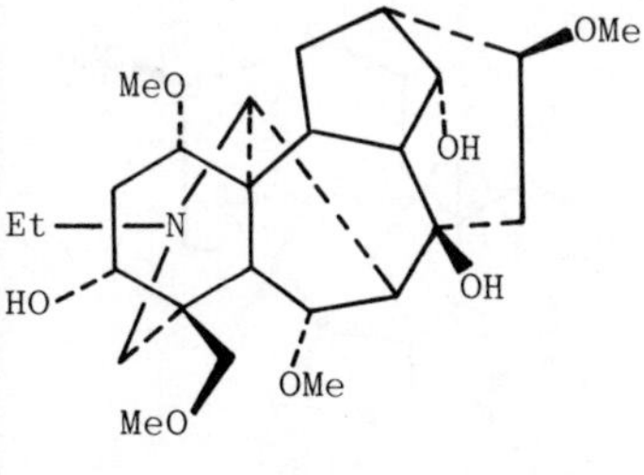

Ezochasmanine (100)

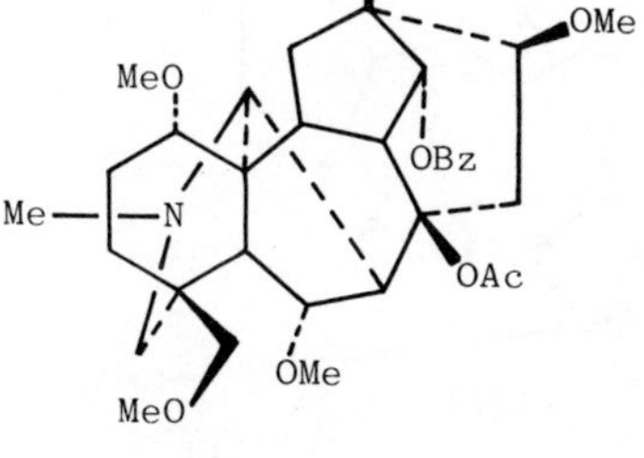

Delphinine (101)

pounds and eight additional bases were isolated by preparative
t.l.c. Two of these, luciculine (95) and the aporphine alkaloid
glaucine, were known from other species. Six new diterpenoid alkaloids were identified primarily from their i.r., ^{1}H and ^{13}C n.m.r.,
and mass spectral data: 1-acetylluciculine (96), $C_{24}H_{35}NO_4$, amorphous; 14-acetylneoline (61), $C_{26}H_{41}NO_7$, amorphous; pyrochasmanine
(97), $C_{25}H_{40}NO_5$, m.p. 124.5-127°C; ezochasmaconitine (98),
$C_{34}H_{47}NO_8$, m.p. 163-165°C; anisoezochasmaconitine (99), $C_{35}H_{49}NO_9$,
m.p. 136-138.5°C; and ezochasmanine (100), $C_{25}H_{41}NO_4$, m.p. 115-
118°C. Ezochasmaconitine (98) and anisoezochasmaconitine (99) are
the first reported examples of diterpenoid alkaloids with an aroyl
group at C(8) and an acetyl group at C(14). Ezochasmanine (100)
corresponds to 3-α-hydroxychasmanine.

Alkaloids of *Atragene sibirica* L. - Chloroform extraction of the
roots of this species (Siberian clematis; Ranunculaceae) followed by
isolation of the bases and column chromatographic separations with
silica and alumina afforded five alkaloids.[56] Two of these were
identified as delphinine (101) and aconitine (14) from their physicochemical and i.r. spectral data and by direct comparisons.

Alkaloids of *Consolida orientalis*. - The structure (102) for 18-
hydroxy-14-O-methylgadesine, $C_{24}H_{37}NO_7$, m.p. 110-114°C, a new alkaloid from this species, was assigned from the i.r., ^{1}H n.m.r., and
mass spectral data and was confirmed by a single-crystal X-ray crystallographic analysis (R = 0.043).[57] *Consolida orientalis* Gay
subsp. *orientalis* is synonymous with *Delphinium orientale* Gay;
incl. *D. hispanicum* Wilk. Delsoline (8), delcosine (103), and
gigactonine (104) were isolated from this plant.

(102)

Delcosine (103) R^1= H, R^2= Me
Gigactonine (104) R^1= Me, R^2= H

Alkaloids of *Delphinium cardiopetalum* DC. - Gonzalez and co-workers[58] have published studies of the alkaloids isolated from the whole plants of *D. cardiopetalum* by ethanol extraction and chromatography on alumina. They isolated hetisinone (105) and a new base, 13-acetylhetisinone (106), $C_{22}H_{27}NO_4$, m.p. 219-220°C. Hydrogenation of (106) with Pd on carbon gave (107). Acetylation of (106) with acetic anhydride/pyridine afforded (108), while hydrolysis with base gave hetisinone (105). Oxidation of (106) with Cornforth reagent produced the diketoacetate (109). From these transformations and the u.v., i.r., ^{1}H n.m.r., and mass spectral data, structure (106) was assigned.

Hetisinone (105) $R^1= \alpha$-OH,H; $R^2=$ H; $R^3=$ CH_2

13-Acetylhetisinone (106) $R^1= \alpha$-OH,H; $R^2=$ Ac; $R^3=$ CH_2

(107) $R^1= \alpha$-OH,H; $R^2=$ H; $R^3=$ Me,H

(108) $R^1= \alpha$-OAc,H; $R^2=$ Ac; $R^3=$ CH_2

(109) $R^1=$ O; $R^2=$ Ac; $R^3=$ CH_2

Alkaloids of *Delphinium dictyocarpum* DC. - Three reports on the alkaloids of *D. dictyocarpum* have appeared.[59,60,61] The structure of dictysine, $C_{21}H_{33}NO_3$, m.p. 184-186°C, has been shown to be (110) from ^{13}C n.m.r. and mass spectral data[59] and from an X-ray crystallographic structure determination.[60]

Two additional bases were isolated from these extracts: acetonyl dictysine (111), $C_{24}H_{37}NO_3$, m.p. 151-153°C, and acetonyl dehydrodictysine (112), $C_{24}H_{32}NO_3$, m.p. 143-145°C. These structures were assigned from the spectral data and chemical transformations. Oxidation of (111) with chromic anhydride afforded (112). Hydrolysis of (112) with 20% aqueous sulfuric acid gave (113). Acetylation of (113) with acetic anhydride produced the monoacetate (114), whereas treatment of (113) with acetyl chloride gave (115). How-

Dictysine (110) $R^1 = \beta$-OH,H; $R^2 = R^3 = H$

Dehydrodictysine (113) $R^1 = O$; $R^2 = R^3 = H$

(114) $R^1 = O$; $R^2 = H$; $R^3 = Ac$

(115) $R^1 = O$; $R^2 = R^3 = Ac$

(111) R = β-OH,H

(112) R = O

ever, (111) and (112) are readily formed on treatment of (110) and (113), respectively, with acetone. Therefore, they were presumed to be artifacts formed during the isolation of the alkaloid fraction, and dictysine (110) and dehydrodictysine (113) must be the naturally occurring bases. In another study of the alkaloids of the aerial parts of *D. dictyocarpum* DC., Salimov and Yunusov[61] reported the isolation of a new base, 14-benzoyldictyocarpine, $C_{33}H_{43}NO_9$, m.p. 143-145°C (116), whose structure was assigned from the i.r., ^1H n.m.r., and mass spectral data.

3 Chemical Studies

Transformation Products from Lycoctonine. - Edwards[62] has reported an unexpected rearrangement during deamination of 4-amino-4-des(oxymethylene)-anhydrolycoctonam (117). In studies of the chemistry of lycoctonine (3), treatment of its derivative lycoctonamic acid (118) with very dilute sulfuric acid afforded the pinacolic dehydration product (119). This acid was converted to (117) by Curtius or Hofmann degradation. When (117) was treated with nitrous acid, the hydroxyketolactam (120) and the aldehydolactam acid (121) were obtained as the main products in 50 and 20% yields, respectively. Since difficulties were encountered in the interpretation of other chemical and spectral evidence, these structures were assigned on the basis of X-ray crystallographic analyses of derivatives. Reduction of (121) followed by acetylation gave the acetoxy-acid (122), whose structure was determined by an X-ray analysis. Oxidation of

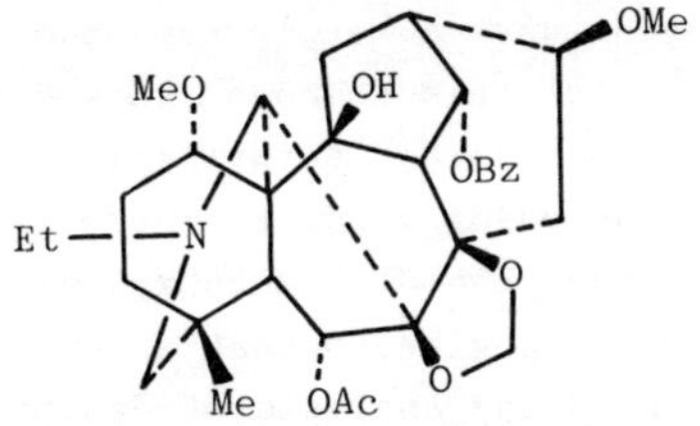

14-Benzoyldictyocarpine (116)

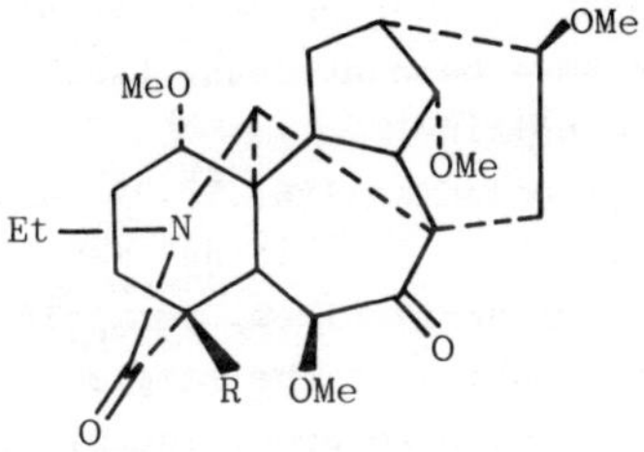

(117) R = NH$_2$
(119) R = COOH

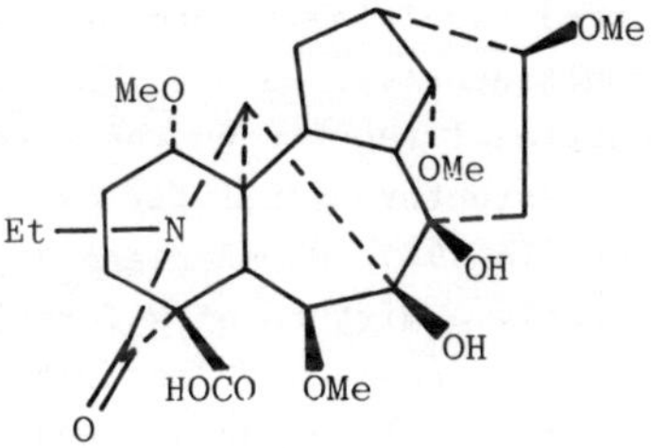

(118)

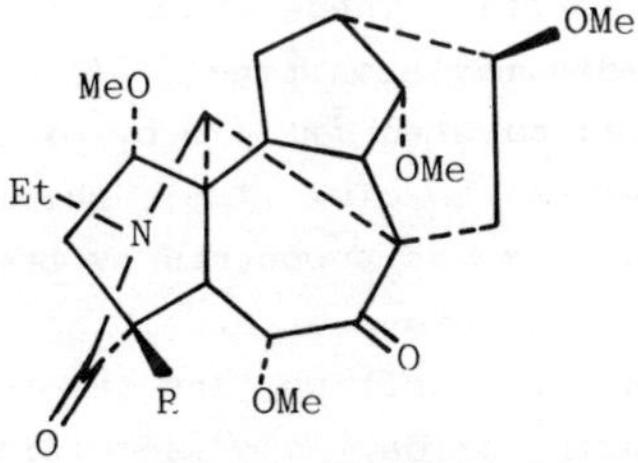

(2) R = COOH
(120) R = CH$_2$OH

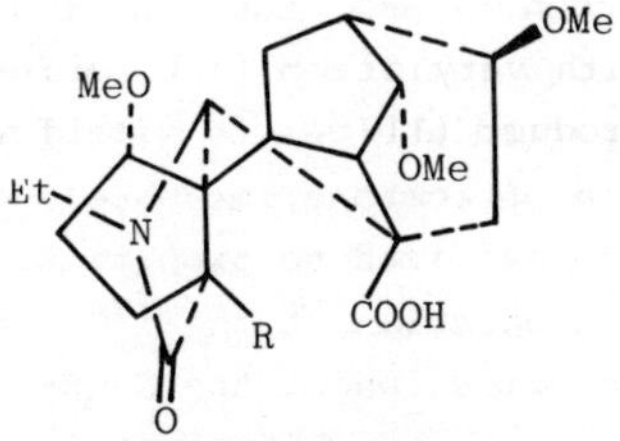

(121) R = CHO
(122) R = CH$_2$OAc

(120) with Jones' reagent gave a ketolactam acid as the main product. An X-ray analysis of this compound showed it to have the structure (2). In addition to the surprising formation of the primary alcohol (120) as the major product, these studies revealed the unexpected information that the C(1) methoxyl group was in an α-configuration. This was opposite to that assumed for lycoctonine on the basis of the earlier X-ray analysis of des(oxymethylene)lycoctonine hydroiodide and suggested that all of the subsequent structure assignments based on this were incorrect (cf. this review, vol. 12, 1982, p. 250). Possible mechanisms for these rearrangements were discussed.

The Origin of Oxonitine. - The solution to a perplexing problem, some seventy years old, has been reported.[63] In 1912 Carr[64] reported oxonitine as one of the permanganate oxidation products of aconitine (14). Since that time, there have been at least twelve additional reports on the chemistry of oxonitine.

Structure (123) for oxonitine was proposed in 1960.[65,66] However, in 1971, Wiesner and Jay[67] suggested that "oxonitine" was probably a mixture of N-formyl and N-acetyl derivatives, resulting from the presence of mesaconitine as an impurity in the samples of aconitine. In a recent study,[63] oxidation of pure aconitine with potassium permanganate in either acetone/acetic acid or in methanol/acetone at 25°C for 5 hrs afforded oxonitine. Spectral data, including ^{13}C n.m.r. analyses, confirmed structure (123) for oxonitine.

Under similar conditions, mesaconitine (10) was resistant to oxidation. However, at 50°C for 48 hrs, mesaconitine gave oxonitine in 75% yield.

The mechanism for the formation of (123) was established by carbon-13 and deuterium labeling experiments. These studies concluded that oxonitine is formed via N-desethylaconitine (124). This intermediate reacts with formaldehyde (generated *in situ* by oxidation of acetone, methanol, or acetaldehyde) to give (125), which is then oxidized to oxonitine. This mechanism was supported by the incorporation of the ^{13}C label on the N-CHO group of oxonitine during oxidation of the N-CH_2- $^{13}CH_3$ group of aconitine, and the formation of oxonitine containing N-CDO during oxidation of aconitine in hexadeuterated acetone.

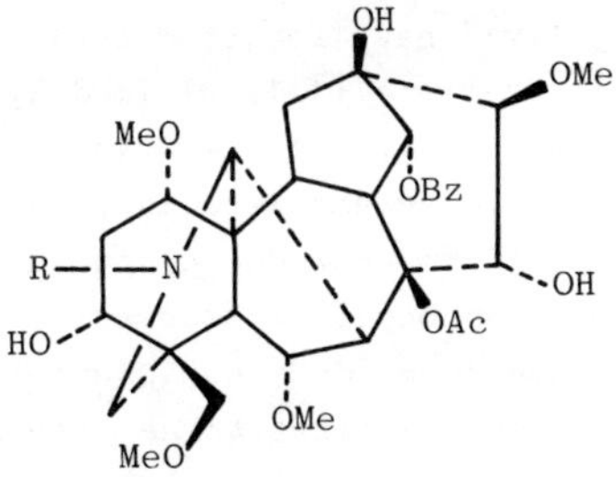

Oxonitine (123) R = CHO
(124) R = H
(125) R = CH$_2$OH

Diterpenoid Alkaloid Synthetic Studies. - Banerjee and coworkers[68] have reported on their synthesis of the ketoester (126) from the octalin (127). Key intermediates in this 14-step route were the decalins (128) and (129). Decalin (129) was converted to (130) by u.v. photolysis followed by deketalization. LAH reduction of (130) followed by oxidation with pyridinium chlorochromate and mesitylation gave (126).

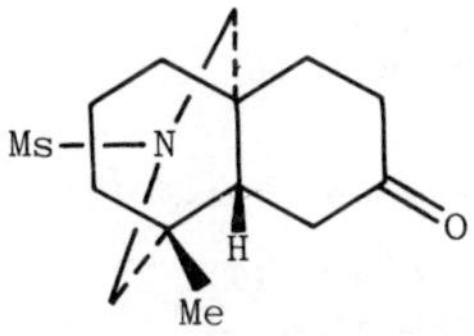

(126)
(127) 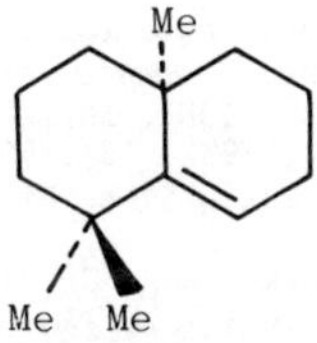
(128) 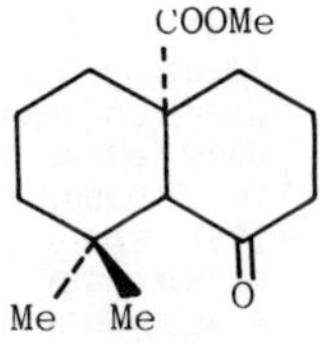

(Ms = methanesulphonyl)

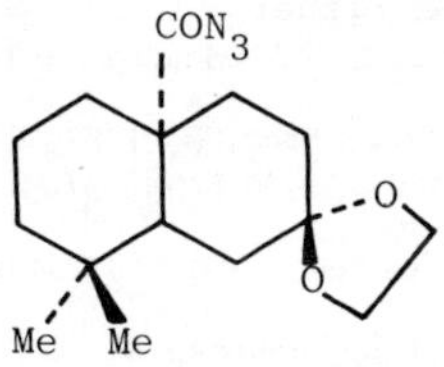

(129)
(130)

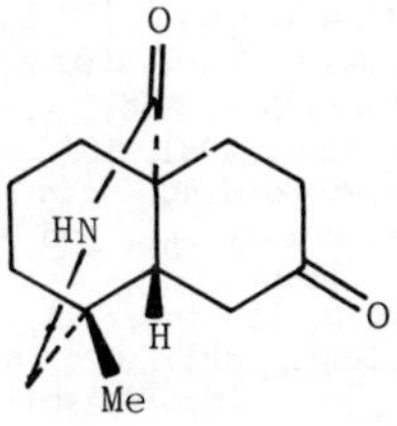

Cornforth and Pengelly[69] have reported their failure to verify the synthesis of the intermediate (131) claimed by Chatterjee.[70] (cf. this Report, 11, 1981, 222). In the reinvestigation, the first three stages of the synthetic scheme were thoroughly examined. None of the original report could be reproduced, including the chemical nature of the products or the physical properties of the products claimed. Consequently, they concluded that Chatterjee did not obtain the reported products during the first three stages or any subsequent stage.

(131)

Acknowledgment: We express our appreciation to Dr. B. S. Joshi for reading the manuscript and making useful suggestions.

References

1   Yuanlong Zhu and Renhong Zhu, *Heterocycles*, 1982, **17**, 607.
2   Yuanlong Zhu and Renhong Zhu, *Zhongyao Tongbao*, 1981, **6**(6), 38.
3   Wang Fengpeng, *Yaoxue Xuebao*, 1981, **16**(12), 943.
4   Wang Fengpeng, *Youji Huaxue*, 1982, (3), 161.
5   Yu Li Dong, Wei Zhou Chen, Guang Sheng Ding, *Chung-kuo Yao Li Hsueh Pao*, 1981, **2**(3), 173.
6   K. Murayama, Japan Kokai Tokkyo Koho JP 81 120,620, 22 Sept. 1981.
7   A.B. Khan and H.M. Taiyab, *Indian J. Pharm. Sci.*, 1981, **43**(3), 120.
8   Tianli Gao, Fulong Liao, Yansong Wang, and Hong Zhuang, *Zhonghua Xinxueguanbing Zazhi*, 1981, **9**(3), 223.
9   K. Murayama, Japan Kokai Tokkyo Koho JP 82 58,627, 08 April 1982.
10 H. Saito, T. Ueyama, N. Naka, J. Yagi, and T. Okamoto, *Chem. Pharm. Bull.*, 1982, **30**(5), 1844.
11 S. Sakai, *Gendai Toyo Igaku*, 1981 2(3), 50.
12 S. Sawada, T. Hasegawa, F. Debata, Y. Konishi, R. Tokura, and Y. Ichinohe, *Bull. Kyoto Univ. Educ. Ser. B*, 1980, **57**, 11.
13 S.W. Pelletier, N.V. Mody, K.I. Varughese, J.A. Maddry, and H. K. Desai, *J. Am. Chem. Soc.*, 1981, **103**, 6536.
14 K.A. Kerr, P.W. Codding, *Acta Crystallogr., Sect. B*, 1982, B**38**(4), 1237.
15 M. Cygler, M. Przybylska, and O.E. Edwards, *Acta Crystallogr., Sect. B*, 1982, B**38**(5), 1500.
16 V.N. Aiyar, P.W. Codding, K.A. Kerr, M.H. Benn, and A.J. Jones, 1981, *Tetrahedron Lett.*, 1981, **22**, 483.
17 B.T. Salimov, M.G. Zhamierashvili, and M.S. Yunusov, *Khim. Prir. Soedin.*, 1981, (5), 621.
18 M.S. Yunusov and S. Yu. Yunusov, *Khim. Prir. Soedin.*, 1970, 334.
19 A.S. Narzullaev, M.S. Yunusov, and S. Yu. Yunusov, *Khim. Prir. Soedin.*, 1973, 497.

20. M. G. Zhamierashvili, V. A. Tel'nov, M. S. Yunusov, and S. Yu. Yunusov, *Khim. Prir. Soedin.*, 1980, 663.
21. H. Hikino, C. Konno, H. Watanabe, and O. Ishikawa, *J. Chromatogr.*, 1981, **211**(1), 123.
22. I. Kitagawa, M. Yoshikawa, Zhao Long Chen, K. Kobayashi, *Chem. Pharm. Bull.*, 1982, **30**(2), 758.
23. C. Konno, M. Shirasaka, H. Hikino, *J. Nat. Prod.*, 1982, **45**(2), 128.
24. S. W. Pelletier, N. V. Mody, K. I. Varughese, Chen Szu-Ying, *Heterocycles*, 1982, **18**, 47.
25. Szuying Chen, Yuqing Liu, Jicheng Wang, *Zum Nan Zhi Wu Yan Jiu*, 1982, **4**(1), 73.
26. Feng-Peng Wang and Qi-Cheng Fang, *Planta Med.*, 1981, **42**(4), 375.
27. Feng-Peng Wang and Chi-Cheng Fang, *Yao Hsueh T'ung Pao*, 1981, **16**(2), 49.
28. Chongren Yang, Dezu Wang, Dagang Wu, Xiaojiang Hao, and Jun Zhou, *Huaxue Xuebao*, 1981, **39**(5), 445.
29. S. W. Pelletier and Z. Djarmati, *J. Am. Chem. Soc.*, 1976, **98**, 2626.
30. Bao-Ren Chen, Yi-Fang Yang, Ru-Mei Tian, Chang-Pan Zhang, You-Zing Xiao, and Ming-Xun Liu, *Yao Hsueh Hsueh Pao*, 1981, **16**(1), 70.
31. Shanhao Jiang, Yuanlong Zhu, Renhong Zhu, *Yaoxue Tongbao*, 1981, **16**(9), 55.
32. Shanhao Jiang, Yuanlong Zhu, and Zhu Renhong, *Yaoxue Xuebao*, 1982, **17**(4), 283.
33. Zi-Ging Zhu, Tong-Lian Chen, Yu-Qing Lin, and Kuei-Tao Zhang, *Lan-Chou Ta Hsueh Hsueh Pao, Tzu Jan K'o Hsueh Pan*, 1980, (3), 135.
34. Xingruo Chang, Hongcheng Wang, Limin Lu, Yuanlong Zhu, and Renhong Zhu, *Yaoxue Xuebao*, 1981, **16**(6), 474.
35. Yuqing Liu, and Guitao Chang, *Yaoxue Tongbao*, 1982, **17**(4), 243.
36. H. Takayama, S. Hasegawa, S. Sakai, J. Haginiwa, and T. Okamoto, *Chem. Pharm. Bull.*, 1981, **29**(10), 3078.
37. H. Takayama, S. Hasegawa, S. Sakai, J. Haginiwa, and T. Okamoto, *Yakugaku Zasshi*, 1982, **102**(6), 525.
38. Dihua Chen and Weiliang Song, *Yaoxue Xuebao*, 1981, **16**(10), 748.
39. M. N. Sultankhodzhaev, M. S. Yunusov, and S. Yu. Yunusov, *Khim. Prir. Soedin.*, 1982, (2), 265.
40. Jing-Han Liu, Hung-Cheng Wang, Yao-Liang Kao, and Jen-Hung Chu, *Chung Ts'ao Yao*, 1981, **12**(3), 1.
41. V.N. Plugar, Ya.V. Rashkes, M.G. Zhamierashvili, V.A. Tel'nov, M.S. Yunusov, and S. Yu. Yunusov, *Khim. Prir. Soedin.*, 1982, (1), 80.
42. E.F. Ametova, M.S. Yunusov, V.E. Bannikova, N.D. Abdullaev, and V.A. Tel'nov, *Khim. Prir. Soedin.*, 1981, (4), 466.
43. Hongcheng Wang, Yaoliang Gao, Rensheng Xu, and Renhong Zhu, *Huaxue Xuebao*, 1981, **39**(7-8-9), 869.
44. Fengpeng Wang, *Acta Pharmaceutica Sinica*, 1981, **16**, 950.
45. M.N. Sultankhodzhaev, M.S. Yunusov, and S. Yu. Yunusov, *Khim Prir. Soedin.*, 1973, **9**, 127.
46. M.N. Sultankhodzhaev, L.V. Beshitaishvili, M.S. Yunusov, M.R. Yagudaev, and S. Yu. Yunusov, *Khim. Prir. Soedin.*, 1980, (5), 665.
47. S.W. Pelletier, N.V. Mody, and Chen Szu-Ying, *Heterocycles*, 1982, **19**, 1523.
48. N.V. Mody, S.W. Pelletier, and Szu-Ying Chen, *Heterocycles*, 1982, **17**, 91.

49 Hongcheng Wang, Dazhu Zhu, Zhiyuan Zhao, and Renhong Zhu, *Acta Chimica Sinica*, 1980, **38**, 475.
50 A. Katz and E. Staehelin, *Tetrahedron Lett.*, 1982, **23**(11), 1155.
51 A. Katz and E. Staehelin, *Helv. Chim. Acta*, 1982, **65**(1), 286.
52 Bi-Yu Wei, Hsien-Wu Kung, Chih-Yuan Chao, Hung-Cheng Wang, and Jen-Hung Chu, *Chung Yao T'ung Pao*, 1981, **6**(2), 26.
53 H. Bando, Y. Kanaiwa, K. Wada, T. Mori, and T. Amiya, *Heterocycles*, 1981, **16**(10), 1723.
54 Z. Karimov and M.G. Zhamierashvili, *Khim. Prir. Soedin*, 1981, (3), 335.
55 H. Takayama, A. Tokita, M. Ito, S. Sakai, F. Kurosaki, and T. Okamoto, *Yakugaku Zasshi*, 1982, **102**(3), 245.
56 E.A. Krasnov and V.S. Bokova, *Khim. Prir. Soedin.*, 1981, (6), 806.
57 A.G. Gonzalez, G. de la Fuente, O. Munguia, and K. Henrick, *Tetrahedron Lett.*, 1981, **22**(48), 4843.
58 A.G. Gonzalez, G. de la Fuente, and M. Reina, *An. Quim., Ser. C*, 1981, **77**(2), 171.
59 B.T. Salimov, B. Tashkhodzhaev, and M.S. Yunusov, *Khim. Prir. Soedin.*, 1982, (1), 86.
60 B. Tashkhodzhaev, *Khim. Prir. Soedin.*, 1981, (4), 230.
61 B.T. Salimov, and M.S. Yunusov, *Khim. Prir. Soedin.*, 1981, (4), 530.
62 O.E. Edwards, *Can. J. Chem.*, 1981, **59**(21), 3039.
63 S.W. Pelletier, J. A. Glinski, and N.V. Mody, *J. Am. Chem. Soc.*, 1982, **104**, 4676.
64 F.H. Carr, *J. Chem. Soc.*, 1912, **101**, 2241.
65 W.A. Jacobs and S.W. Pelletier, *Chem. Ind. (London)*, 1960, 591.
66 R.B. Turner, J.P. Jeshke, and M.S. Gibson, *J. Am. Chem. Soc.*, 1960, **82**, 5182.
67 K. Wiesner and L. Jay, *Experientia*, 1971, **27**, 758.
68 A.K. Banerjee, P.C. Caraballo, H.S. Hurtado, M.C. Carrasco, and C. Rivas, *Tetrahedron*, 1981, **37**(16), 2749.
69 J. Cornforth and T. Pengelly, *Tetrahedron Lett.*, 1982, **23**(22), 2213.
70 S. Chatterjee, *Tetrahedron Lett.*, 1979, 3249.

15
Steroidal Alkaloids

BY D. M. HARRISON

A comprehensive review on enzymic transformations of alkaloids deals, in part, with the steroidal alkaloids.[1] The c.d. spectra of N-salicylidene derivatives of some steroidal amines have been discussed.[2]

1 Alkaloids of the Apocynaceae

The phytochemistry and pharmacology of Holarrhena antidysenterica have been reviewed with emphasis on the alkaloid content of the species.[3]

Paravallaridine (1) has been utilised as the starting point for the preparation of a number of steroidal quaternary ammonium salts which are exemplified by the bis-methiodides (2a) to (2d). These compounds all possessed curare-like activity.[4]

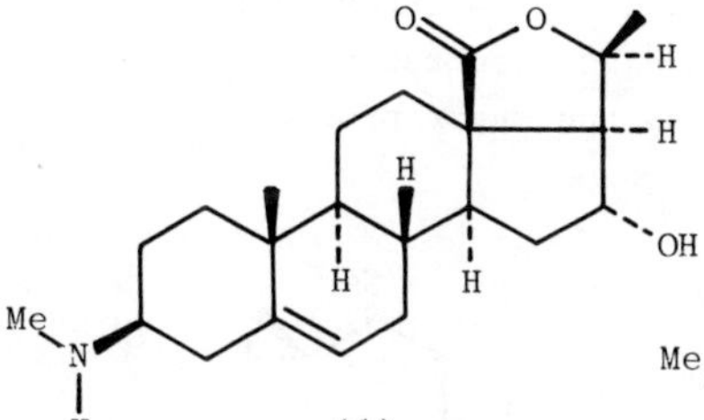

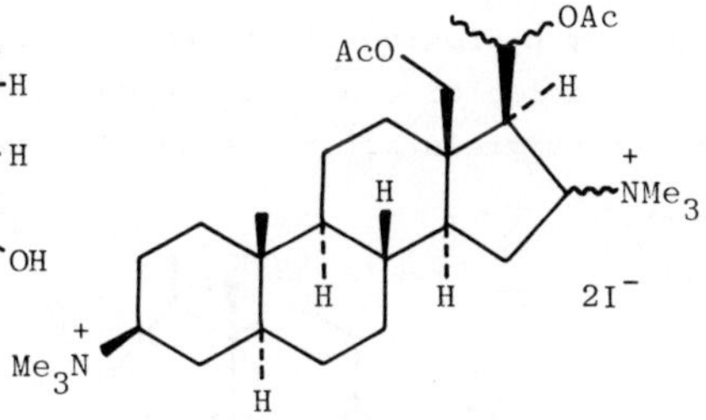

(2) a; 16α,20R
b; 16α,20S
c; 16β,20R
d; 16β,20S

Holamine (3) undergoes a "backbone" rearrangement to yield iso-holamine (4) on treatment with sulphuric acid. The mechanism of this rearrangement was elucidated by ^{2}H and ^{13}C n.m.r. studies on the labelled isoholamine which was formed when holamine was treated with deuteriated sulphuric acid.[5]

The quaternary ammonium salts (5a) and (5b), and the common product (6) of Hofmann elimination, were equilibrated on heating with

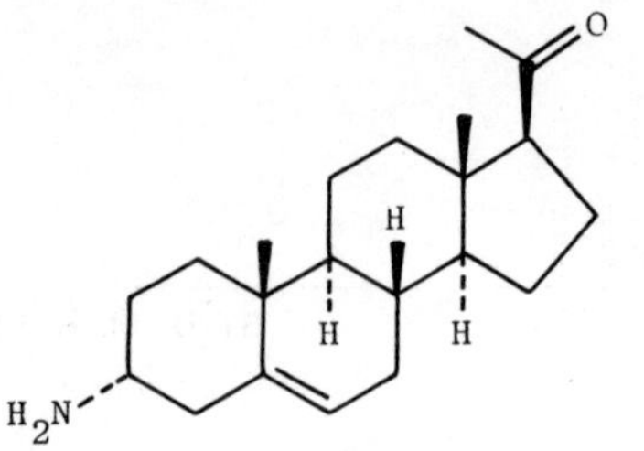

(3)

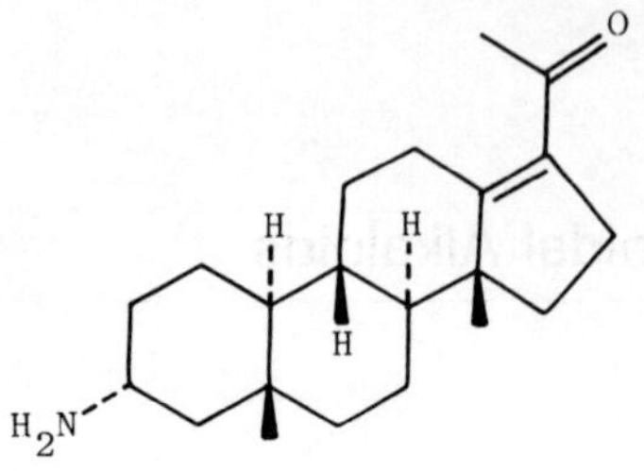

(4)

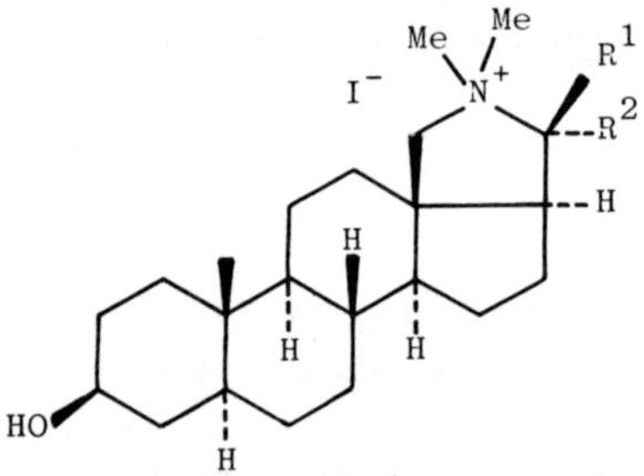

(5) a; R¹ = Me, R² = H
    b; R¹ = H, R² = Me

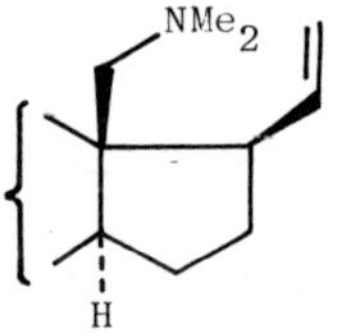

(6)

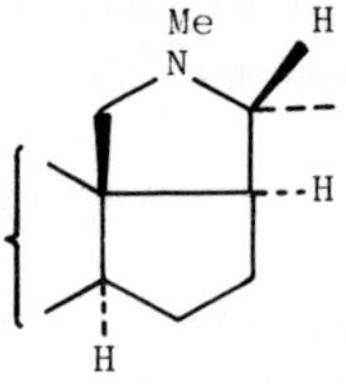

(7)

potassium hydroxide in ethane-1,2-diol. Slow irreversible demethylation of the ammonium salts also occurred under these conditions to yield heteroconanol (7) as the major product.[6]

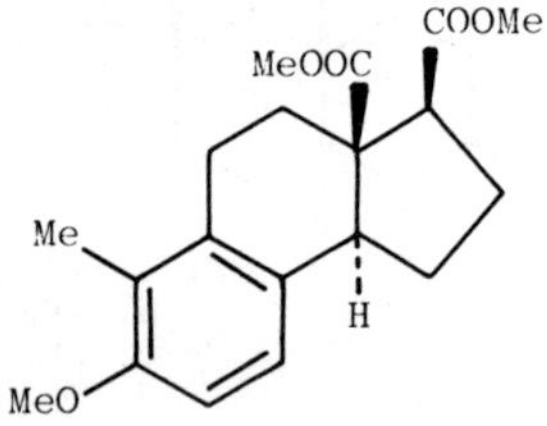

(8)

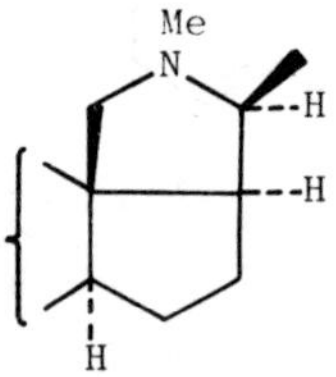

(9)

The racemic diester (8) has been converted into the racemic tertiary amine (9) in a study which is aimed at the synthesis of (dl)-conessine.[7]

The imminium salt (10) gave the oxaziridinium salt (11) on treatment with p-nitroperbenzoic acid, but was inert to hydrogen peroxide. The reactions of the enamine (12) with these two reagents have been described also.[8]

Mass spectroscopic studies on diaminopregnanes have been reported.[9]

2 *Buxus* Alkaloids

The distribution of cycloprotobuxine D (13a) in various parts of <u>Buxus microphylla</u> has been described.[10] The same species has yielded also cycloprotobuxine C (13b), cyclovirobuxine C (13c), cycloprotobuxine A (13d), and cyclovirobuxine D (13e).[11] The latter two bases were isolated also from <u>B. sempervirens</u> together with cyclobuxine D and an unidentified alkaloid.[12]

(13) a; $R^1_2 = R^2_2 = H, Me; R^3 = H$

b; $R^1_2 = H, Me; R^2 = Me; R^3 = H$

c; $R^1_2 = H, Me; R^2 = Me; R^3 = OH$

d; $R^1 = R^2 = Me; R^3 = H$

e; $R^1_2 = R^2_2 = H, Me; R^3 = OH$

3 Solanum Alkaloids

The Solanum steroidal alkaloids have been the subject of an excellent and comprehensive review[13] which contains also some material relevant to Section 4.

A single-crystal X-ray diffraction study has shown that solamaladine possesses the pyrroline structure (14) rather than the isomeric structure (15) which had been assigned previously.[14]

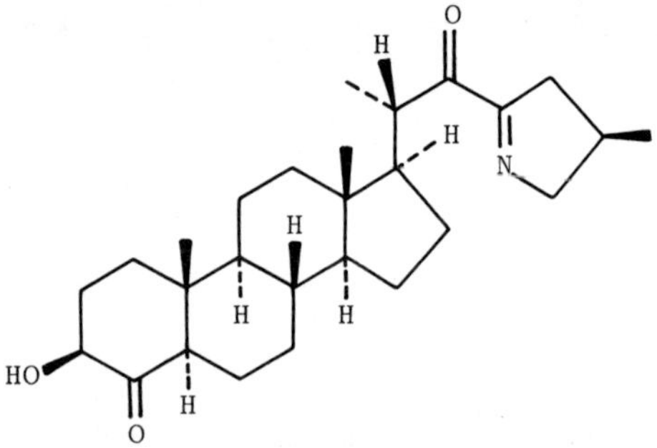

(14)

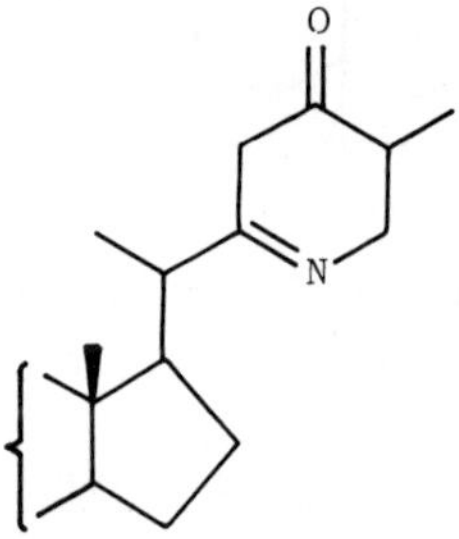

(15)

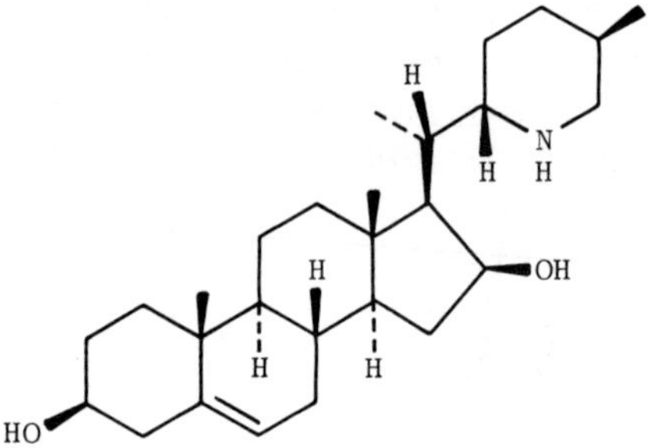

(16)

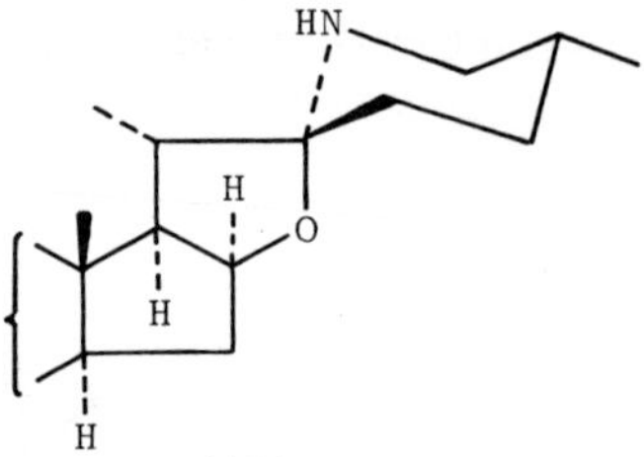

(17)

It was reported earlier that 16β-hydroxy-22,26-epiminocholestane derivatives such as solaverbascine (16) were oxidised smoothly by manganese dioxide to yield spirosolane derivatives such as solasodine (17) (cf. Vol. 11, p.229). Details of this work have now been published.[15] The electrochemical oxidation,[16] epimerisation,[16] and methoxylation[17] of solasodine have been reported.

Approaches to the synthesis of 21,27-bis-nor-solanidine have been described.[18]

The determination of Solanum alkaloids by g.l.c. and by t.l.c.

has been reported.[19]

Thirty two Solanum species have been investigated for their alkaloid content. Solasodine was isolated for the first time from seven of these species.[20] Solasodine has been isolated from S. sisymbrifolium and estimated by a rapid colourimetric method.[21] The maximum solasodine content of both berries and leaves in this species was 7.2%.[22] A new method has been reported for the quantitative determination of solasodine in berries of S. khasianum.[23] Solasodine has been isolated from ripe berries of S. viarum (3.2% yield),[24] and from cell cultures of S. jasminoides[25] and S. verbascifolium.[26]

The glycoalkaloid content of sixteen wild, tuber-bearing Solanum species has been investigated; the solasodine glycosides solamargine and solasonine were identified for the first time in S. berthaultii.[27] The glycoalkaloid content of varieties of the latter species showed no correlation with resistance of the plant to insect infection.[28]

Solamargine and solasonine have been isolated from berries of S. khasianum[29] and S. nigrum,[30] and from flowers of S. laciniatum.[31] The same glycoalkaloids, together with solanine, were detected in extracts of S. erianthum; acid-catalysed hydrolysis of the extract from fruits of this plant gave solasodine in 5.9% overall yield.[32]

A t.l.c. procedure has been described for the quantitative determination of solanine and chaconine, which are glycosides of solanidine (18a).[33] Changes in the solanine and chaconine content of potatoes during growth and storage of the plant have been studied.[34] The glycoalkaloid content of potatoes was increased by application of gibberellic acid.[35]

Potato glycoalkaloids are hydrolysed by rumen microorganisms to yield the aglycon solanidine (18a); much of the latter subsequently undergoes microbiological reduction to 5,6-dihydrosolanidine.[36] The effect of solanine on the growth of potato sprouts in vitro has been studied.[37] β_2-Chaconine (18b) has been prepared by the enzyme-catalysed hydrolysis of potato berries, which were rich in chaconine.[38]

A procedure has been described for the determination of tomatine in tomato plants.[39] The biological and biochemical effects of tomatine[40] and of other Solanum and Veratrum alkaloids [41] have been studied in a variety of biological systems.

The genetic origin of the solamarines in potato cultivars has been investigated.[42]

4 *Fritillaria* and *Veratrum* Alkaloids

Aerial parts of Rhinopetalum stenantherum (*Fritillaria* sp.) have yielded three glycosides of solanidine (18a), namely β_2-chaconine (18b) and the new glycoalkaloids stenantine and stenantidine.[43] Stenantine gave solanidine, D-glucose (two equivalents), and L-rhamnose (one equivalent), on acid-catalysed hydrolysis. The relationship between the two new glycoalkaloids was revealed by partial hydrolysis of stenantine, which gave stenantidine together with β_2-chaconine (18b) and γ-chaconine (18c). The nature of the trisaccharide unit in stenantine was clarified by complete methylation of the glycoside, followed by acid-catalysed hydrolysis, which gave 2,3,4,6-tetra-O-methyl D-glucose, 2,3,4-tri-O-methyl L-rhamnose, and 2,3-di-O-methyl D-glucose. Stenantine and stenantidine were assigned the structures (18d) and (18e), respectively, mainly on the basis of these observations, which were supplemented with spectroscopic data.[43]

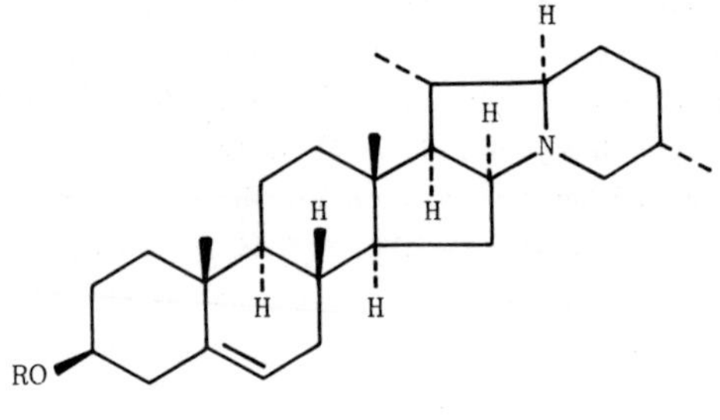

(18) a; R = H

b; R = α-L-rhamnopyranosyl-(1→4)-β-D-glucopyranosyl

c; R = β-D-glucopyranosyl

d; R = β-D-glucopyranosyl-(1→6)-[α-L-rhamnopyranosyl-(1→4)]-β-D-glucopyranosyl

e; R = β-D-glucopyranosyl-(1→6)-β-D-glucopyranosyl

f; R = α-L-rhamnopyranosyl-(1→2)-β-D-glucopyranosyl

g; R = α-L-rhamnopyranosyl-(1→2)-[β-D-glucopyranosyl-(1→4)]-β-D-glucopyranosyl

Methanol extracts of aerial parts of Fritillaria thunbergii have yielded verticine, verticinone, and β_1-chaconine (18f), together with a new glycoside of solanidine (18a) and a glycoside of hapepunine (19a).[44] The structure (18g) of the solanidine glycoside was deduced from the results of standard degradation procedures and spectroscopic studies. The hapepunine glycoside (19b) was assigned the structure which is indicated mainly on the basis of comparison of its ^{13}C n.m.r. spectrum with that of β_1-chaconine. An important feature of the structure investigation of both new glycoalkaloids was the extensive use and interpretation of field-desorption mass spectra.[44]

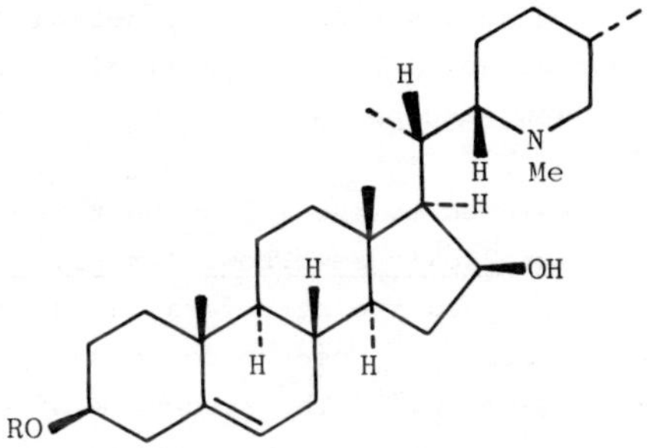

(19) a; R = H

b; R = α-L-rhamnopyranosyl-(1→2)-β-D-glucopyranosyl

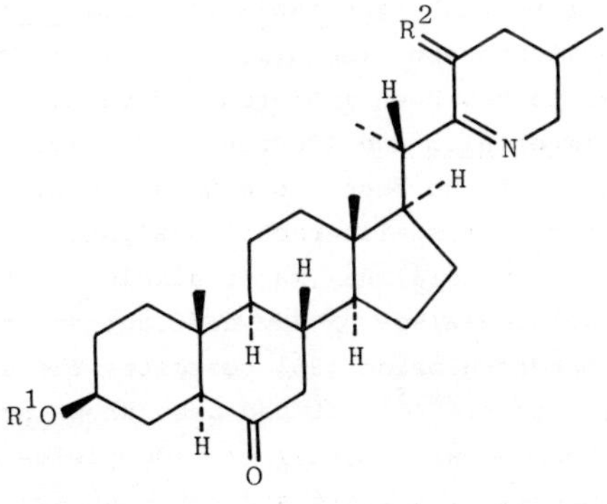

(20) a; R^1=H, R^2= O

b; R^1= β-D-glucopyranosyl, R^2= O

c; R^1= H, R^2= H_2

The alkaloid content of Petilium raddeana (Fritillaria sp.) has been investigated further. Bulbs of the plant yielded imperialine, imperialone, peimisine, petiline, petilidine, and two new bases, petizine (20a) and petizinine (20b).[45] The structures of petizine and petizinine were deduced from spectroscopic study of the bases and of their derivatives, and by hydrolysis of petizinine, which gave the aglycon, petizine. Furthermore, oxidation of the known alkaloid petiline (20c) with manganese dioxide furnished the new base petizine (20a).[45] However, since the structure of petiline[46] is not yet secure,[13] further evidence for the structures proposed for petizine and petizinine would be desirable. In partial contrast to the isolation studies discussed above, aerial parts of P. raddeana yielded imperialine, petiline, petilidine, petilinine, petisine (= petizine), a new base petisidine, and an isomer of imperialine N-oxide.[47]

Structures have been assigned to the following three alkaloids, which were isolated from Korolkowia sewerzowi (Fritillaria sp.). Sevedin N-oxide (21) was assigned the structure and stereochemistry which is indicated, on the basis of spectroscopic study and chemical correlation with sevedin.[48] The structure of sewerzine (22) was deduced solely on the basis of spectroscopic studies on the base, its diacetate, and on the derived dione.[49] Seweridine was shown to possess the interesting structure (23), which is a modified 11,12-seco-(25R)-5α-cevanine, by X-ray crystallography on the hydrochloride salt of the base.[50]

Kaneko has recently reported[51] the isolation from F. thunbergii and the structure determination of isobaimonidine (24) (cf. Vol. 12, p.287). The same alkaloid has been isolated independently by another group from F. imperialis and (fortunately) given the same name; the structure (24) of the base was elucidated by spectroscopic studies and by X-ray crystallographic analysis.[52] In the course of the latter work, imperialine (major alkaloid), verticine, and verticinone were isolated also. X-Ray diffraction studies have shown that verticinone hydrochloride (25) possesses the anticipated trans-fused E/F ring junction.[52,53] It had been shown previously that treatment of verticinone with methyl bromide yields a quaternary salt which possesses a cis-fused E/F ring junction.[54]

The crude drug "protoveratrine", from Veratrum album, consists of two main compounds, namely protoveratrines A (26a) and B (26b). A third component, protoveratrine C (26c), has now been isolated.[55] The structure of this new alkaloid was demonstrated as follows:

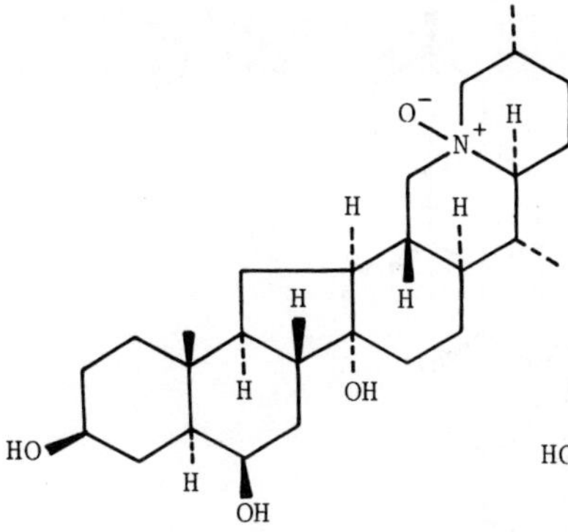

Steroidal Alkaloids

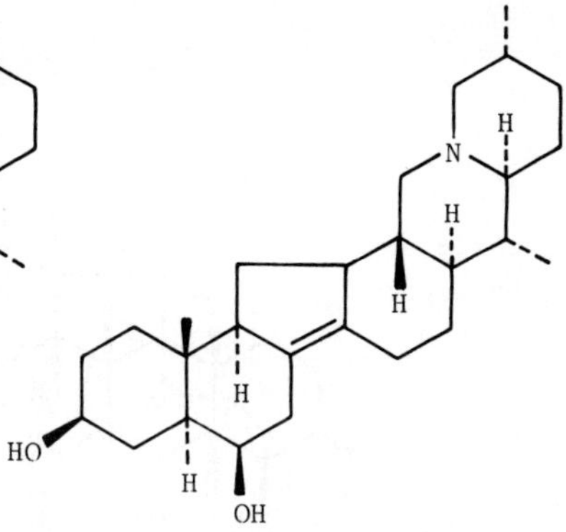

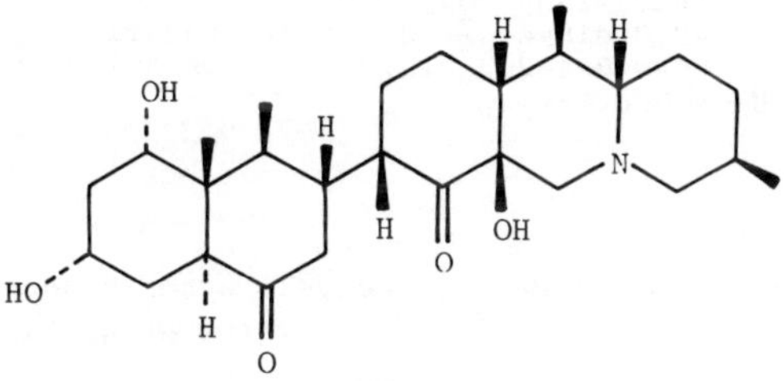

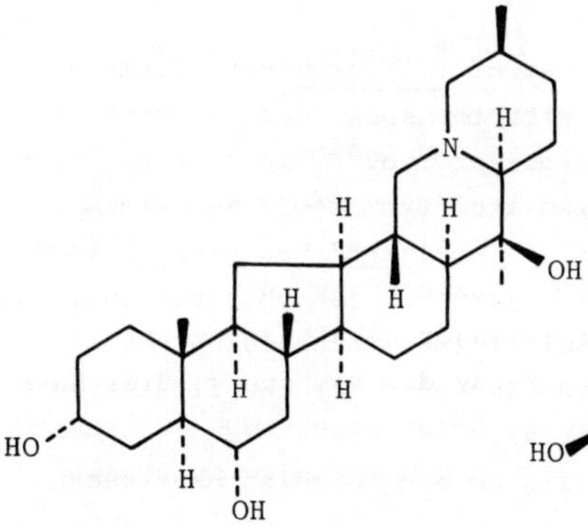

(21)

(22)

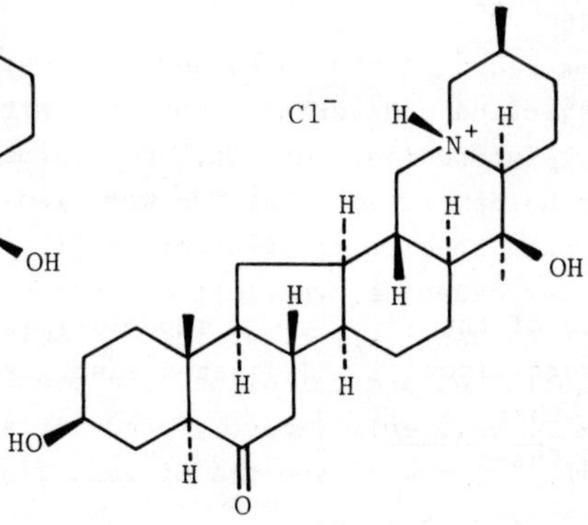

(23)

(24)

(25)

(26) a; R = (+)-(2S)-2-hydroxy-2-methylbutanoyl
    b; R = (+)-2,3-dihydroxy-2-methylbutanoyl
    c; R = (−)-(2S,3R)-2,3-dihydroxy-2-methylbutanoyl
    d; R = H

(a) mass spectrometry and elemental analysis were in accord with the formula $C_{41}H_{63}NO_{15}$; (b) treatment of (26c) with sodium metaperiodate and then with aqueous ammonia gave the known base desatrine (26d); finally, (c) X-ray crystallographic analysis confirmed the structure and relative stereochemistry of protoveratrine C (26c).[55]

Extracts of the roots and rhizomes of V. nigrum have yielded rubijervine and veramine together with the known constituents veratroyl-zygadenine, jervine, and isorubijervine.[56] In another study, four unidentified alkaloids were isolated from roots and stems of the same species.[57] Extraction of V. lobelianum has yielded veratroyl-zygadenine, veralosine, isorubijervine, isorubijervosine, and three unidentified alkaloids.[58] Solanidine, rhinolidine, rhinolinine, and rhinoline have been isolated for the first time from F. valujevii (above-ground parts).[59]

The ^{13}C n.m.r. spectra of veratridine and cevadine have been completely assigned.[60]

Photolysis of the hypoiodite derived from N-acetyljervine (27) gave the dimeric compound (28) as the major product, together with small amounts of the iodides (29) and (30), the cyclic ether (31), and lactone (32).[61] In a preliminary communication (cf. Vol. 5, p.263) the authors had misassigned the structures of several of these compounds (cf. Vol. 7, p.294) and so the early work should be disregarded.

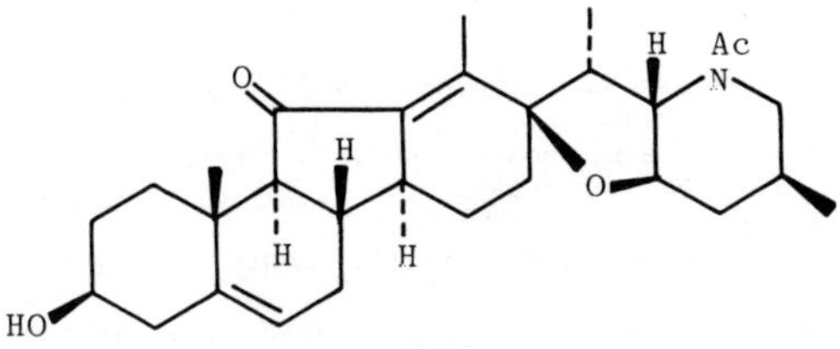

(27)

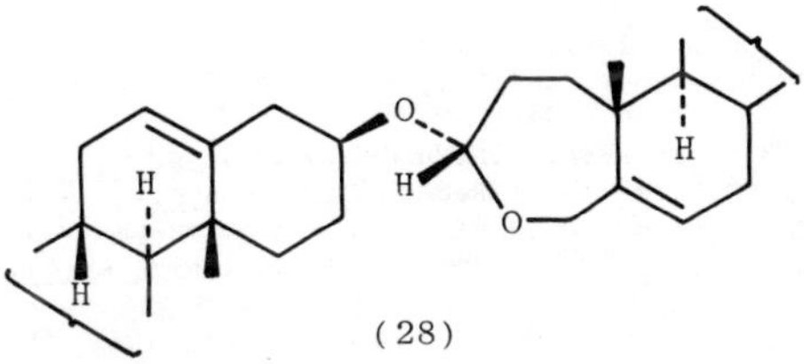

(28) 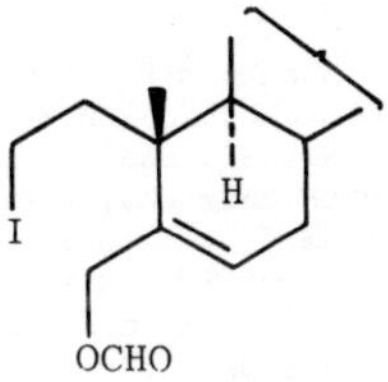 (29)

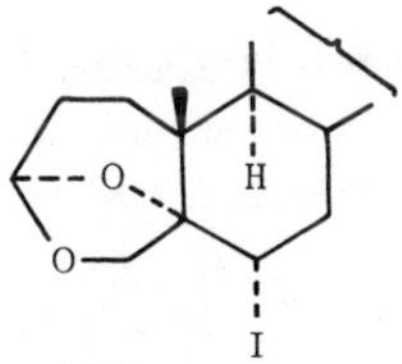

(30) 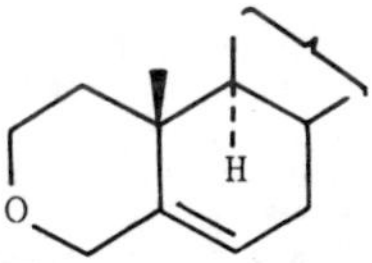 (31) 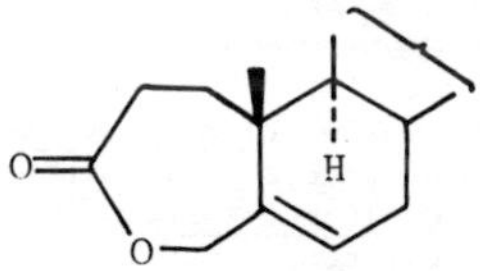 (32)

5 Miscellaneous Steroidal Alkaloids

The chemistry of batrachotoxin and of analogous bases has been summarised in a review on the alkaloids of neotropical poison frogs.[62]

References

1 H.L.Holland, in 'The Alkaloids', ed. R.H.F.Manske and R.G.A.Rodrigo, Academic Press, New York, 1981, Vol. 18, p.323.
2 H.E.Smith, C.A.Taylor, A.F.McDonagh, and Fu-Ming Chen, J. Org. Chem., 1982, 47, 2525.
3 G.N.Chaturvedi, K.P.Singh, and J.P.Gupta, Indian Med. Gaz., 1981, 179 (Chem. Abstr., 1982, 96, 57 612).
4 J.LeMen, G.Buffard, J.F.Desconclois, A.Jehanno, J.Provost, R.Tiberghien, P.Forgacs, F.Roquet, F.Godard, and M.Aurousseau, Eur. J. Med. Chem.-Chim. Ther., 1982, 17, 43 (Chem. Abstr., 1982, 97, 6628).
5 F.Frappier, W.E.Hull, and G.Lukacs, J. Org. Chem., 1981, 46, 4314.
6 G.Van de Woude, M.Biesemans, L.Van Hove, and J.Mertens, Bull. Soc. Chim.

Belg., 1982, 91, 67.
7   M.Mukherjee and L.M.Mukherjee, Indian J.Chem., Sect. B, 1981, 20, 1019
    (Chem. Abstr., 1982, 97, 56 125).
8   P.Milliet, A.Picot, and X.Lusinchi, Tetrahedron, 1981, 37, 4201.
9   H.Budzikiewicz, W.Ockels, and A.C.Campbell, Org. Mass Spectrom., 1982, 17, 107.
10 Yu-Yuan Tang, Chen-An Teng, and Ping-Wen Liang, Chung Ts'ao Yao, 1981, 12,
    No. 3, 43 (Chem. Abstr., 1981, 95, 121 035).
11 Bing-Wen Liang, Xiao-Qing Hou, Chen-An Deng, You-Yuan Tang, Xue-Bin Wang,
    Wei-Yu Yang, Ming-Hua Ciu, and Zhi-Quan Wang, Yao Hsueh T'ung Pao, 1981, 16,
    No. 4, 3 (Chem. Abstr., 1981, 95, 156 427).
12 B.U.Khodzhaev and S.Yu.Yunusov, Khim. Prir. Soedin., 1982, 125 (Chem. Abstr.,
    1982, 97, 3525).
13 H.Ripperger and K.Schreiber, in 'The Alkaloids', ed. R.H.F.Manske and
    R.G.A.Rodrigo, Academic Press, New York, 1981, Vol. 19, 81.
14 A.Usubillaga, V.Zabel, and W.H.Watson, Acta Crystallogr., Sect. B, 1982, 38, 966.
15 G.Adam and H.T.Huong, J. Prakt. Chem., 1981, 323, 839.
16 O.N.Chechina, M.Zh.Zhurinov, Kh.A.Aslanov, and F.Ishbaev, Zh. Obshch. Khim.,
    1982, 52, 163 (Chem. Abstr., 1982, 97, 56 124).
17 M.Zh.Zhurinov, K.E.Karimsakov, M.Ya.Fioshin, Z.M.Muldakhmetov, and O.N.
    Chechina, Deposited Document, 1980, VINITI 4418-80 (Chem. Abstr., 1982, 96,
    123 105); K.E.Karimsakov, M.Zh.Zhurinov, M.Ya.Fioshin, Z.M.Muldakhmetov,
    and O.N.Chechina, Deposited Document, 1980, VINITI 4419-80 (Chem. Abstr.,
    1982, 96, 104 597).
18 D.Miljković and K.Gasi, Chem. Chron., 1980, 9, 325 (Chem. Abstr., 1981, 95,
    187 519); D.A.Miljković and K.M.Gasi, Glas. Hem. Drus. Beograd, 1981, 46,
    No. 6, 263 (Chem. Abstr.; 1982, 96, 143 154); ibid., 1982, 47, No. 3, 19
    (Chem. Abstr., 1982, 97, 39 214).
19 M.Basterrechea, J.L.Mopa, F.Llanes, F.Coll, and V.Verez, Rev. Cubana Farm.,
    1981, 15, 247 (Chem. Abstr., 1982, 97, 3048).
20 R. Carle, Plant Syst. Evol., 1981, 138, 61 (Chem. Abstr., 1981, 95, 200 595).
21 S.C.Pandeya, G.V.Sarababu, and A.B.Bhatt, Indian J. Exp. Biol., 1981, 19,
    1207 (Chem. Abstr., 1982, 96, 82 172).
22 S.C.Pandeya, S.G.V.Babu, and A.B.Bhatt, Planta Med., 1981, 42, 409.
23 U.Gupta and P.K.Basu, Indian J.Exp. Biol., 1981, 19, 1205 (Chem. Abstr., 1982,
    96, 65 127).
24 K.P.Bhusari, V.V.Parashar, A.R.Pingle, and V.R.Dnyansagar, Indian Drugs,
    1981, 19, 5 (Chem. Abstr., 1982, 96, 149 229).
25 S.C.Jain and S.L.Sahoo, Agric. Biol. Chem., 1981, 45, 2909 (Chem. Abstr., 1982,
    96, 65 765).
26 S.C.Jain and S.L.Sahoo, Pharmazie, 1981, 36, 714; Chem. Pharm. Bull., 1981,
    29, 1765.
27 P.Gregory, S.L.Sinden, S.F.Osman, W.M.Tingey, and D.A.Chessin, J. Agric.
    Food Chem., 1981, 29, 1212 (Chem. Abstr., 1981, 95, 183 859).
28 W.M.Tingey and S.L.Sinden, Am. Potato J., 1982, 59, 95 (Chem. Abstr., 1982, 96,
    177 980).
29 B.Bezbaruah, Planta Med., 1981, 43, 77.
30 R.Saijo, K.Murakami, T.Nohara, T.Tomimatsu, A.Sato, and K.Matsuoka,
    Yakugaku Zasshi, 1982, 102, 300 (Chem. Abstr., 1982, 97, 20 704).
31 M.M.Shabana and T.S.El-Alfy, Egypt. J. Pharm. Sci., 1978 (Published 1980),
    19, 337 (Chem. Abstr., 1981, 95, 3392).
32 E.Moreira, C.Cecy, T.Nakashima, J.R.Cavazzani, O.G.Miguel, and R.Krambeck,
    Trib. Farm., 1980, 48, 61 (Chem. Abstr., 1981, 95, 183.867).
33 R.Jellema, E.T.Elema, and T.M.Malingre, J. Chromatogr., 1981, 210, 121.
34 L.K.Klyshev and A.Zh.Dosymbaeva, Izv. Akad. Nauk Kaz. S.S.R., Ser. Biol.,
    1981, No. 2, 14 (Chem. Abstr., 1981, 95, 58 181).
35 Zaib-un-Nisa Abdullah and R.Ahmad, Z. Acker-Pflanzenbau, 1981, 150, 417
    (Chem. Abstr., 1982, 96, 137 838).
36 R.R.King and R.E.McQueen, J. Agric. Food Chem., 1981, 29, 1101 (Chem. Abstr.,
    1981, 95, 111 433).

37 N.M.Talukder and C.Paupardin, C.R. Seances Acad. Sci., Ser. 3, 1981, 293, 549.
38 M.A.Filadelfi and A.Zitnak, Phytochemistry, 1982, 21, 250.
39 K.Drost-Karbowska, Z.Kowalewski, E.Nadolna, and M.Szaufer-Hajdrych, Acta Pol. Pharm., 1980, 37, 663 (Chem. Abstr., 1981, 95, 128 676).
40 B.C.Campbell and S.S.Duffey, J. Chem. Ecol., 1981, 7, 927 (Chem. Abstr., 1982 96, 66 051); H.R.Hohl, P.Stoessel, and H.Haechler, Ann. Phytopathol., 1980, 12, 353 (Chem.Abstr., 1982, 96, 82 867); C.A.Smith and W.E.MacHardy, Phytopathology, 1982, 72, 415 (Chem. Abstr., 1982, 96, 178 113).
41 W.D.Nes, P.K.Hanners, G.A.Bean, and G.W.Patterson, Phytopathology, 1982, 72, 447 (Chem. Abstr., 1982, 96, 196 646).
42 S.L.Sinden and L.L.Sanford, Am. Potato J., 1981, 58, 305 (Chem. Abstr., 1981, 95, 77 028).
43 K.Samikov, Ya.V.Rashkes, and S.Yu.Yunusov, Khim. Prir. Soedin., 1981, 349 (Chem. Abstr., 1982, 96, 20 313).
44 J.Kitajima, T.Komori, T.Kawasaki, and H.R.Schulten, Phytochemistry, 1982, 21, 187.
45 I.Nakhatov, A.Nabiev, and R.Shakirov, Khim. Prir. Soedin., 1981, 616 (Chem. Abstr., 1982, 96, 65 672).
46 R.N.Nuriddinov, B.Babaev, and S.Yu.Yunusov, Khim. Prir. Soedin., 1969, 5, 604 (Chem. Abstr., 1970, 73, 15 049).
47 A.Nabiev, R.Shakirov, and U.T.Shakirova, Khim. Prir. Soedin., 1981, 405.
48 K.Samikov, V.V.Kul'kova, and R.Shakirov, Khim. Prir. Soedin., 1981, 529 (Chem. Abstr., 1982, 96, 123 044).
49 K.Samikov and R.Shakirov, Khim. Prir. Soedin., 1981, 252 (Chem. Abstr., 1981 95, 115 815).
50 S.M.Nasirov, L.G.Kuzmina, K.Samikov, R.Shakirov, D.U.Abdullaeva, Yu.T.Struchkov, and S.Yu.Yunusov, Khim. Prir. Soedin., 1981, 342 (Chem. Abstr., 1982, 96, 20 312; the structure which is depicted in this abstract is incorrect).
51 K.Kaneko, N.Naruse, K.Haruki, and H.Mitsuhashi, Chem. Pharm. Bull., 1980, 28, 1345.
52 I.Masterová, V.Kettmann, J.Majer, and J.Tomko, Arch. Pharm. (Weinheim, Ger.), 1982, 315, 157 (Chem. Abstr., 1982, 96, 196 495; the structure which is depicted in this abstract is incorrect).
53 V.Kettmann, I.Mašterová, and J.Tomko, Acta Crystallogr., Sect. B, 1982, 38, 978.
54 S.Itô, Y.Fukazawa, T.Okuda, and Y.Iitaka, Tetrahedron Lett., 1968, 5373.
55 A.K.Saksena and A.T.McPhail, Tetrahedron Lett., 1982, 23, 811.
56 N.V.Bondarenko, Khim. Prir. Soedin., 1981, 527 (Chem. Abstr., 1982, 96, 65 653).
57 Zhong-Jian Jia, Xi-Fong Hu, Zhen-Mign Xia, and Zi-Qing Zhu, Lan-chou Ta Hsueh Hsueh Pao, Tzu Jan K'o Hsueh Pan, 1981, No. 1, 80 (Chem. Abstr., 1981, 94, 214 465).
58 E.M.Taskhanova and R.Shakirov, Khim. Prir. Soedin., 1981, 404 (Chem. Abstr., 1981, 95, 147 138).
59 K.Samikov and R.Shakirov, Khim. Prir. Soedin., 1981, 530 (Chem. Abstr.,1982, 96, 65 654).
60 D.De Marcano, B.Méndez, H.Parada, and A.Rojas, Org. Magn. Reson., 1981, 16, 314.
61 H.Suginome, K.Kato, and T.Masamune, Bull. Chem. Soc. Jpn., 1981, 54, 3042.
62 J.W.Daly, in 'Progress in the Chemistry of Organic Natural Products', ed. W.Herz, H.Grisebach, and G.W.Kirby, Springer Verlag, Vienna/New York, 1982, Vol. 41, p.205.

16
Miscellaneous Alkaloids

BY J. R. LEWIS

1 Muscarine Alkaloids

A stereospecific synthesis of muscarines and allomuscarines of the D- and L- series has been achieved using 2-deoxy-2-ribose as starting material.[1] Treatment of the ditoluoyl ester (1) with methanolic HCl gave a dimethyl ester which was converted to the dimethylamide (2) by heating to 90°C with dimethylamine.

Subsequent selective tosylation, LiAlH$_4$ reduction, and quaternisation with MeI gave D-(-)-(1$\underline{R}$, 3$\underline{S}$, 4$\underline{R}$)-muscarine iodide (3).

2 Imidazole Alkaloids

A review on imidazoles, covering their chemical and physical properties as well as their formation, has been published.[2] Another novel synthesis of (+)-pilocarpine (5) starting from L-histidine (cf. Vol. 11, p.238) involves its conversion, with inversion, into methyl ($\underline{R}$)-2-bromo-3-(1-methyl-5-imidazolyl)-propionate (4); treatment with dibenzyl ethylmalonate caused reinversion to give (5), which can be converted into (+)-pilocarpine and (+)-isopilocarpine by known methods.[3] The stem bark of Cynometra hankei contains the known imidazoles cynometrine (6; R^1 = Me, R^2 = H) and cynodine (6; R^1 = Me, R^2 = COPh) and the new alkaloids $\underline{N}$-1-demethylcynometrine (6; R^1 = R^2 = H) and $\underline{N}$-1-demethylcynodine (6; R^1 = H, R^2 = COPh);[4] the absolute configuration of the alkaloids is as shown and that of isocynometrine (from Cynometra lujae)[5] is (7). $\underline{N}$-Cinnamoylhistidine has been isolated from the leaves of Lycium cestroides.[6] The betaine clithioneine (8) has been identified as a constituent of the fruiting bodies of the mushroom Clitocybe acromelalga.[7]

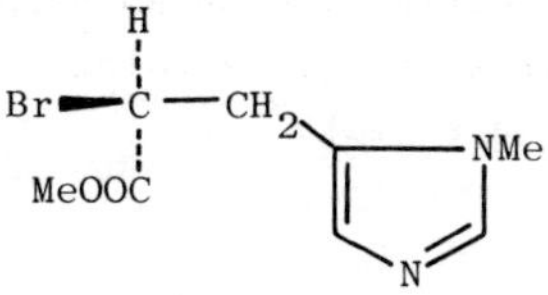

(4)

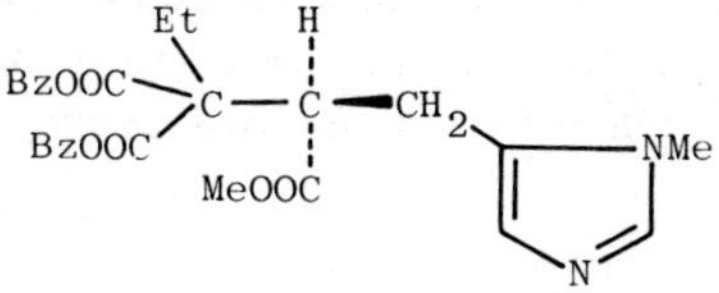

(5)

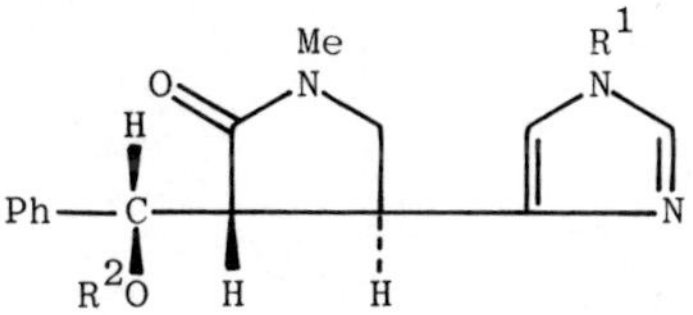

(6)

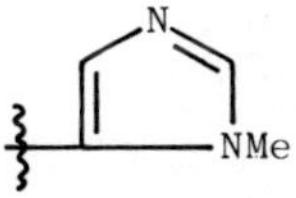

(7)

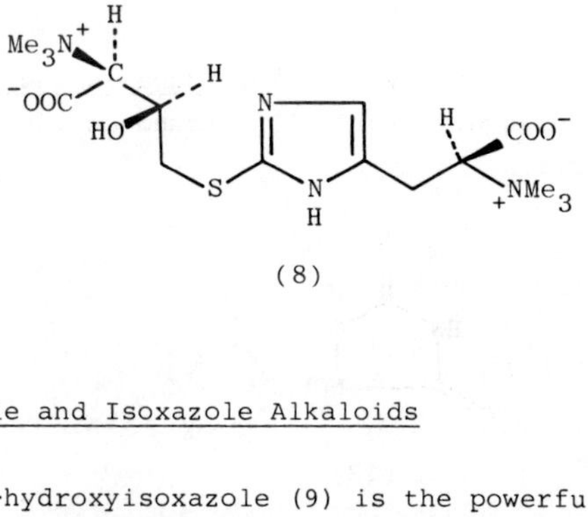

(8)

3 Oxazole and Isoxazole Alkaloids

Triumferol or 4-hydroxyisoxazole (9) is the powerful seed germination inhibitor isolated from the African medicinal plant Triumfetta rhomboidea. Confirmation of its structure and its activity was obtained by synthesis through conversion of ethyl isoxazole-3-carboxylate into its acid chloride followed by reaction with MeMgBr and oxidation of the tertiary alcohol with 30% H_2O_2.[8] The naturally occurring isoxazole and antitumour agent AT-125 (10) has been stereospecifically synthesised from L-glutamic acid by photochemical chlorination to give L-threo- and L-erythro-β-chloroglutamic acids (11). After separation, protection, and cyclization, selective γ

ring-opening with potassium phthalimide, treatment with hydroxyl-
amine, and cyclization gave the L-<u>threo</u> anhydride (12). After
protection, treatment with excess dichlorotris(dimethylamino)phos-
phorane in boiling THF gave the doubly protected isoxazoleacetic
acid, which with thioanisole in CF_3COOH gave (10).[9]

The anthelmintically active plant <u>Quisqualis indica</u> contains its
active principle quisqualic acid (13) (found as its potassium salt)
located in the kernals.[10]

4 Peptide Alkaloids

Simple amides continue to be isolated from the Piperaceae family;
<u>N</u>-<u>trans</u>-feruloylpiperidine, feruperine (14), and its 4,5-dihydro-
derivative have been isolated for the first time from natural
sources (<u>Piper nigrum</u>).[11] Isopiperlonguminine (15; <u>Z,Z</u>; R = H),
corcovadine (15; <u>E,E</u>; R = OAc), and isocorcovadine (15; <u>Z,Z</u>;
R = OAc) have been obtained from <u>Ottonia corcovadensis</u>,[12] and (<u>E,E</u>)-
<u>N</u>-isobutyl-9-piperonyl-nona-2,4-dienoic amide (16) from <u>Ottonia</u>

anisum.[13] Two pungent principles found in the pericarps of Zanthoxylum ailanthoides are (2E, 4E, 8Z, 10E, 12E)-N-isobutyl-2,4,8,10,12-tetradecapentaenamide (γ-sanshoöl) and hydroxy-γ-sanshoöl (the 2'-hydroxy derivative).[14]

A short, convenient, stereoselective synthesis of pepper-derived alkaloids has been carried out by condensing piperonal with the ylide from methyl (E)-4-(diethylphosphono)-2-butenoate to give methyl (E,E)-5-(1,3-benzodioxol-5-yl)-2,4-pentadienoate, which on methoxide-catalysed aminolysis (piperidine, pyrrolidine, etc.) gave the required alkaloid.[15]

Tyramine in conjunction with 'cis' ferulic acid gives the amide found in the roots of bell pepper (Capsicum annuum var. grossum)[16] and with dotriacontanoic acid it gives acalyphamide (17), a constituent of the leaves and twigs of Acalypha indica;[17] also present in the former plant is the benzofuran grossamide (18), which also contains a tyramine moiety.[16] Marmeline (19) has been obtained from the unripe fruits of Aegle marmelos,[18] auranamide (20) from the seeds of Piper aurantiacum,[19] together with aurantiamide (21; R = H) and its acetate (21; R = Ac).[20] N-Benzoyl-L-ornithine (22; R = H) and N^δ-benzoyl-L-γ-hydroxyornithine (22; R = OH) have been isolated from Vicia pseudo-orobus.[21] Intermolecular coupling of two piplatine molecules gives the cyclobutane derivative (23) found in Piper tuberculatum.[22]

Zizyphine A (25; R = Me) and zizyphine B (25; R = H) have been synthesised in a multistep procedure via the intermediate (24)[23] and their dihydro-derivatives have also been synthesised in a 16- and 15- step procedure.[24] (±)-Lunarine (26; X^1 = NH, X^2 = CH_2) and (±)-lunaridine (26; X^1 = CH_2, X^2 = NH) have been synthesised by cyclocondensation via the thiazolidinethione technique, using the appropriate diamine.[25] Fujita's group have also synthesised parabactin A (27)[26] and the codonocarpines (28)[27] using this technique. A review on the method has also been published.[28] Brevigellin (29) has been isolated from the culture of Penicillium brevicompactum.[29] N(1)-Acetyl-N(1)-deoxymayfoline (30; R = Ac) has been found in Maytenus buxifolia together with mayfoline (30; R = OH).[30]

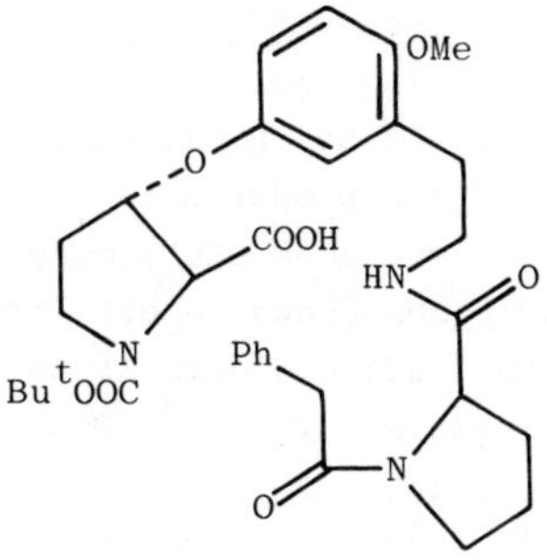

(24)

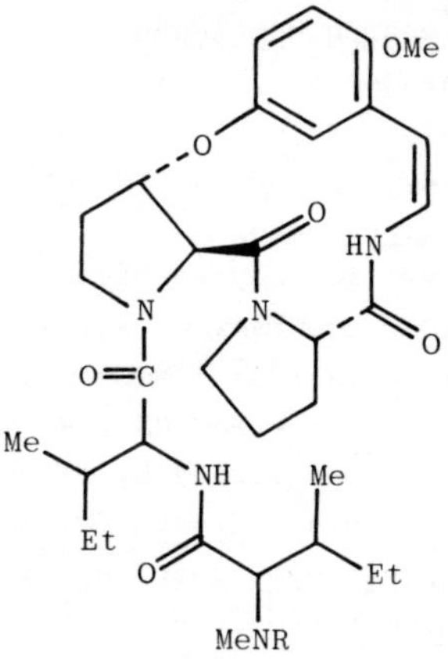

(25)

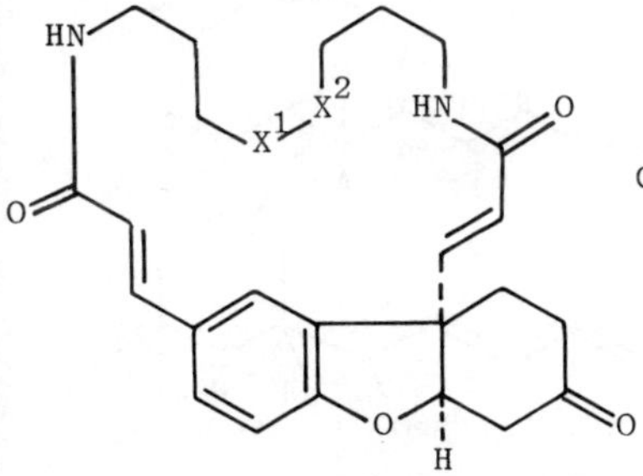

(26)

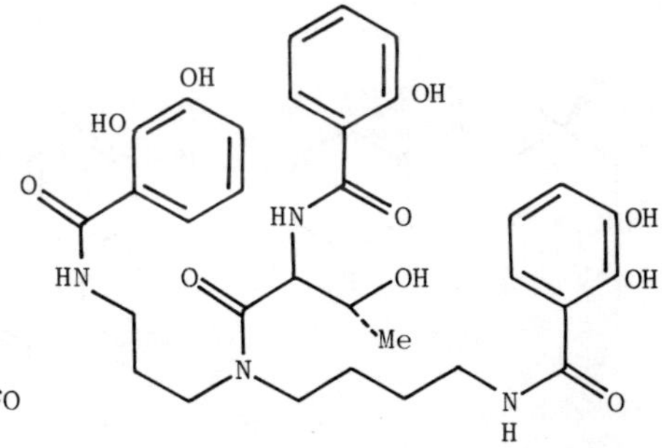

(27)

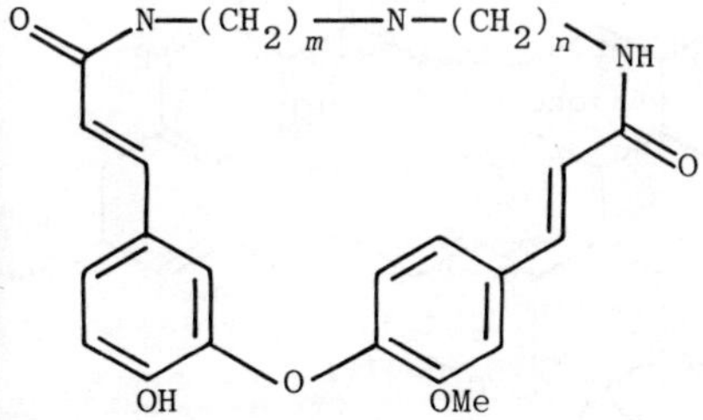

(28) $m = 3, n = 4$ or $m = 4, n = 3$

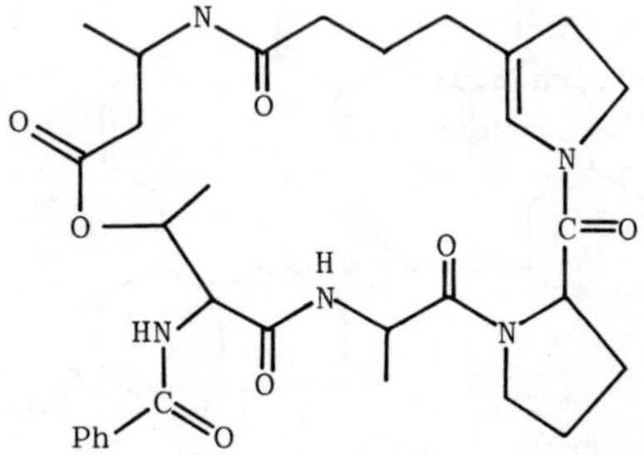

(29)

The search continues for newer and better sources of maytansinoids (the ansa macrolides with anti-tumour activity); larger amounts of maytansine are contained in the leaves of <u>Maytenus confertiflorus</u>[31] while maytanprine is present in its stem.[32] The small upland tree <u>Maytenus rothiana</u>, found near Bombay, appears to be the richest source yet for maytansoids.[33] <u>Trewia nudiflora</u> produces the ansa macrolides trewiasine (31; R^1 = CHMe$_2$, R^2 = Me, R^3 = OMe), dehydrotrewiasine (31; R^1 = MeC=CH$_2$, R^2 = Me, R^3 = OMe), and demethyltrewiasine (31; R^1 = MeC=CH$_2$, R^2 = H, R^3 = OMe), all differing from the maytansine family by possessing a 15-methoxy group.[34] Other ansamacrolides isolated of late are herbamycin B (32) from <u>Streptomyces hygroscopicus</u>[35]; ansatrienin A (33) and ansatrienin B, the hydroquinone counterpart of (33) isolated from <u>Streptomyces collinus</u>, possess a number of different substituents on the ansa ring as compared with the ansamycins.[36]

5 Alkaloid-containing Sources and Unclassified Alkaloids

5.1 *Amanita muscaria*.- A yellow pigment isolated from the fly agaric mushroom has been identified as the dihydroazepine (34) and its synthesis via a biomimetic route involved the conversion of 3-(2-carboxy-5-pyridyl)alanine to the glutaconic dialdehyde (35)[37] and thence to (34).

5.2 *Coix lachryma-jobi*.- The roots of this plant contain the benzoxazolinone (36)[38] as well as its 2-O-glucosyl derivative.

5.3 *Dulacia guianensis*.- A new α-aminotropone, manicoline A (37), has been isolated from the root bark of this tree.[39]

5.4 *Streptomyces aureofaciens*.- A new hypotensive vasodilator (WS-1228A) isolated from this fungus has been identified as the hydrazine-oxime (38).[40]

5.5 *Zanthoxylum arborescens*.- The leaves of this Rutaceous plant contain the cis-piperazine compound (39), the trans-formulation being rejected due to optical activity being associated with the natural product.[41]

References

1 S.Pochet and T.Huynh-Dinh, J. Org. Chem., 1982, 47, 193.
2 M.R.Grimmett in 'Advances in Heterocyclic Chemistry', ed. A.R.Katritzky and A.J.Boulton, Vol. 27, 1980, Academic Press.
3 A.Noordam, L.Maat, and H.C.Bayerman, J.R.Neth. Chem. Soc., 1981, 100, 441 (Chem. Abstr., 1982, 96, 104 569).
4 P.G.Waterman and D.F.Faulkner, Phytochemistry, 1981, 20, 2765.
5 A.Chiaroni, C.Riche, L.Tshissambou, and F.Khuong-Huu, J. Chem. Res.(S), 1981, 182.
6 J.L.Cabrera and H.J.Juliani, An. Asoc. Quim. Argent., 1981, 69, 357 (Chem. Abstr., 1982, 96, 17 254).
7 K.Konno, H.Shirahama, and T.Matsumoto, Tetrahedron Lett., 1981, 22, 1617.
8 T.Kusumi, C.C.Chang, M.Wheeler, I.Kubo, K.Nakanishi, and H.Naoki, Tetrahedron Lett., 1981, 22, 3451.
9 R.B.Silverman and M.W.Holladay, J. Am. Chem. Soc., 1981, 103, 7357.
10 R.Zhang and B.Guan, Zhongcaoyao, 1981, 12, 40 (Chem. Abstr., 1982, 96, 40 796).
11 R.Inatani, N.Nakatani, and H.Fuwa, Agric. Biol. Chem., 1981, 45, 667.
12 S.S.Costa and W.B.Mors, Phytochemistry, 1981, 20, 1305.
13 A.M.Giesbrecht, M.A.DeAlvarenga, O.R.Gottlieb, and H.R.Gottlieb, Planta Medica, 1981, 43, 375.
14 I.Yasudo, K.Takeya, and H.Itokawa, Chem. Pharm. Bull., 1981, 29, 1791.
15 R.A.Olsen and G.O.Spessard, J. Agric. Food Chem., 1981, 29, 942 (Chem. Abstr., 1981, 95, 115 811).
16 T.Yoshihara, K.Yamaguchi, S.Takamatsu, and S.Sakamura, Agric. Biol. Chem., 1981, 45, 2593 (Chem. Abstr., 1982, 96, 48 958).
17 S.K.Talapatra, Indian J. Chem., Section B, 1981, 20, 974.
18 B.R.Sharma, R.K.Rattan, and P.Sharma, Phytochemistry, 1981, 20, 2606.
19 A.Banerji and R.Ray, Indian J. Chem., Section B, 1981, 20, 597.
20 A.Banerji and R.Ray, Phytochemistry, 1981, 20, 2217.
21 S.Hatanaka, S.Kaneko, K.Saito, and Y.Ishida, Phytochemistry, 1981, 20, 2291.
22 R.Braz Filho, M.P. De Souza and M.E.O.Mattos, Phytochemistry, 1981, 20, 345.
23 U.Schmidt, A.Lieberknecht, H.Boekens, and H.Griesser, Angew. Chem., 1981, 93, 1121.
24 U.Schmidt, H.Boekens, A.Lieberknecht, and H.Griesser, Tetrahedron Lett., 1981, 22, 4949.
25 Y.Nagao, S.Takao, T.Miyasaka, and E.Fujita, J. Chem. Soc., Chem. Commun., 1981, 286.
26 Y.Nagao, T.Miyasaka, Y.Hagiwara, and E. Fujita, Tennen Yuki Kagobutsu Toronkai Koen Yoshishu 24th, 1981, 474 (Chem. Abstr., 1982, 96, 143 122).
27 Y.Nagao, K.Seno, and E.Fujita, Tetrahedron Lett., 1980, 21, 4931.
28 N.Yoshimitsu and E.Fujita, Heterocycles, 1982, 17, 537.
29 N.J.McCorkindale and R.L.Baxter, Tetrahedron, 1981, 37, 1795.
30 M.Diaz and H.Ripperger, Phytochemistry, 1982, 21, 255.
31 X.-F.Wang, R.-F.Wei, J.-Y.Chen, D.-Q.Jiang, Yao Hsueh Hsueh Pao, 1981, 16, 59 (Chem. Abstr., 1981, 95, 138 482).
32 H.-F.Wang, C.-Y.Cheng, Y.-G.Wei, and T.-C.Chiang, Yao Hsueh Tung Pao, 1980, 15, 44 (Chem. Abstr., 1981, 95, 67 849).
33 D.F.Nettleton Jnr., D.M.Balitz, M.Brown, J.E.Moseley, and R.W.Myllymaki, J. Nat. Prod., 1981, 44, 340.
34 R.G.Powell, D.Weisleder, and C.R.Smith, J. Org. Chem., 1981, 46, 4398.
35 A.Nakagawa, N.Sadakane, and S.Omura, Nippon Kagaku Kaishi, 1981, 892 (Chem. Abstr., 1981, 95, 24 781).

36 M.Damberg, P.Russ, and A.Zeeck, Tetrahedron Lett., 1982, 23, 59.
37 H.Barth, G.Burger, H.Doepp, M.Kobayashi, and H.Musso, Liebigs Ann. Chem., 1981, 2164.
38 N.Shigematsu, I.Kouno, and N.Kawano, Yakugaku Zasshi, 1981, 101, 1156 (Chem. Abstr., 1982, 96, 103 944).
39 J.Polonsky, J.Varenne, T.Prangé, C.Pascard, H.Jacquemin, and A.Fournet, J. Chem. Soc., Chem. Commun., 1981, 731.
40 H.Tanaka, K.Yoshida, Y.Itoh, and H.Imanaka, Tetrahedron Lett., 1981, 22, 3421.
41 J.A.Grina and F.R.Stermitz, Tetrahedron Lett., 1981, 22, 5257.